ROUTLEDGE LIBRARY EDITIONS: GEOLOGY

Volume 27

SPACE AND TIME IN GEOMORPHOLOGY

SPACE AND TIME IN GEOMORPHOLOGY

Binghamton Geomorphology Symposium 12

Edited by
COLIN E. THORN

Routledge
Taylor & Francis Group

LONDON AND NEW YORK

First published in 1982 by George Allen & Unwin Ltd

This edition first published in 2020
by Routledge
2 Park Square, Milton Park, Abingdon, Oxon OX14 4RN

and by Routledge
52 Vanderbilt Avenue, New York, NY 10017

Routledge is an imprint of the Taylor & Francis Group, an informa business

British Library Cataloguing in Publication Data
A catalogue record for this book is available from the British Library

ISBN: 978-0-367-18559-6 (Set)
ISBN: 978-0-429-19681-2 (Set) (ebk)
ISBN: 978-0-367-27667-6 (Volume 27) (hbk)
ISBN: 978-0-367-27818-2 (Volume 27) (pbk)
ISBN: 978-0-429-29797-7 (Volume 27) (ebk)

Publisher's Note
The publisher has gone to great lengths to ensure the quality of this reprint but
points out that some imperfections in the original copies may be apparent.

Disclaimer
The publisher has made every effort to trace copyright holders and would welcome
correspondence from those they have been unable to trace.

Space and Time in Geomorphology

Edited by Colin E. Thorn

Department of Geography, University of Illinois at Urbana-Champaign

London
GEORGE ALLEN & UNWIN
Boston   Sydney

GEORGE ALLEN & UNWIN LTD.
40 Museum Street, London WC1A 1LU

LIBRARY OF CONGRESS CATALOGING IN PUBLICATION DATA
SPACE AND TIME IN GEOMORPHOLOGY
   Bibliography: P.
   Includes Index.
   1. Geomorphology–Congresses.
   I. Thorn, Colin E.
   GB400.2.S67   551.4   81.21740

ISBN 0-04-551056-3   AACR2

BRITISH LIBRARY CATALOGUING IN PUBLICATION DATA
Space and time in geomorphology.—(The 'Binghamton'
   symposia in geomorphology; no. 12)
   1. Geomorphology–Congresses
   I. Thorn, Colin E. II. Series
   551.4   GB400.2

ISBN 0-04-551056-3

Produced by
Publishers Creative Services Inc., New York
and printed in Great Britain by Mackays of Chatham Ltd

Preface

Uniformitarianism and the ergodic hypothesis are fundamental concepts in contemporary geomorphology. In the case of uniformitarianism there has been a protracted debate over the true usefulness of the term (e.g. Gould 1965). In truth, the worst that can be said about it is that it is cacophonous and anachronistic. However, the ergodic hypothesis has rarely been put to the test (Savigear 1952 is one of a few classic exceptions), although in a generalized form it is inextricably woven into the fabric of contemporary geomorphology.

As geomorphic theory has developed, both uniformitarianism and the ergodic hypothesis have been recast. Magnitude and frequency (Wolman & Miller 1960) updated simplistic uniformitarianism, only to be modified itself by the notions of 'healing time' (Wolman & Gerson 1978), episodic erosion (Schumm 1977) and the entire range of threshold-related concepts (Coates & Vitek 1980). Church and Mark (1980) have seriously questioned ergodic principles (space – time substitution) in their comparison of static and dynamic allometry. Indeed, Church (1980) has gone right to the heart of the matter by suggesting that process geomorphology provides little or no insight into landscape evolution. Such a perspective appears readily defensible and places even greater onus on stratigraphic interpretation for studies of landscape evolution. Unfortunately, the weak linkage between the alluvial record and interfluvial geomorphology (Costa 1975, Caine 1974), plus an insightful reappraisal of the stratigraphic record itself (Ager 1973) suggest an equally pressing need for careful reconsideration of traditional concepts on these fronts as well.

Clearly, geomorphology stands at something of a Rubicon. Faced with a dearth of data, it is essential to press forward on another front, namely, that of establishing a more cogent and comprehensive paradigm. There is certainly nothing new in this notion, as it has been stated eloquently by Davis (1899) and Chorley (1962). A meaningful grasp of landscape evolution would seem to depend upon: (a) understanding the present spatial distribution of processes and process rates, especially the linkage between erosional and depositional components; (b) comparison of spatial versus temporal change; and (c) careful appraisal of the character and composition of the stratigraphic record. The size of the task is herculean, but one obvious starting point is a theoretical examination of the characterisitics of available data sets for the specific purpose of shaping future field measurement studies more purposefully.

This symposium was conceived with many of the above points in mind. It was inspired by the success and significance of the ninth 'Binghamton' symposium on thresholds in geomorphology (Coates & Vitek 1980) and was consciously designed to complement the initial focus on across-threshold variability with one upon between-threshold variability. Each participant was invited to use a data set to address between threshold variability in either a spatial or temporal context or, if possible, both.

The paper by Derek C. Ford and John J. Drake makes a singularly appropriate introduction because in it they suggest that karst erosion has no intrinsic thresholds and therefore the entire span of Holocene time constitutes a meta-stable karst erosion phase. Integration of a variety of measurement techniques permits variations in limestone erosion and speleothem deposition to be examined over a range of timescales. However, temporal variability can be successfully evaluated only in association with a sound grasp of spatial variability.

Equilibrium concepts are central to geomorphology, and John B. Thornes sets out to examine both the behavior and stability of equilibria. He is able to demonstrate that autocorrelative properties permit recognition of basic process structures in temporal series, even when a large portion of the data sequence is missing. Estimation of spatial and temporal variability using the 'normal generated distribution' (ngd) is developed and demonstrated by H. Charles Romesburg and Jerome V. DeGraff. The ngd is particularly flexible and its estimation is simpler than that of the beta-binomial distribution. The ergodic hypothesis is tested mathematically by Richard G. Craig, using data from the Appalachian Mountains of Pennsylvania. In addition to tending to confirm the ergodic principle, the model suggests that stable landforms cannot occur, but that there is an oscillating effect about a preferred value.

Michael J. Bovis examines the spatial coherence of soil loss and soil loss controls in alpine tundra, sub-alpine and montane forests of the Colorado Front Range, while Shiu-hung Luk combines field and laboratory experiments to investigate within-site and between-site variability of sediment yield from artificial rainfall. The common use of the coefficient of variation provides a useful comparative link between the two papers.

T. Nelson Caine's paper on the spatial variability of surficial soil movement in three, geographically disparate, alpine zones emphasizes the importance of sampling spacing. Data from all sites suggest an absence of statistical independence when measurements are <12 m apart. In the

Canadian Rocky Mountains debris accretion, debris shift, rockfall and flood/debris flow are not currently producing gross changes in slope morphology according to James S. Gardner. In fact, most of the Holocene has been a period of relative quiescence and thus stands in sharp contrast with that of late Wisconsin deglaciation. A decade of observations and a total of 4500 measurements have provided Ian A. Campbell and John L. Honsaker with a powerful data base from which to appraise badland development in Dinosaur Provincial Park, Alberta. One of the most interesting questions to arise from this paper is the extent to which badlands are truly viable high-speed surrogates for other landscapes.

It is appropriate that Antony R. Orme's study of a beach shore zone should reflect the dynamism of such an environment by focussing upon short-interval sampling over a single summer season. The results highlight rather nicely the presence of multiple equilibria types within close spatial proximity to each other. William L. Graf integrates his own studies of arroyo development in the Henry Mountains, Utah, with those of renowned earlier workers. His synthesis of temporal development and spatial location in a network provides a comprehensive explanation of current patterns of arroyo behavior.

The interdependence between geomorphologists and Quaternary stratigraphers is underscored by W. Hilton Johnson's overview of the present interaction between geomorphic and stratigraphic models. His comments on the representative qualities of present environments *vis-à-vis* the Holocene, let alone the entire Quaternary, serve as a timely reminder of the problems confronted when trying to use the present as an analog for the geologic past. Wayne Wendland's paper shows those not already familiar with the idea that climatologists were among the first to recognize the significance of step functions in paleo-reconstructions. He is able to identify 13 periods of apparently rapid change in sedimentation in the past 9000 years and, in turn, to relate these to airstream frequencies.

Soil geomorphology may lay justifiable claim to being a potential keystone in linking process geomorphology and stratigraphy. Leon R. Follmer uses the geomorphic surface associated with the Sangamon soil to elucidate some important principles in the use of paleosols for geomorphic reconstructions. Scott F. Burns and Philip J. Tonkin use the theoretical framework of Butler's K-cycles as an appropriate structure for development of a Synthetic Alpine Slope model for soil development in the ridge-top tundra province of the Colorado Front Range. A sharp break between Lithic Haploxeralfs on slope crests and shoulders, versus proto-Vertisols at slope toes, was identified on San Clemente Island, California by Daniel R. Muhs, and it appears to be a general characteristic in Mediterranean environments.

In closing, it seems reasonable to suggest that the ninth and twelfth

'Binghamton' symposia provide persuasive evidence that much more careful theoretical appraisals must precede field installations if attempts to overcome scale-linkage problems in geomorphology are to attain fruition. For too long field geomorphologists have viewed theory as an anathema. However, general acceptance of the symbiotic nature of theory and field study is a necessary precursor to full flowering and maturation of geomorphology.

Colin E. Thorn
University of Illinois
March 1981

REFERENCES

Ager, D. V. 1973. *The nature of the stratigraphic record.* New York: Wiley.

Caine, T. N. 1974. The geomorphic processes of the alpine environment. In *Arctic and alpine environments,* J. D. Ives & R. G. Barry (eds), 721–48. London: Methuen.

Chorley, R. J. 1962. *Geomorphology and general systems theory.* U.S. Geol. Surv. Prof. Paper 500-B.

Church, M. 1980. Records of recent geomorphological events. In *Timescales in geomorphology,* R. A. Cullingford, D. A. Davidson & J. Lewin (eds), 13–29. New York: Wiley.

Church, M. and D. M. Mark 1980. On size and scale in geomorphology. *Prog. Phys. Geog.* **4,** 342–90.

Coates, D. R. and J. D. Vitek 1980. *Thresholds in geomorphology.* London: George Allen & Unwin.

Costa, J. E. 1975. Effects of agriculture on erosion and sedimentation in the Piedmont province, Maryland. *Geol Soc. Am. Bull.* **86,** 1281–6.

Davis, W. M. 1899. The geographical cycle. *Geog. J.* **14,** 481–504.

Gould, S. J. 1965. Is uniformitarianism necessary? *Am. J. Sci.* **263,** 223–8.

Savigear, R. A. G. 1952. Some observations on slope development in south Wales. *Trans Inst. Br. Geogs* **18,** 31–51.

Schumm, S. A. 1977. *The fluvial system.* New York: Wiley-Interscience.

Wolman, M. G. and J. P. Miller 1960. Magnitude and frequency of forces in geomorphic processes. *J. Geol.* **68,** 54–74.

Wolman, M. B. and R. Gerson 1978. Relative scales of time and effectiveness of climate in watershed geomorphology. *Earth Surf. Proc.* **3,** 189–208.

ACKNOWLEDGEMENTS

The opportunity to organize this symposium really stems from a decade of insightful hard work by Marie Morisawa and Donald R. Coates in creating and sustaining the Binghamton Geomorphology Symposia Series. In this instance credit must also go equally to Stanley A. Schumm for provocative and stimulating research on the nature of geomorphic processes. More immediately I would like to acknowledge the financial assistance of the National Science Foundation and the University of Illinois at Urbana through the good offices of the Head of the Department of Geography, Arthur Getis. Foremost, I wish to thank the contributing authors for 16 excellent papers, and Susan L. Seidel for handling numerous typing chores during preparation of the manuscript.

Colin E. Thorn

Contents

Preface   v

1 The spatial variation of soil loss and soil loss controls
  Michael J. Bovis   1

2 Soil–geomorphic models and the spatial distribution and
  development of alpine soils
  Scott F. Burns and Philip J. Tonkin   25

3 The spatial variability of surficial soil movement rates in alpine
  environments
  Nel Caine   45

4 Variability in badlands erosion: problems of scale and threshold
  identification
  Ian A. Campbell and John L. Honsaker   59

5 The ergodic principle in erosional models
  Richard G. Craig   81

6 The geomorphology of the Sangamon surface: its spatial and
  temporal attributes
  Leon R. Follmer   117

7 Spatial and temporal variations in karst solution rates: the structure
  of variability
  D. C. Ford and J. J. Drake   147

8 Alpine mass-wasting in contemporary time: some examples from the
  Canadian Rocky Mountains
  James S. Gardner   171

9 Spatial variation of fluvial processes in semi-arid lands
  William L. Graf   *193*

10 Interrelationships among geomorphic interpretations of the
   stratigraphic record, process geomorphology and geomorphic
   models
   W. Hilton Johnson   *219*

11 Variability of rainwash erosion within small sample areas
   Shiu-hung Luk   *243*

12 The influence of topography on the spatial variability of soils in
   Mediterranean climates
   Daniel R. Muhs   *269*

13 Temporal variability of a summer shorezone
   Antony R. Orme   *285*

14 Using the normal generated distribution to analyze spatial and
   temporal variability in geomorphic processes
   H. Charles Romesburg and Jerome V. DeGraff   *315*

15 Problems in the identification of stability and structure from temporal
   data series
   John Thornes   *327*

16 Geomorphic responses to climatic forcing during the Holocene
   Wayne M. Wendland   *355*

   Index   *373*

Contributors

Michael J. Bovis
Department of Geography, University of British Columbia, Vancouver, British Columbia V6T 1W5, Canada

Scott F. Burns
Department of Soil Science, Lincoln College, Canterbury, New Zealand

Nel Caine
Institute of Arctic and Alpine Research, University of Colorado, Boulder, Colorado 80309, USA

Ian A. Campbell
Department of Geography, University of Alberta, Edmonton, Alberta T6G 2H4, Canada

Richard G. Craig
Department of Geology, Kent State University, Kent, Ohio 44242, USA

Jerome V. DeGraff
USDA Forest Service, Sierra National Forest, 1130 'O' Street, Fresno, California 84322, USA

J. J. Drake
Department of Geography, McMaster University, Hamilton, Ontario, L8S 4K1, Canada

Leon R. Follmer
Illinois State Geological Survey, Natural Resources Building, University of Illinois, Urbana, Illinois 61801, USA

D. C. Ford
Department of Geography, McMaster University, Hamilton, Ontario, L8S 4K1, Canada

James S. Gardner
Department of Geography, Faculty of Environmental Studies, University of Waterloo, Waterloo, Ontario, N2L 3GL, Canada

William L. Graf
Department of Geography, Arizona State University, Tempe, Arizona 85281, USA

John L. Honsaker
Department of Geography, University of Alberta, Edmonton, Alberta T6G 2H4, Canada

W. Hilton Johnson
Department of Geology, University of Illinois at Urbana-Champaign, Urbana, Illinois 61801, USA

Shiu-hung Luk
Department of Geography, Erindale College, University of Toronto, Mississanga, Ontario L5L IC6, Canada

Daniel R. Muhs
Department of Geography, University of Wisconsin, Madison, Wisconsin 53706, USA

Antony R. Orme
Department of Geography, University of California, Los Angeles, California 90024, USA

H. Charles Romesburg
Department of Forestry and Outdoor Recreation and Computer Center, Utah State University, Logan, Utah 84322, USA

John Thornes
Department of Geography, London School of Economics and Political Science, Houghton Street, London WC2A 2AE, UK

Philip J. Tonkin
Department of Soil Science, Lincoln College, Canterbury, New Zealand

Wayne M. Wendland
Climatology Section, Illinois State Water Survey, Box 5050, Station A, Champaign, Illinois 61820, USA

Space and Time in Geomorphology

1

The spatial variation of soil loss and soil loss controls

Michael J. Bovis

INTRODUCTION

The problem of measuring directly the quantity of clastic material eroded from an area over a specified period has been approached at a wide range of scales by workers in the geomorphic and agricultural engineering communities over the past 40 years. The techniques of measurement have varied appreciably according to the size of the study area. At the catchment scale (usually 1 km² or larger) a measure of sediment produced from all sources may be obtained from in-channel records of bedload and suspended load. It is by no means certain that most of the sediment will be generated from the erosion of upstream areas: Zone I in the terminology of Schumm (1977, p. 3). In southwest British Columbia, for example, the main stems of the Fraser and Lillooet River systems carry very high suspended loads; yet many tributaries, including those occupied by glaciers, yield relatively small amounts of material (Slaymaker & McPherson 1977, Table 3). Scour and lateral migration of the main channels through unconsolidated late-glacial and Postglacial deposits accounts in large measure for the high main-stem turbidities. In this instance, in-channel data provide a substantial overestimate of primary denudation rates within the respective catchments.

At a much smaller scale and in a totally different climatic and hydrologic setting, the study of Leopold *et al.* (1966) indicated a high sediment production by sheetwash in Schumm's Zone I and modest contributions from channel extension and migration; in fact most of the arroyo beds experienced net aggradation over the period of study. As the size of the catchment decreases, the volume of floodplain and low terrace materials decreases and a fairly close relationship would be expected between sediment production rates on hillslopes and sediment delivery rates in channels. A remarkable illustration of this is documented by Hadley and Lusby (1967) in a very small, homogeneous catchment of 12 acres (0.05 km^2) in western Colorado. Reservoir survey after a storm event indicated a sediment volume of 0.09 ac-ft (5588 t/km^2). Hillslope measurements from erosion pins indicated 0.11 ac-ft (4572 t/km^2), within a factor of 1.2 of the preceding measurement. Normally, diversity of terrain conditions within catchments as small as 1 km^2 would prevent such close agreement between erosion estimates derived from quite different methods of measurement. A lack of closely coupled responses between the slope and channel systems may be a concern, as already indicated. Even if responses are closely coupled, there remains the problem of assessing the partial contributions of various sub-areas within the catchment to the total sediment output.

Two types of installation have been used by geomorphologists to resolve this problem: (1) those designed to measure the flux of material across a unit width of contour using Gerlach troughs or fractional-acre plots (Costin *et al.* 1960, Hayward 1969, Campbell 1970, Bovis 1978); (2) those designed to measure net surface lowering directly with erosion pins or microtopographic surveying (Schumm 1964, Leopold *et al.* 1966, Campbell 1974). Painted markers or tracers have also been used; however this technique does not measure soil loss since in many cases tracers are substantially larger than the mean grain diameter of soil within the experimental plot. Whichever method is used, the scale of study shifts from the catchment scale (typically 10^6 m^2 and above) to the plot scale (10^0–10^2 m^2). The actual area sampled is a very small fraction of the total catchment; therefore the number of plots and their precise locations can exert a strong influence on the results of the study (Boughton 1967).

To reduce bias and improve areal coverage some studies have first defined 'strata' or relatively homogeneous sub-areas within which soil loss is expected to be less variable than in the study area as a whole. Strata have been defined from cover type alone or from a combination of variables commonly cited as soil loss controls: for example, slope angle, grain-size characteristics and proportion of total plot area without vegetation. Strata provide a framework for stratified, random sampling; however, truly random plot locations may be difficult to obtain in the field, particularly when topographic maps are used to determine the sample points, or where logistical constraints exist (Bovis 1978). Sites may be more precisely positioned when largescale air photographs or maps are used (Hayward 1969, Campbell 1970). In most cases, all that can be reasonably claimed is a 'representative' sample of sites but this may prove difficult to substantiate. Provided that sites have not been deliberately selected, some level of statistical analysis is probably justified.

Individual data points (plots) are almost certain to yield a biased estimate of the mean erosion rate in a particular terrain unit; therefore replication of plots is desirable since the local mean and variance can be computed. Comparison of local and zonal variations of soil loss may then be possible (Hayward 1969, Bovis 1978). Alternatively, a tentative order-of-magnitude ranking of several terrain units may be possible in terms of annual soil loss (Bovis & Thorn 1981).

The topic of spatial validity of soil loss data has been neglected in geomorphic studies and there are instances in the literature in which erosion rates are published with little or no areal control. This, in turn, has led to comparisons of computed rates between different environments with scant questioning as to the validity of the individual estimates. Nevertheless, in many instances plot-scale studies are the only reliable source of information concerning primary denudation rates.

SAMPLING DESIGN AND STATISTICAL ANALYSIS

In the non-agricultural areas normally investigated by geomorphologists there are relatively few data on the spatial variability of soil loss, and the variables which control it, over relatively small, homogeneous terrain units (less than 0.5 km^2). In many studies of soil loss or tracer movement, the explanation of the observed movement in terms of site controls and climatic inputs usually takes precedence over the study of spatial variation. Since process mechanisms can be studied at virtually any site, the precise location of a plot is often not critical. For example, there is very little reference to areal sampling design in the recent 'process' studies by Bryan *et al.* (1978), Imeson *et al.* (1980) and Pearce (1973).

The first requirement in an areal study of soil loss is a method of stratifying the study area. In many cases, detailed ground surveys will not be available and spatial strata must be based on information from small-scale maps which produce very generalized boundaries relative to studies at plot scales. Site groups which are similar in terms of cover type, bedrock, slope angle and soil type at the map scale may be heterogeneous when characteristics are measured at the plot scale (Bovis 1978). On the basis of this experience, there is no substitute for a thorough preliminary survey of slope, soils and vegetation factors over the proposed study area, which argues strongly in favour of small catchments (1 km^2 or smaller). Cover type has been identified as an important control of soil loss in so many studies that it is probably the most suitable variable for defining strata. There exists the possibility of statistical confounding of cover type with other factors such as slope angle, slope length and soil erodibility, which appear in standard soil loss equations. These are possible sources of appreciable 'within-groups' variance of soil loss if cover type alone is used to define strata. Multivariate cluster analysis of control variables measured at the plot scale produces a complex stratification and may result in clusters with non-contiguous plots. This is illustrated by the fact that dry alpine tundra sites showed greater similarity with dry montane forest sites than with tundra meadow sites (Bovis 1978).

Stratification of study plots allows statistical comparison of mean soil loss rates from different terrain units using one-way analysis of variance:

$$X_{ij} = \mu + \alpha_i + e_{ij} \qquad (1.1)$$

in which μ is the grand mean across all plots, α_i the mean soil loss of the ith stratum and e_{ij} the deviation of the jth plot value from the ith mean. As Hoel (1971, p. 296) points out, a stratified, random design will also provide a more precise estimate of the grand mean than will simple random sampling. Although the design in Equation 1.1 has a long history of use in the agricultural literature, its use in erosion studies in geomorphology is by no means widespread. Nested designs such as that presented by Slaymaker (1972) are also amenable to this type of analysis.

Replication of plots enables the spatial validity of the areal sampling to be assessed. An appropriate analysis-of-variance model is:

$$X_{ijk} = \mu + \alpha_i + \beta_{ij} + e_{ijk} \qquad (1.2)$$

where group i contains j sites, at each of which k replicate plots exist. In both models it is assumed that samples are randomly generated within strata and that replicates are also randomly positioned within a 'replicate sample space'. As noted earlier, the assumption of randomness cannot usually be met at all sites. The approach used by Bovis (1978) was first to establish a representative sample of ten sites within each stratum. Then at randomly generated pairs of sites in each stratum, pairs of replicate plots were positioned randomly from the original plot location. Each plot is assumed to function without interference from adjacent plots so that independence of data points can be claimed; otherwise, standard tests of significance cannot be used. The use of representative sites instead of strictly random locations should be borne in mind when statistical results are presented, as Moseley and O'Loughlin (1980) point out. An alternate approach is to avoid the problem of unsatisfied statistical assumptions and compute a mean rate for each terrain unit from which an areally weighted, mean denudation rate can be derived.

Judgments as to adequate sample size require that the variance of soil loss in each sub-area be known in advance. The required accuracy of the project must also be specified. Ideally, a pilot study could be conducted to obtain rough estimates of this parameter; sample sizes could then be adjusted to ensure a roughly constant standard error of the mean in each group. Such an approach is obviously time consuming. There is now sufficient evidence that sample distributions of soil loss and tracer displacement tend to the log-normal (Caine 1968, Bovis 1978). This is due to the low probability of uphill displacement coupled with the possibility of very large downhill displacements or yields at the most active sites. (However, Kwaad (1977, Table VI) indicates that uphill transport by rainsplash may approach 50% of downhill splash on slopes flatter than about 17 degrees.) Since samples are often not large enough to warrant a goodness-of-fit test to log-normality, the coefficient of variation is used instead (Table 1.1). There is a consistent tendency to right skewness over a range of environments since all coefficient values are higher than 0.5 (Agterberg 1974, p. 211). The implication is that average rates of movement will be seriously biased unless the geometric mean is reported. The studies by Costin *et al.* (1960), Hayward (1969), Caine (1976), Bovis (1978) and Bovis and Thorn (1981) all attempted to obtain a representative sample of sites from a range of cover types and it is interesting to note that relatively high coefficients resulted. All of these studies showed that the cover types associated with highest soil loss occupied a small percentage of each study area. It is likely, therefore, that a small sample of random or representative sites will underestimate both the mean and the variance of soil loss. On the other hand, a small sample of subjectively chosen sites – the norm if one looks at the past 25 years of hillslope research – will probably overestimate the mean and underestimate the variance, since the overriding tendency has been to select relatively active sites. Reduction cf the variance would occur since relatively stable terrain has traditionally been ignored. The bias toward active sites is explained by the need for a favorable signal-to-noise ratio when measurement techniques are relatively crude, and the desirability of obtaining significant changes over what is usually a short period of study.

SPATIAL COHERENCE OF SOIL LOSS AND SOIL LOSS CONTROLS

The large body of literature which exists on the spatial coherence of rainfall is not matched by similar studies of soil loss variation. In the agricultural sciences this is of lesser significance than in geomorphology, since usually the intention is to monitor the erosional effects of different plot treatments. In addition, agricultural terrain conditions are much more spatially uniform than the natural slopes investigated by geomorphologists. The samples in the study reported here are small, but allow estimates to be made of the relative variability of soil loss within different environments. Attention is focused first at the local scale by comparing the responses of replicate plots over areas of approximately 2500 m^2.

Local variation

Ten Gerlach troughs were established in each of the following zones of the east slope of the Front Range: montane forest (Group C), sub-alpine forest (Group D) and alpine tundra (Group E) (Fig. 1.1). The logistical difficulties encountered during the sampling have been described elsewhere and will not be reiterated here (Bovis 1978). At two sites in each stratum, two replicate plots were set up within a 50 m x 50 m replicate sample space, using a co-ordinate origin 4 m along the contour from the original plot. The original sites were established in July 1971; replicates in May 1972. The study was terminated in June 1973. Sites yielding relatively large amounts of material were visited at weekly intervals during the period of maximum soil loss (June through September). Otherwise, sites were visited monthly during the snow-free season.

Soil loss totals, grain-size parameters of eroded soil and precipitation data are reported for replicated sites in Tables 1.2 through 1.7. No grain size data are reported for samples smaller than 0.5 g. Precipitation was measured with small wedge-shaped storage gauges (Tru-chek type). A short calibration run against the standard eight-inch gauge showed no significant variation from the line of equal values, reinforcing the conclusions of Huff (1955).

TABLE 1.1
Coefficient of variation of erosion rates.

Study	Environment	Method of measurement	Sample size	Coefficient of variation	Duration of study
Costin et al. (1960)	alpine, sub-alpine, montane	fractional acre plots	56	2.5	3 years
Leopold et al. (1966)	semi-arid arroyo	erosion pins	45	1.0	5 years
Hayward (1969)	alpine tundra	fractional acre plots	20	1.8	1 year
Campbell (1970)	semi-arid badland	fractional acre plots	9	1.3	3½ months
Caine (1976)	alpine tundra	tracers	91	3.5	4 years
Kwaad (1977)	temperate forest	splash boards	11	1.1	1 year
Bovis (1978)	alpine, sub-alpine, montane	Gerlach troughs	29	2.7	2 years
Bovis & Thorn (1981)	alpine tundra	Gerlach troughs and sediment traps	44	1.9	3 months

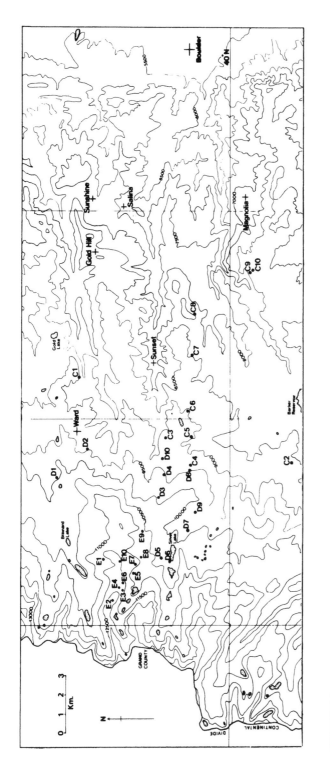

FIGURE 1.1
The study area, Colorado Front Range. Area below 7500 ft not sampled. Area to southwest of Silver Lake excluded for logistical reasons.

A summary of spatial coherence of the four variables at each replicated site is given by the multiple correlation coefficients in Table 1.8. The correlation statistic measures the association between fluctuations of each variable through time across each replicate sample space. Significant correlation therefore implies close proportionality between responses, not necessarily equality. Although the total amounts of soil eroded above each trough differ significantly in some cases, for example the C 8 and D 7 groups, there is moderately good correlation between soil loss fluctuations. Correlations for precipitation are significant, with the important distinction that precipitation values are more similar in an absolute sense than soil loss values. With the exception of the mean grain diameter in the E 6 group (dry alpine tundra), there is no association between fluctuations in the size parameters of eroded soil within each group. The $\bar{X}_e$ parameter in Tables 1.2 through 1.7 shows that most plots consistently

TABLE 1.2
Variation of soil loss and precipitation at montane forest site C 2.

Plot	Date	Soil loss (g)	$\bar{X}_e{}^a$	$S_e{}^b$	Precip- itation (cm)	Date	Soil loss (g)	$\bar{X}_e{}^a$	$S_e{}^b$	Precip- itation (cm)
C 2	4 July	3.9	0.3	2.6	5.2	13 Sept.	23.3	−1.3	1.9	2.7
C 2A	1972	8.3	−0.9	1.5	9.3		32.5	−1.5	1.3	2.7
C 2B		113.2	0.0	1.5	6.9		43.7	−1.5	1.5	3.3
C 2	2 August	4.4	0.2	2.4	3.0	21 Sept.	2.8	−0.9	1.7	<0.1
C 2A		0.2	*	*	3.4		1.2	−0.5	1.7	<0.1
C 2B		57.0	0.2	1.5	3.3		3.0	−0.1	1.8	<0.1
C 2	8 August	14.8	−1.5	2.0	0.7	29 Sept.	0.7	−0.7	2.2	0.1
C 2A		0.1	*	*	0.8		0.4	*	*	0.5
C 2B		24.3	−0.3	1.7	0.8		0.6	0.2	1.5	0.1
C 2	15 August	11.2	−0.6	1.8	0.9	6 Oct.	0.4	*	*	<0.1
C 2A		11.1	−1.1	1.9	1.4		0.3	*	*	<0.1
C 2B		43.9	−0.5	1.6	1.3		1.0	−1.3	1.2	<0.1
C 2	23 August	1.5	0.1	2.7	0.5	17 Oct.	0.2	*	*	0.1
C 2A		0.9	0.4	1.5	1.0		<0.1	*	*	0.3
C 2B		4.9	0.1	1.5	1.0		<0.1	*	*	0.3
C 2	30 August	0.8	1.0	2.0	1.2	27 Oct.	2.1	−2.1	1.4	1.1
C 2A		1.3	0.5	2.0	0.3		1.9	−1.3	0.9	1.5
C 2B		7.6	−0.9	2.0	0.3		5.9	−0.8	1.2	1.4
C 2	5 Sept.	1.4	−0.4	2.1	0.9	12 June	1.1	−1.2	1.5	*
C 2A		12.9	−1.8	1.1	1.7	1973	20.4	−0.3	1.8	*
C 2B		4.6	0.6	1.4	2.0		20.5	−0.6	1.4	*

* No data.
 a Phi-mean grain diameter of eroded soil sample.
b Inclusive graphic standard deviation of eroded soil sample (phi-units).

yield material larger than the mean grain diameter of the plot material (compare with $\bar{X}_p$ values in Table 1.9). This observation, coupled with the poor correlations of the size parameters, suggests that the relatively mobile coarse fraction in the range 2 mm to 8 mm is non-uniformly distributed across many plots. Coarse particles which are one or two phi-units larger than X_p are larger than the mean roughness elements on the granular slopes.

The weighted-mean grain diameter for eroded soil over the period June 1972–June 1973 is computed from:

$$\bar{X}'_e = \Sigma \bar{x}_{ei} w_i / \Sigma w_i \tag{1.3}$$

TABLE 1.3
Variation of soil loss and precipitation at montane forest site C 8.

Plot	Date	Soil loss (g)	$\bar{X}_e{}^a$	$S_e{}^b$	Precip-itation (cm)	Date	Soil loss (g)	$\bar{X}_e{}^a$	$S_e{}^b$	Precip-itation (cm)
C 8	29 June	66.2	−0.6	1.8	2.7	21 Sept.	4.2	−0.1	1.6	0.1
C 8A	1972	21.4	−1.8	1.4	1.1		2.3	−0.5	1.5	0.2
C 8B		5.5	−0.1	1.1	4.3		0.9	0.7	2.0	0.2
C 8	1 August	276.7	1.0	1.8	4.7	28 Sept.	2.1	−0.8	1.5	0.2
C 8A		42.2	−0.7	1.7	4.2		0.6	−0.3	1.1	<0.1
C 8B		13.7	0.2	1.8	3.2		0.3	*	*	<0.1
C 8	8 August	6.8	0.4	2.0	0.3	6 Oct.	10.2	−1.7	1.5	<0.1
C 8A		4.1	−1.5	1.0	0.1		0.4	*	*	<0.1
C 8B		1.1	0.1	1.4	0.1		0.3	*	*	<0.1
C 8	15 August	26.9	0.3	1.9	1.6	26 Oct.	8.9	−1.8	1.5	1.4
C 8A		12.1	−1.2	1.1	1.6		3.2	−1.9	1.4	2.5
C 8B		9.9	−0.2	1.5	1.6		0.5	*	*	0.6
C 8	22 August	6.2	−1.9	1.8	0.6	14 Nov.	2.2	−0.9	1.2	*
C 8A		3.3	−2.0	1.3	0.6		<0.1	*	*	*
C 8B		0.9	0.1	1.2	0.6		<0.1	*	*	*
C 8	29 August	2.9	−0.2	1.4	2.0	8 May	10.5	−1.3	1.3	*
C 8A		3.2	−0.9	1.2	2.1	1973	18.4	−2.3	1.1	*
C 8B		0.8	0.2	1.4	2.0		0.3	*	*	*
C 8	5 Sept.	5.9	−0.6	1.5	1.9	11 June	18.7	−0.7	1.4	4.4
C 8A		4.1	−1.8	1.3	1.7		4.2	−1.5	1.0	2.8
C 8B		1.0	−0.1	1.3	1.8		0.7	0.5	0.9	3.0
C 8	13 Sept.	31.8	−1.0	2.4	1.8	13 June	28.3	0.4	2.1	1.3
C 8A		26.6	−1.0	1.5	1.7		3.3	−1.0	1.4	1.3
C 8B		4.4	−0.5	1.7	1.8		1.3	−0.5	1.8	1.4

* No data.
[a] Phi-mean grain diameter of eroded soil sample.
[b] Inclusive graphic standard deviation of eroded soil sample (phi-units).

where $\bar{x}_{ei}$ and w_i are, respectively, the mean grain diameter and the dry weight of mineral material eroded during period i. This computation is restricted to plots at which six or more data points are available with at least 0.5 g of eroded soil. Plot D 7 and all of the D 10 group are thereby eliminated (Table 1.9). The difference between $\bar{X}_p$ and the weighted mean for eroded soil correlates quite well with S_p, the sorting coefficient of plot soil ($r = 0.51$, yielding $t = 2.35$ for 16 degrees of freedom, significant at the 5% level). This suggests that the greatest divergence between mean sizes of eroded and plot soil occurs where plot soil is most poorly sorted.

The absolute amounts of material yielded through time across each sample space are compared in Table 1.10 by one-way analysis of variance. Data points for a given plot are considered to be independent of one another in two senses. First, plots do not interact with one another in space; distances of separation of several meters assure this. Secondly,

TABLE 1.4
Variation of soil loss and precipitation at sub-alpine forest site D 7.

Plot	Date	Soil loss (g)	$\bar{X}_e$ [a]	S_e [b]	Precip- itation (cm)	Date	Soil loss (g)	$\bar{X}_e$ [a]	S_e [b]	Precip- itation (cm)
D 7	31 July	0.3	*	*	3.0	20 Sept.	0.1	*	*	<0.1
D 7A	1972	7.3	1.8	2.7	1.4		0.8	1.2	2.3	<0.1
D 7B		7.5	−2.6	1.3	1.5		0.1	*	*	<0.1
D 7	7 August	0.5	1.7	2.0	1.8	27 Sept.	<0.1	*	*	<0.1
D 7A		3.5	1.5	2.7	1.3		0.6	0.5	2.0	<0.1
D 7B		2.2	−1.4	1.7	2.0		<0.1	*	*	<0.1
D 7	14 August	0.5	1.0	2.5	0.6	6 Oct.	0.6	−1.9	1.0	0.1
D 7A		2.7	−0.4	1.5	0.1		0.3	*	*	0.1
D 7B		0.3	*	*	0.1		0.4	*	*	0.6
D 7	21 August	<0.1	*	*	1.7	16 Oct.	<0.1	*	*	0.6
D 7A		9.1	−3.1	2.2	1.9		2.0	−2.1	1.5	0.5
D 7B		2.5	−2.8	1.1	2.0		1.2	−2.3	0.4	0.8
D 7	28 August	<0.1	*	*	2.7	24 Oct.	<0.1	*	*	0.9
D 7A		1.9	0.9	2.7	2.7		1.3	−0.6	2.4	1.5
D 7B		2.9	−2.3	0.4	3.2		<0.1	*	*	1.3
D 7	4 Sept.	<0.1	*	*	2.8	18 June	<0.1	*	*	*
D 7A		1.9	0.4	2.3	2.7	1973	4.3	−2.7	1.1	*
D 7B		1.2	−1.6	1.2	3.2		1.1	−2.3	0.3	*
D 7	12 Sept.	<0.1	*	*	0.6					
D 7A		1.8	0.2	2.2	0.9					
D 7B		0.8	−0.0	2.0	0.4					

* No data.
[a] Phi-mean grain diameter of eroded soil sample.
[b] Inclusive graphic standard deviation of eroded soil sample (phi-units).

successive values in time are independent in that they are not constrained by the value of soil loss recorded in the preceding time period. The fact that the major forcing variable, rainfall energy, has a pronounced seasonal cycle does not imply intrinsic dependence between soil loss yields through time. It is possible that the antecedent positions of relatively coarse particles relative to the trough edge could produce some degree of intrinsic dependence; however, this effect was not considered in the study. Data points occur at irregular intervals through time but all plots in a given sample space were visited at the same time. The null hypothesis tested in each case is that the time-averaged responses at each site are identical. All analyses are based on $\log_e$ transformations. Significant differences exist between plots in the C 8, D 7 and D 10 groups and no significant differences at C 2, E 6 and E 9. The relatively uniform response at alpine tundra groups E 6 and E 9 is considered to be due to the 'uniformly heterogeneous' dry alpine tundra surface, and the absence of variable arboreal canopy closure such as exists at montane forest group C 8 and sub-alpine forest groups D 7 and D 10.

The spatial variability of plot characteristics in Table 1.9 is expressed by a simple coefficient of variation statistic in view of the small sample sizes (Table 1.11). Slope angle, sorting coefficient and mean infiltration rate exhibit similar relative variability over the sample spaces. Averaged over all groups, the relative variability of organic content, silt-clay content

TABLE 1.5
Variation of soil loss and precipitation at sub-alpine forest site D 10.

Plot	Date	Soil loss (g)	$\bar{X}_e{}^a$	$S_e{}^b$	Precip- itation (cm)	Date	Soil loss (g)	$\bar{X}_e{}^a$	$S_e{}^b$	Precip- itation (cm)
D 10	3 June 1972	0.3	*	*	*	28 Sept.	0.2	*	*	*
D 10A		0.9	2.1	2.3	0.9		1.3	2.0	2.7	2.0
D 10B		0.8	*	*	1.1		0.5	*	*	6.0
D 10	29 June	0.2	*	*	2.8	26 Oct.	<0.1	*	*	1.6
D 10A		1.3	2.7	2.6	2.9		1.3	<0.0	2.0	2.0
D 10B		0.9	*	*	2.4		0.1	*	*	1.9
D 10	31 July	0.2	*	*	2.9	13 June	0.1	*	*	*
D 10A		1.5	*	*	1.6	1973	0.5	*	*	*
D 10B		0.4	*	*	2.4		0.2	*	*	*
D 10	18 August	0.3	*	*	4.3					
D 10A		1.2	1.4	2.4	1.7					
D 10B		0.3	*	*	2.4					

* No data.
[a] Phi-mean grain diameter of eroded soil sample.
[b] Inclusive graphic standard deviation of eroded soil sample (phi-units).

and bare soil area are similar. The mean grain diameter is most variable, with a strong tendency for highest values where soil losses are high. In view of the small samples, each coefficient is probably an underestimate. Much higher values were noted in the larger-scale sampling carried out by Carson (1967). The soil loss coefficients are generally consistent with the results in Table 1.10.

Variation within site clusters

As reported in Bovis (1978), the similarity-clustering routine used to classify study sites at the beginning of the study used nominal scale data derived from small-scale maps. The analysis yielded three rather loosely

TABLE 1.6
Variation of soil loss and precipitation at alpine tundra site E 6.

Plot	Date	Soil loss (g)	$\bar{X}_e{}^a$	$S_e{}^b$	Precipitation (cm)	Date	Soil loss (g)	$\bar{X}_e{}^a$	$S_e{}^b$	Precipitation (cm)
E 6	6 July	6.6	−1.2	1.0	4.0	20 Sept.	0.5	−0.0	1.2	0.5
E 6A	1972	16.6	−2.0	1.2	2.9		4.7	−1.8	1.2	0.3
E 6B		6.3	−1.6	0.8	1.9		1.1	−0.3	1.3	0.9
E 6	28 July	2.3	−0.9	1.2	1.8	27 Sept.	0.5	−0.5	1.0	0
E 6A		5.0	−1.0	1.1	1.3		3.8	−1.9	1.2	0
E 6B		0.7	−0.8	0.8	3.0		0.7	−1.2	1.1	0
E 6	7 August	1.2	−0.1	1.1	1.0	5 Oct.	1.1	−0.4	0.9	1.0
E 6A		4.0	−1.0	1.1	1.3		5.0	−1.6	1.2	0.5
E 6B		0.6	−0.9	1.2	0.4		1.6	−1.0	1.1	0.4
E 6	14 August	15.4	−1.7	1.0	0.1	16 Oct.	1.1	−2.0	1.1	0.8
E 6A		10.7	−1.9	1.1	0.1		8.0	−1.8	0.7	1.0
E 6B		8.1	−2.2	0.9	0.1		0.2	*	*	0.5
E 6	21 August	6.5	−1.1	1.3	2.2	24 Oct.	1.1	−1.5	1.0	1.3
E 6A		3.5	−1.2	1.1	1.7		2.3	−1.8	0.6	1.0
E 6B		1.4	−1.3	0.9	1.1		0.0	*	*	1.0
E 6	28 August	4.4	−1.5	0.9	1.7	13 Nov.	12.1	−2.8	0.8	*
E 6A		2.9	−1.1	0.8	1.8		1.8	−1.8	0.8	*
E 6B		4.8	−1.7	0.6	1.5		0.2	*	*	*
E 6	4 Sept.	6.7	−1.7	1.2	3.8	17 June	21.4	−2.0	0.9	*
E 6A		1.0	−0.0	1.3	4.2	1973	68.3	−1.8	1.1	*
E 6B		8.8	−2.4	0.9	4.2		20.0	−1.8	0.9	*
E 6	12 Sept.	4.4	−1.3	1.0	1.5					
E 6A		19.7	−2.5	1.1	2.3					
E 6B		8.4	−2.0	1.0	2.3					

*   No data.
[a]   Phi-mean grain diameter of eroded soil sample.
[b]   Inclusive graphic standard deviation of eroded soil sample (phi-units).

TABLE 1.7
Variation of soil loss and precipitation at alpine tundra site E 9.

Plot	Date	Soil loss (g)	$\bar{X}_p{}^a$	$S_p{}^b$	Precip- itation (cm)	Date	Soil loss (g)	$\bar{X}_p{}^a$	$S_p{}^b$	Precip- itation (cm)
E 9	7 July	18.1	−2.5	1.3	5.9	20 Sept.	0.4	*	*	0.5
E 9A	1972	46.3	−4.1	0.4	3.6		0.1	*	*	0.5
E 9B		57.6	−4.3	0.7	4.7		0.3	*	*	0.5
E 9	29 July	21.5	−2.3	2.0	1.9	26 Sept.	0.1	*	*	0
E 9A		6.3	−2.4	1.3	1.0		0	*	*	0
E 9B		13.0	−3.0	1.8	0.8		0.1	*	*	0
E 9	7 August	9.3	−2.1	1.9	1.4	5 Oct.	0.1	*	*	0.5
E 9A		1.2	−2.3	1.6	1.3		0	*	*	0
E 9B		0.9	0.8	1.6	0.6		0.2	*	*	0.8
E 9	14 August	48.3	−2.5	1.1	0.5	16 Oct.	0.0	*	*	0.8
E 9A		7.0	−2.4	1.3	0.8		0	*	*	1.1
E 9B		17.3	−2.9	2.2	0.7		0.1	*	*	1.0
E 9	21 August	3.7	1.5	3.5	2.0	26 Oct.	0.3	*	*	1.1
E 9A		4.4	−3.3	0.7	2.0		0.1	*	*	1.3
E 9B		3.4	1.4	2.3	1.8		0.2	*	*	1.4
E 9	28 August	5.2	0.2	1.5	3.0	17 June	5.9	−3.1	0.7	*
E 9A		2.3	−2.5	1.3	3.0	1973	2.2	−1.5	1.3	*
E 9B		3.3	0.2	2.1	2.8		2.6	0.3	1.3	*
E 9	4 Sept.	3.2	−1.2	2.0	3.3					
E 9A		0.3	0.6	1.4	3.3					
E 9B		1.2	1.2	1.6	3.3					
E 9	12 Sept.	2.1	0.5	1.5	1.3					
E 9A		0.9	−0.1	1.8	1.1					
E 9B		1.7	0.8	1.7	1.0					

* No data.
[a] Phi-mean grain diameter of eroded soil sample.
[b] Inclusive graphic standard deviation of eroded soil sample (phi-units).

TABLE 1.8
Multiple correlations.

Plot group	Soil loss	N	$\bar{X}_e$	N	S_e	N	Precip- itation	N
C 2	0.61*	13	0.66	9	0.18	9	0.96**	11
C 8	0.87**	15	0.51	11	0.42	11	0.96**	12
D 7	0.65**	11	0.34[a]	8	0.47[a]	8	0.94**	10
D 10	0.36	7	no data		no data		0.12	5
E 6	0.62*	15	0.78**	12	0.11	12	0.84**	13
E 9	0.87**	14	0.41	9	0.40	9	0.92**	13

* Significant at the 10% level.
** Significant at the 5% level.
[a] Simple correlation coefficient.

knit clusters of sites corresponding roughly with the boundaries of the montane forest, sub-alpine forest and alpine tundra cover zones of Marr (1967). These were eventually reduced to groups C, D and E in this study. Important variation within the montane group occurs between Ponderosa pine–bunch grass communities on south-facing slopes, with high soil loss (C 1, C 2, C 8) and Douglas fir–lodgepole pine–aspen communities on north-facing slopes with much lower soil losses (C 3, C 4, C 6, C 7). Within the alpine tundra zone, important soil loss variation occurs between dry tundra areas (E 5, E 6, E 7, E 9) and tundra meadow communities (E 1, E 2, E 3, E 4, E 8, E 10). Sites were subsequently regrouped using principal-components analysis of plot characteristics. Each plot is then defined in the score space by component scores, from which the Euclidean distances between plots are known. Site clusters can then be formed by collapsing the distance matrix by the method described in Clark (1974). The coherence of each major cluster depicted in Figure 1.2 is measurable by the *F*-

TABLE 1.9
Characteristics of plot soil and eroded soil, replicated sites.

Plot	Slope angle (degrees)	Per cent bare soil	$\bar{X}_p{}^a$	$S_p{}^b$	Soil loss (g) 72–73	$\bar{X}'_e$	Per cent silt/ clay	Per cent organic	Mean infiltration (cm/hr)
C 2	12	22	1.0	3.3	68.6	−1.0	19.7	12.3	4.9
C 2A	10	25	0.2	2.0	91.6	−1.1	7.1	4.9	4.2
C 2B	10	25	0.3	2.2	330.3	−0.3	9.7	2.5	3.7
C 8	19	25	0.3	2.6	517.4	0.2	8.5	4.6	6.6
C 8A	20	32	−0.4	2.1	149.0	−1.3	4.7	4.0	6.3
C 8B	12	24	0.5	2.5	41.5	−0.0	9.6	7.5	5.6
D 7	17	4	0.7	3.5	2.8	*	17.5	27.8	4.3
D 7A	21	8	0.8	3.9	37.4	−0.6	21.2	21.0	4.1
D 7B	12	4	1.8	4.6	20.1	−2.2	31.0	50.0	4.0
D 10	13	1	1.6	3.5	0.9	*	28.4	29.3	3.9
D 10A	11	1	1.6	3.4	7.1	*	26.9	30.4	3.4
D 10B	12	2	2.4	4.3	2.4	*	34.4	25.8	5.1
E 6	13	47	−1.3	2.0	85.2	−1.7	4.9	4.7	5.7
E 6A	11	33	−1.1	2.2	156.8	−1.8	6.3	3.6	6.1
E 6B	12	31	−0.3	2.8	62.8	−1.9	11.7	7.3	5.0
E 9	17	39	−0.4	3.1	118.1	−2.1	12.7	6.9	5.7
E 9A	16	11	1.3	3.5	70.9	−3.6	24.2	8.6	5.2
E 9B	17	27	2.4	3.7	102.2	−3.3	33.4	9.6	no data

$\bar{X}'_e$ = weighted-mean grain diameter of eroded soil, June 1972–June 1973.
$^a\bar{X}_p$ = mean grain diameter of plot soil; $^b S_p$ = inclusive graphic standard deviation of plot soil; both in Φ units and based on the fraction finer than 8 mm.
*Eroded samples generally too small for meaningful grain-size analysis.

statistic, defined by the ratio of between-groups to within-groups variance of scores. There is considerable divergence between clusters defined at the map scale (groups C, D, E) and those derived from component scores. Plots C 4, C 5, C 7, D 5 and E 6B are not included in any of the three major clusters but do not form a coherent sub-cluster. In addition, plot C 2 is separated from replicate plots C 2A and C 2B. These anomalies can be attributed mainly to either very high or very low scores on the silt-clay and organic content variables at these five plots in comparison with plots with which they are similar in many other respects.

Clusters based on plot-scale measurements have a more homogeneous soil loss response than those derived from map-scale data (Table 1.12). The five unclassified plots are excluded from parts B and C of the table. Mean annual soil losses are not significantly different between the montane forest and dry alpine tundra sites in cluster 2. Cluster 1 is composed

TABLE 1.10
One-way analysis of variance of soil loss (replicate groups),
June 1972–June 1973.

Replicate group	Degrees of freedom	F-value	Significance
C 2	2,39	2.58	>10%
C 8	2,46	10.10	< 1%
D 7	2,36	18.83	< 1%
D 10	2,18	10.67	< 1%
E 6	2,41	2.55	>10%
E 9	2,36	0.80	>10%

TABLE 1.11
Coefficient of variation, plot characteristics and eroded soil (replicate groups).

Replicate group	Slope angle (degrees)	Per cent bare soil	$\bar{X}_p{}^a$	$S_p{}^a$	Soil loss (g) 72–73	Per cent silt/clay	Per cent organic	Mean infiltration rate (cm/hr)
C 2	0.11	0.07	0.87	0.28	0.89	0.55	0.78	0.14
C 8	0.26	0.16	3.54	0.11	1.06	0.34	0.35	0.08
D 7	0.27	0.43	0.55	0.14	0.86	0.30	0.46	0.04
D 10	0.08	0.43	0.25	0.13	0.93	0.13	0.08	0.21
E 6	0.08	0.24	0.59	0.18	0.48	0.47	0.37	0.10
E 9	0.03	0.55	1.28	0.09	0.25	0.44	0.16	0.06
Means	0.14	0.31	1.18	0.16	0.75	0.36	0.37	0.11

[a]$\bar{X}_p$ = mean grain diameter of plot soil; [a]S_p = inclusive graphic standard deviation of plot soil; both in Φ units and based on the fraction finer than 8 mm.

mainly of Group D sites and indicates a greater homogeneity within the original sub-alpine forest cluster in comparison with the montane forest and alpine tundra clusters.

Estimates of the percentage of the total study area occupied by each major cover type (Table 1.13) are based on the maps in Krebs (1973) and Komarkova and Webber (1978). The 30 sites are representative of about 74% of all major cover types recognized by Krebs. A weighted-mean annual soil loss for the study area is 17 g. The relatively small area occupied by actively eroding montane forest sites (3.4%) is consistent with a positively skew sample distribution of annual soil loss. By comparison, the weighted-mean summer soil loss from the alpine tundra is 22 g (Bovis & Thorn 1981, Table 6). From this it is concluded that the annual weighted-mean soil loss in the alpine tundra is approximately 1.7 times greater than the weighted regional average, since summer soil loss is 70–80% of the total annual soil loss in both alpine and non-alpine areas (Table 1.13).

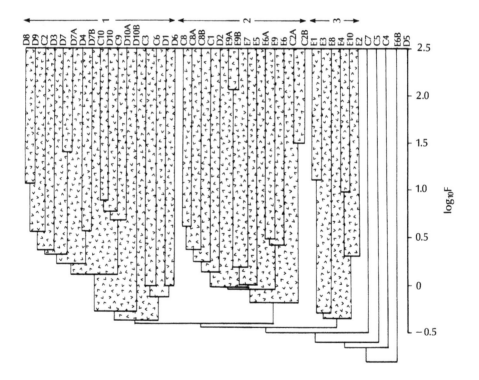

FIGURE 1.2
Component-score clusters.

Selected characteristics of non-replicated plots are listed in Table 1.14. Characteristics in clusters 1, 2 and 3 are generally less variable than in clusters C, D and E (Table 1.15). This is particularly true of the bare soil

TABLE 1.12
One-way analysis of variance, between-cluster soil loss variations, June 1972–June 1973 (log transformation).

Plot clusters	Degrees of freedom	F-value	Significance
A. C, D, E (map-scale)	2,39	3.30	5%
B. 1, 2, 3 (plot-scale)	2,34	23.82	1%
C. Cluster 2 (alpine *vs.* montane)	1,11	2.99	N.S.

TABLE 1.13
Plot clusters based on component scores (see Fig. 1.2).

Cluster number	Plots	Cover types	Geometric mean soil loss 6/72–6/73 (g)	Geometric mean soil loss 6/72–9/72 (g)	Per cent of study area
1	C 2 C 3 C 6 C 9 C 10	dense montane forest communities on north-facing slopes; principally Douglas fir and lodgepole pine			
	D 1 D 3 D 4 D 6 D 7 D 7A D 7B D 8 D 9 D 10 D 10A D 10B	sub-alpine forest; principally Engelmann spruce and sub-alpine fir			
			4.2	3.5	53.1
2 (a)	C 1 C 2A C 2B C 8 C 8A C 8B D 2	Ponderosa pine– bunch grass	178.1	155.6	3.4
2 (b)	E 5 E 6 E 6A E 7 E 9 E 9A E 9B	dry alpine tundra	70.8	42.2	5.7
		(cluster 2 (overall):	116.4	85.2	9.1)
3	E 1 E 2 E 3 E 4 E 8 E 10	tundra meadow	2.1	1.4	11.5
				Total:	73.7

variable, which is most strongly correlated with annual soil loss. The $\overline{X}_p$ parameter is more variable in clusters 1, 2 and 3 and this causes the grand mean coefficient of variation to be higher than in part A of the table. In all other variables there is an overall decrease of about 20% in the relative variability in clusters 1, 2 and 3. The mean coefficients for replicate sites (Table 1.11) are appreciably lower than values in Table 1.15, which points to a progressive increase of the variance of each variable, relative to the mean, as the spatial scale increases. The same effect is noted in the soil loss coefficients and points to a joint decrease in the spatial coherence of both plot variables and soil loss. The soil loss coefficients are generally greater than those of any individual plot variable (Table 1.15). This dis-

TABLE 1.14
Characteristics of plot soil and eroded soil, non-replicated sites.

Plot	Slope angle (degrees)	Per cent bare soil	$\overline{X}_p{}^a$	$S_p{}^a$	Soil loss (g) 1972–73	$\overline{X}'_e$	Per cent silt/ clay	Per cent organic	Mean infiltration (cm/hr)
C 1	22	42	0.4	2.9	932.4	−1.3	12.7	6.2	4.2
C 3	11	1	3.2	3.4	0.3	*	40.2	52.6	7.3
C 4	13	1	1.5	3.5	2.8	*	26.5	35.8	5.8
C 5	6	24	−0.7	1.7	37.4	−1.1	3.0	8.7	5.0
C 6	17	2	2.0	3.3	0.6	*	32.5	48.6	7.0
C 7	22	8	1.4	3.4	5.1	−0.3	22.8	12.4	4.9
C 9	19	2	3.1	3.7	6.1	1.4	39.3	34.7	3.2
C 10	12	10	1.4	3.3	28.2	1.0	23.5	21.4	3.5
D 1	14	1	3.8	3.7	0.8	*	45.4	40.0	5.5
D 2	28	18	−0.3	3.2	73.7	*	13.6	9.3	6.8
D 3	7	9	0.6	3.4	13.2	−1.0	16.3	9.3	4.5
D 4	13	1	0.7	3.8	4.7	−0.7	24.1	33.5	4.3
D 5	19	1	4.6	3.9	0.1	*	61.5	54.9	8.2
D 6	18	1	5.5	3.3	0.4	*	69.6	64.6	6.2
D 8	9	21	2.2	3.2	4.2	*	27.4	21.4	3.7
D 9	9	8	2.3	3.2	5.3	2.3	28.1	13.0	4.0
E 1	8	5	4.7	3.7	0.3	*	57.4	28.2	5.5
E 2	13	18	2.2	3.5	48.0	3.9	30.5	23.3	3.3
E 3	9	1	3.9	4.2	0.6	*	48.4	34.9	4.9
E 4	13	26	−0.5	3.2	8.8	−1.0	13.0	14.0	5.2
E 5	21	25	0.1	3.4	54.3	−1.1	13.4	8.7	2.9
E 7	24	12	3.8	3.4	7.3	−0.6	46.3	9.2	3.9
E 8	16	12	5.5	2.9	1.6	*	73.4	39.9	5.3
E 10	14	11	1.6	3.5	4.8	*	25.1	17.6	4.4

$^a\overline{X}_p$ = mean grain diameter of plot soil; S_p = inclusive graphic standard deviation of plot soil; both in Φ units and based on the fraction finer than 8 mm.
* = Eroded samples generally too small for meaningful grain-size analysis.
$\overline{X}'_e$ = weighted-mean grain diameter of eroded soil, September 1971–June 1973.

crepancy is accounted for by soil loss being a multivariate response to both climatic events and plot variables. In addition, the list of plot variables is not comprehensive.

SOIL LOSS PREDICTION

In view of the wide range of conditions sampled in this study no simple, efficient predictive equation for annual soil loss can be expected. Approximately 65% of the variance of annual soil loss (1972–3) is accounted for by the single term: bare soil per cent times sine of slope angle (Fig. 1.3). The significant relationship on log-log paper ($F = 73.9$ for 1,40 d.f., $r^2 = 0.645$) suggests a power-law relationship:

$$Y = 5.39X^{1.30}$$

where Y is annual soil loss. Separation of the bare soil (X_1) and slope (X_2) terms in the equation:

$$Y = 0.28X_1^{0.67}X_2^{1.36}$$

does not improve the overall predictive power ($F = 75.0$ for 1,40 d.f., $r^2 = 0.659$). Since the tangent and sine functions are similar over the range of slopes in this study, it is worth noting that the exponent of the slope term (1.36) is close to the values found by Zingg (1.4), Musgrave (1.35) and Kirkby (1.35), these results being summarized in Carson and Kirkby (1972,

TABLE 1.15
Coefficient of variation of plot characteristics and soil loss (plot clusters).

Cluster 1. (see Table 1.13)	Slope angle (degrees)	Per cent bare soil	$\bar{X}_p{}^a$	$S_p{}^a$	Soil loss (g) 72–73	Per cent silt/clay	Per cent organic	Mean infiltration rate (cm/hr)
A. C	0.35	0.76	1.16	0.23	1.70	0.68	0.95	0.23
D	0.39	1.15	0.81	0.12	1.65	0.52	0.55	0.28
E	0.30	0.63	1.45	0.19	0.98	0.73	0.75	0.20
Means:	0.35	0.85	1.14	0.18	1.44	0.64	0.75	0.20
B. 1	0.29	1.16	0.63	0.11	1.51	0.42	0.49	0.26
2	0.32	0.38	3.29	0.23	1.31	0.81	0.37	0.24
2(a)	0.40	0.28	2.62	0.18	1.06	0.33	0.41	0.24
2(b)	0.26	0.48	2.80	0.23	0.70	0.76	0.32	0.25
3	0.25	0.74	0.77	0.11	1.74	0.54	0.54	0.17
Means:	0.30	0.61	2.02	0.17	1.26	0.57	0.43	0.23

$^a\bar{X}_p$ = mean grain diameter of plot soil; S_p = inclusive graphic standard deviation of plot soil; both in Φ units and based on the fraction finer than 8 mm.

p. 210). Slope length was also a significant variable in each of these three studies. In the study reported here, the scale of plots relative to the slope profiles on which they occur means that the effect of slope length is masked by local variation in erodibility, controlled mainly by the bare soil variable.

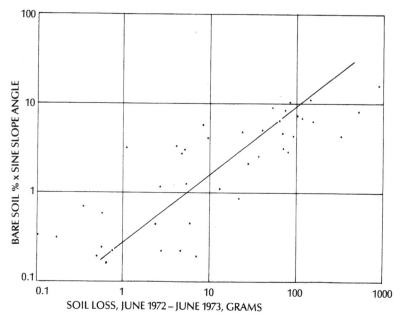

FIGURE 1.3
*Log–log regression, annual soil loss vs. bare soil % x
sine of slope angle.*

TABLE 1.16
Stepwise regression of soil loss on plot variables.

Step	Variable	R^2(%)	Sequential F-value	Significance level	Residual mean-square
1	Bare soil per cent	60.8	62.2	0.1%	0.393
2	Silt-clay per cent	73.9	13.1	0.1%	0.268
3	Sine of slope angle	76.8	2.9	5%	0.245
4	Mean infiltration rate, cm/hr	80.2	3.4	2.5%	0.215

The relative importance of other variables in the statistical prediction of annual soil loss is shown by the stepwise, linear regression results, remembering that the apparent ranking of terms is one which maximizes the regression sum of squares (Table 1.16). Untransformed X-scores are used since some of the variables are not right-skewed. Soil loss scores are $\log_e$ transformed. Silt-clay percentage has a significant inverse correlation with annual soil loss ($r = -0.78$, $t = -7.88$ for 40 d.f.), which reflects the acknowledged greater susceptibility of granular soils to rainsplash transport. The apparent minor status of slope angle and infiltration rate is due to their relatively slight variation over the sample of sites.

DISCUSSION

The single most important problem to be resolved in erosion studies conducted at the plot scale concerns the spatial coherence of erosion rates over relatively small areas. There are still few data concerning the areas to which 'point' estimates of erosion can be realistically extrapolated. This study indicates that extrapolation to areas of approximately 2500 m^2 may be uncertain in forested environments. Better results are obtained in the alpine tundra where variable canopy closure is not a factor. Larger numbers of replicate installations are obviously required in the forested environments.

The study indicates the value of deriving plot clusters from field measurement of attributes, in that the ratio of between-groups to within-groups variance of annual soil loss increases by a factor of seven in comparison with the ratio obtained from clusters based on small-scale maps (compare lines A and B, Table 1.12). The regrouping of plots as component-score clusters also achieves a marked reduction in the spatial variability of most plot characteristics (Table 1.15). Although component-score clusters are based on microscale measurements, they correlate reasonably well with major plant communities, when allowance is made for the obvious twofold division of cluster 2 (Table 1.13). Cover type data as detailed as that now available in Krebs (1973) and Komarkova and Webber (1978) would obviously have benefitted this study in its design stages (1970–71).

In the past ten years there has been a tendency to move away from plot-scale studies on hillslopes; greater effort is now directed at the catchment scale. The problems of areal coverage and signal-to-noise ratio, experienced by many practitioners of plot-scale studies, are generally of much lesser magnitude in catchment studies. The latter have the added advantage of access to a continuous data record and are better able to compare sediment and solute yields. As noted earlier, however, this method measures sediment from all sources within a watershed, and plot studies may provide valuable data on sediment sources as well as areally weighted estimates of primary denudation. The implication is that plots are likely to attain greatest utility within a catchment-scale study, rather than as isolated entities as in this study.

Plots continue to be valuable in relatively small-scale studies of land treatment effects. In the geomorphic literature, the method was used effectively by Caine (1976) to investigate the geomorphic activity of spring snowmelt versus summer rainstorms in alpine areas subjected to winter snowpack augmentation. The value of plot data in testing models of hillslope development is less certain. As noted earlier under soil loss prediction, plots are small relative to the slope profiles on which they occur, so that distance from divide does not emerge as an erosional control. This is to be expected in situations where rainsplash is a dominant agent; however, verification of a distance-independent process would require much shorter slope profiles than existed in this study. Areas with relatively high drainage density (and, therefore, relatively short slope profiles) would seem to offer the greatest scope for small-scale studies of sediment movement through a slope profile.

The complex sampling problems posed by plot-scale studies are insufficient justification for their abandonment in favor of small catchment studies. The implicit equating of sediment yield with primary denudation is probably erroneous in many cases.

ACKNOWLEDGEMENTS

This research was partially funded by two grants from the Penrose Bequest Fund of the Geological Society of America. Logistical support was provided by the Institute of Arctic and Alpine Research, University of Colorado, Boulder. My thanks go to both organizations.

REFERENCES

Agterberg, F. P. 1974. *Geomathematics: mathematical background and geoscience applications.* Amsterdam: Elsevier.

Boughton, W. C. 1967. Plots for evaluating the catchment characteristics affecting soil loss, 1: design of experiments. *N.Z.J. Hydrol.* **6**, 113–19.

Bovis, M. J. 1978. Soil loss in the Colorado Front Range: sampling design and areal variation. *Z. Geomorph.* **29**, 10–21.

Bovis, M. J. and C. E. Thorn 1981. Soil loss variation within a Colorado alpine area. *Earth Surf. Proc. and Landforms,* **6**, 151–63.

Bryan, R. B., A. Yair and W. K. Hodges 1978. Factors controlling the initiation of runoff and piping in Dinosaur Provincial Badlands, Alberta, Canada. *Z. Geomorph.* **29**, 151–68.

Caine, N. 1968. The log-normal distribution and rates of soil movement: an example. *Rev. Geomorph. Dyn.* **18**, 1–7.

Caine, N. 1976. The influence of snow and increased snowfall on contemporary geomorphic processes in alpine areas. In *Ecological impacts of snowpack augmentation in the San Juan Mountains, Colorado*, H. W. Steinhoff & J. D. Ives (eds), 145–200. Final Report No. CSU-FNR-7052-1 to U.S. Bureau of Reclamation, Denver, Colorado.

Campbell, I.A. 1970. Erosion rates in the Steveville Badlands, Alberta. *Can. Geog.* **14**, 202–16.

Campbell, I. A. 1974. Measurement of erosion on badland surfaces. *Z. Geomorph.* **21**, 122–37.

Carson, M. A. 1967. The magnitude of variability in samples of certain geomorphic characteristics drawn from valley-side slopes. *J. Geol.* **75**, 93–100.

Carson, M. A. and M. J. Kirkby 1972. *Hillslope form and process*. Cambridge: Cambridge University Press.

Clark, J. A. 1974. *Mudflows in the San Juan Mountains, Colorado: controls and work*. Unpubl. M.A. thesis, Univ. Colorado, Boulder.

Costin, A. B., D. J. Wimbush and C. Kerr 1960. *Studies in catchment hydrology in the Australian Alps II: surface runoff and soil loss*. Division of Plant Industry. Canberra, Australia: CSIRO

Hadley, R. F. and G. C. Lusby 1967. Runoff and hillslope erosion resulting from a high-intensity thunderstorm near Mack, Colorado. *Water Resources Res.* **3**, 139–43.

Hayward, J. A. 1969. *The use of fractional-acre plots to predict soil loss from a mountain catchment*. Lincoln Papers in Water Resources 7. Lincoln College, Canterbury, N.Z.

Hoel, P. G. 1971. *Introduction to mathematical statistics*. New York: Wiley.

Huff, F. A. 1955. Comparison between standard and small-orifice raingauges. *Am. Geophys. Union Trans* **36**, 689–94.

Imeson, A. C., F. J. Kwaad and H. J. Mücher 1980. Hillslope processes and deposits in forested areas of Luxembourg. In *Timescales in geomorphology*, R. A. Cullingford, D. A. Davidson & J. Lewin (eds), 31–42. New York: Wiley.

Komarkova, V. and P. J. Webber 1978. An alpine vegetation map of Niwot Ridge Colorado. *Arct. Alp. Res.* **10**, 1–29.

Krebs, P. V. 1973. Vegetation. In *Environmental inventory and land-use recommendations for Boulder County, Colorado*. R. F. Madole (ed.). *Occasional Paper* 8, 63–79. Institute of Arctic and Alpine Research, Univ. Colorado, Boulder.

Kwaad, F. J. 1977. Measurement of rainsplash erosion and the formation of colluvium beneath deciduous woodland in the Luxembourg Ardennes. *Earth Surf. Proc.* **2**, 161–73.

Leopold, L. B., W. W. Emmett and R. M. Myrick 1966. *Channel and hillslope processes in a semi-arid area in New Mexico*. U.S. Geol Survey Prof. Paper 352-G.

Marr, J. W. 1967. *Ecosystems of the east slope of the Front Range in Colorado*. Studies in Biology 8. Boulder: Univ. Colorado Press.

Moseley, M. P. and C. O'Loughlin 1980. Slopes and slope processes. *Prog. Phys. Geog.* **4**, 97–106.

Pearce, A. J. 1973. *Mass and energy flux in physical denudation: defoliated areas, Sudbury, Ontario.* Tech. Report 75-1. Montreal: Dept. of Geological Sciences, McGill University.

Schumm, S. A. 1964. Seasonal variation of erosion rates and processes on hillslopes in western Colorado. *Z. Geomorph.* **5**, 214–38.

Schumm, S. A. 1977. *The fluvial system.* New York: Wiley-Interscience.

Slaymaker, O. 1972. Patterns of present sub-aerial erosion and landforms in mid-Wales. *Trans, Inst. Br. Geogs.* **55**, 47–68.

Slaymaker, O. and H. J. McPherson 1977. An overview of geomorphic processes in the Canadian Cordillera. *Z. Geomorph.* **21**, 169–86.

2

Soil–geomorphic models and the spatial distribution and development of alpine soils

Scott F. Burns and Philip J. Tonkin

INTRODUCTION

Geomorphological processes affect spatial and temporal changes to the land surface (Thornes & Brundsen, 1977), and consequently influence the distribution and development of soils. The intimacy of these relationships is illustrated by soil–geomorphic studies (Butler 1959, Ruhe 1969, Daniels *et al.* 1971). Models of soil landscape and soil stratigraphic relationships derived from these studies have been used to infer past and interpret present environments.

This paper discusses the results of a study of the spatial and temporal relationships of soils and landscapes with respect to the environmental factors of topography, climate, vegetation, parent material and time in the alpine environment of the Colorado Front Range. The alpine environment, the ecological zone above treeline, is divided into three geomorphic provinces, and the relative significance of time–space models is evaluated for each province. The Synthetic Alpine Slope model illustrates the use of soil–geomorphic models in interpreting and mapping the spatial distribution and relative development of soils within one of the alpine geomorphic provinces. From this model, processes presently modifying the landscape can be inferred and the results of these processes described in a quantitative manner.

USE OF SOILS BY GEOMORPHOLOGISTS

Geomorphologists and pedologists have an overlapping interest in the results of processes affecting changes to the landscape (Daniels *et al*, 1971). The land surface or soil surface is essentially the same whether it is called a geomorphic surface (Daniels *et al*. 1971), a ground surface (Butler 1959), or a pedomorphic surface (Dan & Yaalon 1968). The geomorphic surface and associated soils together reflect the climatic and geomorphic history of the region in which they evolved, for both were exposed to the processes of subaerial weathering at the same time. The term geomorphic surface will be used in this paper.

In studying these similar objects, with similar goals, geomorphologists and pedologists, with their slightly different orientations, have developed different concepts, each with its own nomenclature. To many geomorphologists, the slope is the basic equilibrium unit, while the catena (Milne 1935), the soil landscape body (Schelling 1970), the soil landscape unit (Huggett 1975), or the soil geomorphic unit (Ruhe 1974) are the pedologist's basic equilibrium units.

Jenny (1941) emphasizes that soils are products of the interaction of the spatial 'state factors' of climate, parent material, organisms and topography, and the temporal factor of time. Daniels *et al*. (1971) stress that distribution and development of both soils and geomorphic surfaces are products of the interaction of these same five 'state factors'.

Geomorphic history of a site after the landform was established is also the history of the soil continuum of that site (Morrison 1978). Soil-stratigraphic studies are used to understand the spatial distribution and textural and mineralogical variation of the soil parent material. History of the past vegetation can be obtained from the study of phytoliths (Lutwick 1969) and plant micro- and macrofossils associated with organic and buried soils. Clues to past history can be inferred from the reconstructed vegetational history and from properties of both present and buried soils. Periods of landscape stability and the shape of the previous topography can be determined from soil-stratigraphic studies and the identification of buried soils (Morrison 1978). Relative ages of the soils can also be estimated using morphological and chemical differentiation of the soil profile. Overall, information from each of the state factors adds to the total knowledge about the system.

Once each factor has been evaluated, an appropriate soil–geomorphic model can be formed that stresses the environmental time–space relationships that predominated while the soil formed (Lavkulich 1969). The model of landscape and soil history, obtained from soil data, allows the geomorphologist to interpret anomalous and/or different site data more successfully. The systems analysis approach helps interpret the past and present environment, and it aids in estimating possible future environmental changes (Ruhe 1975).

SOIL–GEOMORPHIC MODELS

Various soil–geomorphic models have been used around the world. Milne (1935), in his concept of the catena, was one of the first to recognize how landscape evolution can affect soil distribution and properties. Some early models concerned with soils that could not have been developed under the present climate are reviewed by Daniels *et al.* (1971). Butler's (1959) development of the K-cycle model brought attention to the spatial and temporal effects of both erosion and deposition, as recognized by studies of soil stratigraphy and development in the landscape.

Most of the different soil–geomorphic models of the past 20 years have been developed for Quaternary deposits and their associated erosional and depositional geomorphic surfaces. Models have been formulated for coastal regions (Daniels & Gamble 1978), arid regions (Gile *et al.* 1970), volcanic topography (Leslie 1973a, 1973b), forested and unforested slopes (McCraw, 1967, Ruhe 1974, Huggett 1975, Parsons 1978), and mountain environments (Richmond 1962, Birkeland 1967, Tonkin *et al.* 1981). Many models also have been derived for soil–landscape systems in loess, and most are summarized or reviewed by Ruhe (1969, 1974).

Some soil–geomorphic models emphasize temporal relationships, while others stress spatial relationships. In temporal models, such as chronosequences, the landscapes compared have similar climate, vegetation, topography and parent material while the ages of the soils and geomorphic surfaces differ. In the spatial models, the landscapes compared have similar ages, but either climate, vegetation, topography or parent material have varied. Temporal models will always have some spatial variability, and spatial models will always have some temporal variability, as factors can never be totally isolated (Vreeken 1975). For instance, even though a catena is primarily a spatial model of soils, there is some temporal variability on the backslope and toeslope related to the redistribution of solids and solutes within the slope system.

THE K-CYCLE MODEL

The K-cycle model (Butler 1959) is to the pedologist what the Dynamic Metastable Equilibrium model (Schumm 1977) is to the geomorphologist. The unstable period of erosion and deposition in the K-cycle model is equivalent to the geomorphic threshold of the dynamic metastable equilibrium. The K-cycle period of stability when soils form is equivalent to the periods between the thresholds of dynamic metastable equilibrium. Most soil–geomorphic models should be developed within this landscape periodicity model, for it puts soils into a time-space framework.

The fundamental principle of the K-cycle model is the alternation of a phase of instability with a phase of stability. The unstable period of erosion and deposition initiates the K-cycle, with destruction and burial of old surfaces by such events as till production and deposition, mass movement, stream aggradation, dune movement or loess deposition. The stable period follows when soils develop on the new erosion and deposition geomorphic surfaces. These surfaces are called the geomorphic surfaces of one K-cycle. The K_1 geomorphic surface was formed in the first K-cycle back from present, the K_2 geomorphic surface was formed in the second K-cycle back from present, and so on (Fig. 2.1).

Each zone of the K-cycle model is distinguishable by the spatial arrangement of soils and the occurrence of relict, exhumed or buried soils (Ruhe 1975). The **erosion zone** has erosional geomorphic surfaces and soils that show increasing soil morphological and chemical differentiation with increasing geomorphic-surface age. The **deposition zone** is characterized

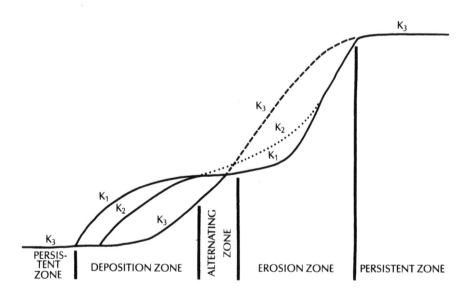

FIGURE 2.1
The K-cycle. K_1, K_2 and K_3 cycle geomorphic surfaces on a hillslope illustrate the relationship of geomorphic surface zones. Names have been modified from Butler (1959).

by chronological sequences of depositional geomorphic surfaces and their associated soils. A key feature of this zone is the occurrence of buried soils within the stratified deposits, denoting previous K-cycles. In the **alternating zone** there is spatial and temporal complexity, where both erosion and deposition may be contemporary processes. Exhumed, truncated, cumulative and polygenetic soil profiles (Birkeland 1974) are present. The **persistent zone** is where either erosional or depositional geomorphic surfaces are of sufficient age that the soil continuum is determined by spatial factors superimposed on the original temporal relationships that governed the distribution of the surfaces. This zone is characterized by soils and/or relict soils.

Though presented as a two-dimensional slope in Figure 2.1, the K-cycle can be represented in all types and sizes of environments. The erosional zone, for instance, could be in the mountains and the depositional zone in the adjoining plains. Soil–geomorphic models based on temporal state factors are of greater significance in the interpretation of soil landscape relationships in the erosion, alternating and deposition zones, whereas soil–geomorphic models formed from spatial state factors are more appropriate to the interpretation of relationships in the persistent zone.

ALPINE GEOMORPHIC PROVINCES

The alpine tundra zone is divided into three geomorphic provinces which are topographically defined, but between which both geomorphic processes and the temporal and spatial pattern of soils differ (Table 2.1, Fig. 2.2). This classification was developed for the Southern Rocky Mountains (Burns 1980), but it should be general enough to be applicable in

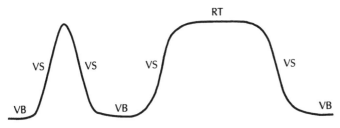

FIGURE 2.2
Alpine geomorphic provinces. *Cross-section of alpine region showing the ridge-top tundra province (RT), valley-side tundra province (VS) and the valley-bottom tundra province (VB).*

other alpine regions with comparable climatic and physiographic conditions. On a macroscale, the three alpine geomorphic provinces as a whole can be interpreted with respect to the K-cycle model.

(a) The *ridge-top tundra province* comprises the broad, rolling ridges (interfluves) that have not been glaciated in late Pleistocene time. This province can be interpreted as a persistent zone which, with the exception of periglacial processes (Benedict 1970), has in recent times been characterized by relative stability. This stability is reflected in the mature soils with deep profiles that characterized this province.

(b) The *valley-side tundra province* is located on the steep slopes of the valley walls surrounding the valley bottoms. This province is analogous to the erosion, alternating and deposition zones of several K-cycles. This province has been undergoing dynamic change, reflected in slope instability and mass wasting throughout the past 10 000 yr as Holocene glaciers receded. This instability is evidenced in the soils and associated geomorphic surfaces which are both temporally and spatially complex, and dominated by young and predominately thin soils intermixed with rock outcrops.

(c) The *valley-bottom tundra province* is primarily present on the cirque floors, but can be present further down valley. This province is equivalent to the major deposition zones of K-cycles during which diamictons of glacial, gravitational and fluvial origins have accumulated on the valley floor. Smaller-scale K-cycles are also recognizable with soils formed on glacially scoured bedrock and in till deposited by Holocene glaciers.

Duration of the stable phases of K-cycles that occur in each province can be measured by a soil's development compared to a steady state. For the Colorado alpine zone, the steady state (Yaalon 1971) is equated to a morphological state of the soil characterized by the development of an

TABLE 2.1
Control of soil variability in the three alpine provinces.

Alpine province	Dominant geomorphic processes	Soil variability control by state factors	K-cycle stable phase, period of duration[a]
ridge-top	periglacial	spatial > temporal	long/medium
valley-side	gravity some glacial	temporal > spatial	short
valley-bottom	glacial fluvial gravity	temporal > spatial	short/medium

[a] 'Long': greater than 15 000 yr; 'medium': 5000 to 15 000 yr; 'short': less than 5000 yr.

A/B/C horizon sequence with a cambic horizon. Where the stable phase of a K-cycle is of sufficient duration that the soil development approaches a steady state, the soil variability is controlled by spatial factors related to topography, such as snow distribution and soil catenary relationships. Conversely, where the stable phase has not had enough time for soil development to achieve a steady state, soil variability is primarily controlled temporally by the ages of the geomorphic surfaces.

'Short' intervals of duration of the stable phases of these alpine K-cycles are considered to be less than 5000 yr; 'medium' lengths are approximately 5000 to 15 000 yr; and 'long' periods are greater than 15 000 yr. The short phases are estimated from glacial soils that have not reached steady state characteristics (Mahaney 1974). Medium-length intervals are characterized by soils that have minimum steady state requirements. Long periods are based on the ages of well-developed glacial soils (Burns 1980).

The lengths of the stable phases of the K-cycles vary in each province. On the ridge-top tundra province, soil variability is primarily spatially controlled because the long and medium K-cycle stable phases have been long enough for the soils to reach a steady state. In the valley-side tundra province where K-cycles are kept short through gravitational processes, soil variability is controlled primarily by temporal causes. In the valley-bottom tundra province, the K-cycle stable phases are mainly of short duration and soil variability is primarily temporally controlled, although some medium length K-cycle stable phases allow for some spatial control.

THE SYNTHETIC ALPINE SLOPE: A SPATIAL SOIL–GEOMORPHIC MODEL FOR THE RIDGE-TOP TUNDRA PROVINCE

Model theory

The purpose of this symposium is to discuss variability between geomorphic thresholds. The soil–geomorphic models for the valley-side and valley-bottom tundra provinces, which are primarily temporal and are therefore strongly influenced by events re-occurring at the geomorphic threshold, are not discussed here. An alpine slope model that concentrates on variability between thresholds has been synthesized for the ridge-top tundra province from observations made during a study of soil distribution and development in the Indian Peaks region of the Colorado Front Range (Burns 1980). The model is based on observed spatial relationships between aspect, topography, seasonal snow accumulation and the distribution of plant communities, alpine loess and soils.

This model, which is a dimensionless measure of topography, is called the **Synthetic Alpine Slope** (SAS). It is named in this manner because it is based on a synthesis of different observations (Caine 1979), it was never hypothesized, and it has not yet been found in its entirety on one slope. Yet, where parts of it are found, the repetition of its characteristics points to its existence. Although similar to the 'mesotopographic unit' of Billings (1973), the SAS contains more specific components.

An understanding of two field relationships led to the derivation of the SAS. First, the direct relationship between alpine vegetation and the soil types beneath them on the ridge-top tundra province, as noted in the literature by Baptie (1968), was established after half of the soil pits of the study had been completed. Second, it had been established that alpine plant communities mimicked winter snow cover (Billings & Bliss 1959, Marr 1967, Komarkova & Webber 1978, Willard 1979). From these two observations, it was concluded that variations in soil characteristics follow topographically controlled variation in snow cover, and that the distribution of the latter indirectly controls the distribution of the soils. The SAS model summarizes this last relationship.

The SAS would be classified as three different types of models by Dijkerman (1974). First, since it is based on observations, it could be called an 'observational model.' Second, it is used as an 'explanatory model' to relate soil distribution and development to yearly snow cover on alpine slopes. Third, it is used as a 'predictive model' in the mapping of alpine soils. Its use as all three types of models has proved quite successful.

Characteristics of the microenvironmental SAS sites

Each of the seven microenvironmental sites of the SAS is defined by an edaphic–topographic–snow cover relationship (Table 2.2, Fig. 2.3).

(a) The *extremely windblown* (EWB) sites are found on the tops of the drainage divides on the ridge crests and knolls. As these sites are extremely dry and windblown, the vegetation is sparse, primarily cushion plants.

(b) The *windblown* (WB) sites, located from the tops to 30% down the drainage divides, are also dry and primarily have mixed vegetation of cushion plants, sedges and herbs.

(c) *Minimal snow cover* (MSC) sites occur in similar positions on the drainage divides as the WB sites, but in the cols and on the large plateaus. They generally have more snow cover in winter than the WB sites and are characterized by thick turf vegetation, primarily kobresia.

(d) *Early-melting snowbank* (EMS) sites are located 30% to 90% of the way down the drainage divides on gentle slopes. These sites are characterized by a great variety of vegetation, and a snowbank generally persists all winter, but usually melts completely by early June.

(e) *Late-melting snowbank* (LMS) sites are present 50% to 90% of the way down the drainage divides in leeward nivation hollows. Normally, the snowbanks melt out in early July and August. Vegetation is sparse, and commonly the sites are quite rocky.

(f) *Semipermanent snowbank* (SPS) sites, which are found 60% to 90% of the way down the drainage divides in nivation hollows, have no vegetation on them, for the snowbanks rarely melt out completely under present climatic conditions. Soils of all three of the snowbank sites have

TABLE 2.2
Site characteristics of the synthetic alpine slope.

Micro environmental sites	Approximate snow-free days per year[a]	Slope angle (degrees)	Mean annual soil temperature (°C at 50 cm)	Associated indicator plant species	Braun–Blanquet groupings of vegetation (Komarkova & Webber 1978)
extremely windblown (EWB)	>300	0-8	+0.5	*Paronychia pulvinata* *Silene acaulis*	Association *Sileno–Paronychietum*
windblown (WB)	225-300	0-10	+1.4	*Carex rupestris* *Dryas octopetala* *Trifolium dasyphyllum*	Association *Potentillo–Caricetum;* Ass. *Trifolietum dasyphyllum;* Ass. *Eritricho–Aretiodis.*
minimal snow cover (MSC)	150-200	0-10	−0.1	*Kobresia myosuroides* *Carex elynoides* *Kobresia myosuroides* with *Acomastylus rossii*	Alliance *Caricion foeneo–elynoidis;* Ass. *Selaginello densae – Kobresietum myosuroidis*
early-melting snowbank (EMS)	100-150	5-15	+0.5	*Acomastylus rossii* *Trifolium parryii* *Deschampsia caespitosa* *Vaccinium* spp.; *Salix* spp.	Order *Trifolio–Deschampsietalia*
late-melting snowbank (LMS)	50-100	10-40[b]	+3.4	*Sibbaldia procumbens* *Carex pyrenaica* *Juncus drummondii*	Order *Sibbaldio–Caricetalia pyrenaicae*
semi-permanent snowbank (SPS)	0	5-40	0	no vegetation	no vegetation
wet meadow (WM)	about 100	0-5	−0.1	*Carex scopulorum* *Pedicularis groenlandica* *Salix* spp. (wet)	Class *Scheuchzerio–Caricetea fuscae;* Class *Betulo–adenstyletea;* Class *Montio–Cardaminetea.*

[a]Burns (1980) and May (1973).
[b]Backslopes.

undergone downslope movement which has produced lobes, terraces and buried soils.

(g) *Wet meadow* (WM) sites are characterized by bog vegetation and are situated below snowbank sites in depressions and on turf-banked terraces and lobes at the bottoms of the drainage divides. Winter snowbanks generally cover these sites, and meltwater from the melting of this snow and from snowbanks upslope contribute to the saturated conditions. Except for the WM sites, snow cover increases downslope. The SAS microenvironmental sites are identified in the field by indicator plant species that have adapted to these different snow covers (Table 2.2).

Similar slope models occur on both windward and leeward slopes (Fig. 2.4). As there is less snow accumulation on windward slopes, LMS and SPS sites generally do not occur there. As a result, the WB and MSC sites extend further down the slope, with the EMS present just above treeline. North- and south-facing slopes form transitions between the windward and leeward slopes. Because all seven SAS sites are located on the leeward slopes, the leeward model (Fig. 2.3) is emphasized in this paper.

Soil characteristics of SAS microenvironmental sites

Soils on the SAS differ physically and chemically, but only the physical characteristics are mentioned here because they are of greatest interest to the geomorphologist. Physically, the SAS soils differ mostly in A horizon characteristics, thicknesses of the eolian deposits in the soils, moisture

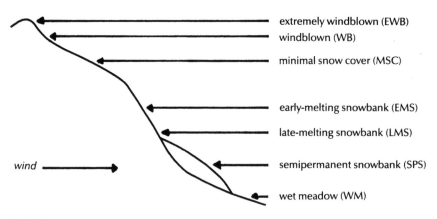

FIGURE 2.3
Synthetic Alpine Slope model. *Microenvironmental sites on the leeward side of the slope.*

regimes, and mean annual soil temperatures. Most soils are deeper than 1 m, and all except those in the SPS nivation hollow show a decrease in clay with depth. Most profiles show an increase in silt with depth. Each SAS microenvironmental site contains one or more soil taxa. Soil data are implicit in the use of soil taxonomic names (Soil Survey Staff 1975) and space limitations preclude extensive descriptions of soil characteristics here.

Eolian-deposited surficial sediments with a texture of loam or finer and a silt mode are called loess. However, these deposits are considered to be a distinctive alpine facies of loess in that they are less sorted and have a more variable fine sand to silt ratio than 'lowland' loess. The variation in sedimentalogical properties can be attributed to the thinness of the loess deposits which subjects them to change due to pedological processes, internal deformation and turbations resulting from freeze–thaw processes, and erosion and re-deposition of these sediments.

Below is a brief description of the dominant soils for each SAS, including their classifications (Soil Survey staff 1975) and relative abundances (in parentheses).

(a) *EWB:* Dystric Cryochrept (90%), Typic Cryumbrept (10%). These poorly developed, well-drained soils have a thin, sandy A horizon over a thin, cambic B horizon. No loess deposits occur on their surfaces.

(b) *WB:* Dystric Cryochrept (80%), Typic Cryumbrept (20%). Poorly developed, well-drained soils present here have thin A horizons that overlie weakly developed but thick cambic B horizons. Patches of loess up to 8 cm thick are common, and some loess is mixed in each A horizon. The A horizons are thicker than common EWB soils, but are not as thick as the A horizons of the MSC sites.

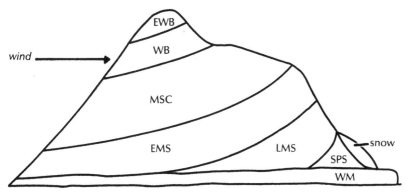

FIGURE 2.4
Synthetic Alpine Slope model on both leeward and windward portions of alpine slopes.

(c) *MSC:* Pergelic Cryumbrept (80%), Dystric Pergelic Cryochrept (20%). These well-drained soils have thick, fine-textured A horizons of eolian origin overlying cambic B horizons and are the best developed soils of the SAS. Except for the soils of the WM units, their A horizons have the highest organic matter content. They have the strongest color development of the SAS soils and the lowest annual soil temperatures.

(d) *EMS:* Typic Cryumbrept (60%), Pachic Cryumbrept (30%), Dystric Cryochrept (10%). The moderately well-drained EMS soils are not as well developed as the soils of the MSC slope positions, but their A horizons are the thickest of the SAS soils of all SAS sites. Pocket gopher activity is greatest here and may be the cause of the overthickened A horizons. Loess is found in the A horizons, but it is not as thick as in the soils of the MSC sites.

(e) *LMS:* Dystric Cryochrept (100%). These poorly developed, moderately well-drained soils have thin A horizons that overlie weakly developed, cambic B horizons. No loess deposition is present on their surfaces. These soils have the warmest mean annual soil temperatures of the SAS.

(f) *SPS:* Lithic Cryorthent (headwall soil), Pergelic Cryoboralf over Pergelic Cryochrept (nivation hollow soil). The headwall soil that forms at the back of the snowbank is the least developed alpine soil and is basically a thin, sandy A horizon over a C horizon. The nivation hollow soil at the base of the snowbank is a sandy soil that has a clay buildup with depth in it that overlies a buried soil of possible Altithermal age (Burns 1979).

(g) *WM:* Pergelic Cryaquept, Humic Pergelic Cryaquept, Histic Pergelic Cryaquept, Pergelic Cryohemist, Pergelic Cryaquoll, Histic Pergelic Cryaquoll (no abundances estimated). These poorly drained soils have either A and/or O horizons of variable thickness overlying gleyed or mottled B and C horizons. They also have cold mean annual soil temperatures with possible permafrost being reported in several localities. These soils contain large amounts of silt and clay which have probably been deposited on higher slopes and snowbanks by wind action and carried downslope by the action of water.

Relative development of the SAS soils

For each of 21 physical and chemical soil characteristics, the dominant soils for each SAS site were compared, ranked for relative soil development, and then the ranks summed (Fig. 2.5, Burns 1980). The rank sums show relative development, with the common soil of the MSC sites being the most developed. This degree of development undoubtedly stems from the fact that MSC sites are the most stable of the SAS sites. In contrast, the other SAS sites are less stable, with the EWB and WB sites being highly affected by wind and water erosion and the EMS, LMS and SPS sites being affected by water erosion and mass wastage. For the upper portion of the SAS, soil development increases as effective wind decreases at the site, and for the lower portion of the SAS, soil development increases as yearly snow cover decreases (Fig. 2.5).

Jenny's (1941) state factors of soil development were investigated for their role in affecting relative soil development and distribution (Burns 1980). The effects of the macroclimate and the lithologic changes in the crystalline bedrock were found to be less significant than other state factors, thus leaving a simpler system to evaluate the interplay of the remaining factors. Time was also judged to be insignificant, for most soils probably have reached a steady state with their microenvironment. Based on the pieces of soil evidence such as soil depths, colors and eolian fines which were compared to the steady state data of Mahaney (1974), it is hypothesized that steady states exist in these soils. A strong correlation between the presence of loess and soil development indicates that eolian fines may have a controlling influence on alpine pedogenesis. Its presence seems to control many soil characteristics, such as amount of fine particles throughout the profile, source of weathering material, and cation additions for plant vigor and chemical reactions. Vegetation and loess together control the thickness of the A horizon and organic matter content. But

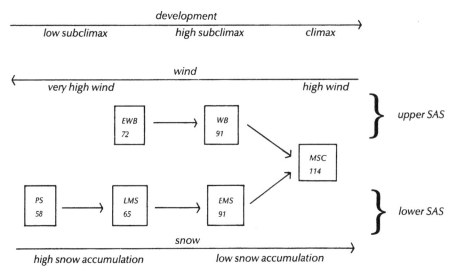

FIGURE 2.5
Successive stages of soil development for SAS soils. Numbers are rank sums of soil development. Maximum rank sum is 147. WM is not listed but it has a value of 100.

overall, topography was determined to be the most important soil-forming factor, for it regulates yearly snow cover and loess distribution, which together control plant communities, soil distribution and soil development.

Subclimax and climax soils of the SAS

In the SAS model for the Indian Peaks, the dominant soil of the MSC site is the most developed and most extensive soil, and therefore it is considered to be the regional climax soil, i.e. the maximum-developed soil for the Front Range alpine climate. The other SAS soils, except for the soils of the WM sites, are considered topographically controlled subclimax soils, i.e. long-persisting soil development stages short of the climax state. These climax and subclimax soils are in a steady state with their respective micro-environments, the latter being controlled by wind, snow and topography. Variations in soils within each SAS site group are a result mainly of changes in vegetational densities. Changes in effective wind for the upper three SAS sites or yearly snow cover for the lower four SAS sites can change soil characteristics and relative soil development of the different subclimax and climax states (Fig. 2.6). Therefore, it is felt that given enough time, the different alpine soils of the SAS will develop to a maximum subclimax stage of development for that site and will remain in a steady state with the microenvironment until the yearly snow cover or wind changes to allow development to a different subclimax or climax level.

Soils in dry alpine areas like the Southern Rocky Mountains do not seem to develop the same characteristics of well-developed soils below treeline and in other mountain ranges. The classical development sequence of soils in dry mountainous areas is Entisol to Inceptisol to Alfisol (Birkeland 1974), and for wet mountainous areas, it is Entisol to Inceptisol to Spodosol (Stevens 1968). The alpine soil development of the Southern Rocky Mountains does not advance past the Inceptisol stage in the climax state, which probably is a result of lack of effective moisture moving through the soil, for they do not develop Bt or eluviated horizons.

Even though the SAS model explains the distribution and development of soil characteristics in the profiles as resulting primarily from spatial factors, there are some temporal components in this model where a K-cycle is noted in the lower four SAS sites. The EWB, WB and MSC soils are all estimated to be at least 15 000 yr old, if not much older. Dated buried soils of 10 400 ± 400 yr BP (Mahaney & Fahey 1980) and 5360 ± 210 yr BP (Burns 1979) have been noted in WM and SPS sites respectively. Other undated buried soils also have been found in LMS and EMS sites. These buried soils that occur in only a few pits show that the lower four SAS units may be younger than the upper three SAS units because of the greater instability of the environment. Yet, it seems these soils have reached a steady state with their microenvironment and therefore a subclimax state. With or without buried soils in the profiles of these lower four SAS sites, soil development is similar, further reinforcing the idea that subclimaxes have been reached for the microenvironment. The ages of the SAS soils are old enough for the soil variability to be explained primarily by spatial factors.

The periods of time between the beginning of K-cycles, or between geomorphic thresholds, is 5000 to 15 000 yr for the lower SAS sites, but is much longer for the upper SAS sites. The MSC sites seem to be very stable sites based on the soil data of deep soils, 7.5YR colors, accumulation of silt and clay in the profile, loess accumulation, and overall development based on the rank sums. Other upper SAS soils show similar soil characteristics to the MSC site soils, but they are not as well developed. Compared to the other two alpine geomorphic provinces, the ridge-top tundra province is the most stable province based on the soil data and has the longest time between geomorphic thresholds.

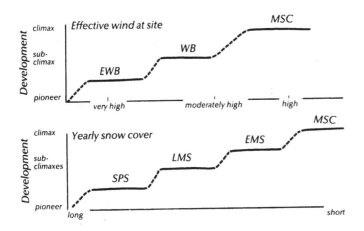

FIGURE 2.6
Steady state levels and soil development on the SAS. Solid lines denote steady state stages and dashed lines denote periods of changing soil characteristics in response to a climatic change. Changes in microclimate can shift soil characteristics from one subclimax to another.

CONCLUSION

Soil data can be important tools that help the alpine geomorphologist assess landform development through supplying data on such things as age, vegetative history, microclimate and parent material. To achieve the most effective interpretation, soil data investigations should include all five state factors of Jenny (1941). The data should be evaluated in the following manner. (a) What is the variability of the soils? (b) Why is there this variability? (c) Once the why is answered, the data can be interpreted using the proper temporal or spatial model.

The alpine environment is complex, and the soils not only vary temporally but spatially, and the researcher must be aware of both possibilities when interpreting the soil data. For instance, care must be taken in interpreting an alpine chronosequence, for it must be checked for any spatial variations as might be recognized under the SAS model. Too many times geomorphologists overemphasize chronosequences when other factor-related models are important in the explanation. The converse is true for SAS interpretations, for temporal variations must be isolated and interpreted. Spatial and temporal variations like these alpine examples also occur in other environments.

Soil–geomorphic models can be used to explain soil variability in the three alpine geomorphic provinces. The K-cycle (Butler 1959) is an important tool in developing soil–geomorphic models, for it provides a time–space framework that facilitates interpretation. Models based on temporal state factors are most effective for soil–geomorphic relationships in the valley-side and valley-bottom tundra provinces because K-cycle stable phases are of short duration. The spatially oriented Synthetic Alpine Slope model has proven to be effective in explaining soil variability in the ridge-top tundra province of Colorado, where K-cycle stable phases are much longer in duration. There, the soils have reached a steady state within their microenvironments.

The Synthetic Alpine Slope model summarizes the spatial pattern of soils that is reflected by the geomorphic positions of the ridge-top tundra province. Pergelic Cryumbrepts, the soils of the **minimal snow cover** sites of the SAS, are the most developed and most abundant soils on the SAS. These soils are the climax soils of the region, with the other SAS soils being at subclimax stages of development when compared to the Pergelic Cryumbrepts. Topography is the most important soil-forming factor, for it controls both snow cover and loess distribution which in turn influence plant communities, soil distribution and the subclimax and climax states of soil development.

The study of alpine soils is important to the geomorphologist. By studying the spatial distribution of soils, one can come to an understanding of the contemporary geomorphic processes and their effect on landscape development within the ridge-top tundra province. These processes are important for they are preparing and preconditioning the landscape very slowly for a change at a new geomorphic threshold. The soil data for the ridge-top tundra province, primarily the MSC sites, point to a stable system with extremely long periods of time between geomorphic thresholds.

ACKNOWLEDGEMENTS

The authors wish to thank Peter Birkeland, Nel Caine and E. J. B. Cutler for their comments on discussions of concepts used in this paper, and Alistair Campbell who diligently reviewed an earlier version of this paper. Logistical support for the field research was supplied by the Mountain Research Station of the University of Colorado and the Institute of Arctic and Alpine Research. The Colorado state office of the Soil Conservation Service was also extremely supportive of the field work. Monetary support for the research came from a NASA grant #PY 06-003-200 to Jack Ives. The Department of Soil Science at Lincoln College was extremely helpful in the production of this paper.

REFERENCES

Baptie, B. 1968. *Ecology of alpine soils of Snow Creek Valley, Banff National Park,* Unpubl. MS thesis, Univ. Calgary.
Benedict, J. B. 1970. Downslope soil movement in a Colorado alpine region: rates, processes, and climatic significance. *Arct. Alp. Res.* **2**, 165–227.
Billings, W. D. 1973. Arctic and alpine vegetations: similarities, differences, and susceptibility to disturbance. *Bioscience* **23**, 697–704.
Billings, W. D. and L. C. Bliss 1959. An alpine snowbank environment and its effect on vegetation, plant development and productivity. *Ecology* **40**, 388–97.
Birkeland, P. W. 1967. Correlation of soils of stratigraphic importance in western Nevada and California, and their relative rates of profile development. In *Quaternary Soils*, R. R. Morrison & H. E. Wright (eds), 71–91. Vol. 9: VII INQUA Congress. Desert Research Institute, Univ. Nevada, Reno.
Birkeland, P. W. 1974. *Pedology, weathering, and geomorphological research.* New York: Oxford University Press.
Burns, S. F. 1979. Buried soils beneath alpine perennial snowbanks may date the end of the Altithermal. *Abstracts with Papers, Rocky Mountain Section, Geol. Soc. Amer.* **11**, 267.
Burns, S. F. 1980. *Alpine soil distribution and development, Indian Peaks, Colorado Front Range.* Unpubl. PhD dissertation, Univ. Colorado, Boulder.
Butler, B. E. 1959. *Periodic phenomena in landscapes as a basis for soil studies.* CSIRO Australia, Soil Pub. 14.

Caine, Nel 1979. The problem of spatial scale in the study of contemporary geomorphic activity on mountain slopes (with special reference to the San Juan Mountains). *Studia Geom. Carpatho-Balcanica* **13**, 5–22.

Dan, J. and D. H. Yaalon 1968. *Pedomorphic forms and pedomorphic surfaces.* Trans Int. Congr. Soil Sci. Adelaide, Australia **IV**, 577–84. Amsterdam: Elsevier.

Daniels, R. B. and E. E. Gamble 1978. Relations between stratigraphy, geomorphology, and soils in coastal plain areas of southeastern USA. *Geoderma* **21**, 41–65.

Daniels, R. B., E. E. Gamble and J. G. Cady 1971. The relation between geomorphology and soil morphology and genesis. *Adv. Agron.* **23**, 51–88.

Dijkerman, J. C. 1974. Pedology as a science: the role of data, models, and theories in the study of natural soil systems. *Geoderma* 11, 73–93.

Gile, L. H., J. W. Hawley and R. B. Grossman 1970. *Distribution and genesis of soils and geomorphic surfaces in a desert region of southern New Mexico.* Madison, Wisconsin: Soil-Geomorphology Field Conference Guidebook, Soil Sci. Soc. Am.

Huggett, R. J. 1975. Soil landscape systems: a model of soil genesis. *Geoderma* **13**, 1–22.

Jenny, Hans 1941. *Factors of soil formation.* New York: McGraw-Hill.

Komarkova, Vera and P. J. Webber 1978. An alpine vegetation map of Niwot Ridge, Colorado. *Arct. Alp. Res.* **10**, 1–29.

Lavkulich, L. M. 1969. Soil dynamics in the interpretation of paleosols. In *Pedology and Quaternary research*, S. Pawluk (ed.), 25–37. Edmonton: University of Alberta Press.

Leslie, D. M. 1973a. Quaternary deposits and surfaces in a volcanic landscape on Otago Peninsula. N.Z. *J. Geol. Geophys.* **16**, 557–66.

Leslie, D. M. 1973b. Relationship between soils and regolith in a volcanic landscape on Otago Peninsula. N.Z. *J. Geol. Geophys.* **16**, 567–74.

Lutwick, L. E. 1969. Identification of phytoliths in soils. In *Pedology and Quaternary research*, S. Pawluk (ed.), 77–82. Edmonton: University of Alberta Press.

Mahaney, W. C. 1974. Soil stratigraphy and genesis of neoglacial deposits in the Arapaho and Henderson Cirques, central Colorado Front Range. In *Quaternary environments*, W. C. Mahaney (ed.), 197–240. Toronto: York University Series in Geography, Geographical Monographs 5.

Mahaney, W. C. and B. D. Fahey 1980. Morphology, composition, and the age of a buried paleosol, Front Range, Colorado, USA. *Geoderma* **23**, 209–18.

Marr, J. W. 1967. Ecosystems of the east slope of the Front Range in Colorado. *Studies in Biology* 8. Boulder: University of Colorado Press.

May, D. C. E. 1973. *Models for predicting composition and production of alpine tundra vegetation from Niwot Ridge, Colorado.* Unpubl. M.A. thesis, Univ. Colorado, Boulder.

McCraw, J. D. 1967. The surface features and soil pattern of the Hamilton Basin. *Earth Sci. J.* **1**, 59–81.

Milne, G. 1935. Some suggested units of classification and mapping, particularly for East African soils. *Soil Res.* **4**, 183–98.

Morrison, R. B. 1978. Quaternary soil stratigraphy–concepts, methods and problems. In *Quaternary soils*, W. C. Mahaney (ed.) 77–107. Third York Quaternary Symposium. Norwich, England: Geo Abstracts.

Parsons, R. G. 1978. Soil-geomorphology relations in mountains of Oregon, USA. *Geoderma* **21**, 25–39.

Richmond, G. M. 1962. *Quaternary Stratigraphy of the La Sal Mountains, Utah.* U.S. Geol. Survey Prof. Paper 324.

Ruhe, R. V. 1969. *Quaternary landscapes in Iowa.* Ames: Iowa State University Press.

Ruhe, R. V. 1974. Holocene environments and soil geomorphology in Midwestern United States. *Quat. Res.* **4**, 487–95.

Ruhe, R. V. 1975. *Geomorphology.* Boston: Houghton Mifflin.

Schelling, J. 1970. Soil genesis, soil classification, and soil survey. *Geoderma* **4**, 165–93.

Schumm, S. A. 1977. *The fluvial system.* New York: Wiley.

Soil Survey staff, 1975. *Soil taxonomy–a basic system of soil classification for making and interpreting soil surveys.* U.S. Dept. of Agric., Soil Conservation Service, Agric. Handbook 436.

Stevens, P. R. 1968. *A chronosequence of soils near the Franz Joseph Glacier.* Unpublished PhD dissertation, Lincoln College, New Zealand.

Thornes, J. B. and D. Brunsden 1977. *Geomorphology and time.* London: Methuen.

Tonkin, P. J., J. B. J. Harrison, I. E. Whitehouse and A. S. Campbell 1981. Methods for assessing late Pleistocene and Holocene erosion history in glaciated mountain drainage basins. In *Erosion and sediment transport in Pacific rim steeplands symposium*, T. R. H. Davies & A. J. Pearce (eds), 527–40. Int. Assn, Sci. Hydd., Publ. 132.

Vreeken, W. J. 1975. Principal kind of chronosequences and their significance in soil history. *J. Soil Sci.* **26**, 378–93.

Willard, B. E. 1979. Plant sociology of alpine tundra, Trail Ridge, Rocky Mountain National Park, Colorado. *Q. J. Colorado Sch. Mines* **74**, 1–119.

Yaalon, D. H. 1971. Soil forming processes in time and space. In *Paleopedology–origin, nature, and dating of paleosols.* D. H. Yaalon, (ed.), 29–39, Int. Soil Sci. Soc. Jerusalem: Israel University Press.

3

The spatial variability of surficial soil movement rates in alpine environments

Nel Caine

INTRODUCTION

The consideration of spatial scale has a long history in geomorphology, particularly with reference to the mapping and classification of landforms (e.g. Tricart 1952, Haggett *et al.* 1965, Derbyshire 1976). Far less work has been done on the problem of spatial scales in the study and modelling of geomorphic processes, although one accidental result of a realist approach to the subject has been a marked reduction in the space- and timescales of interest to the geomorphologist (Chorley 1978). There are, of course, exceptions, for example in the work of Schumm and Lichty (1965) which concerned the way in which cause-and-effect links change with the scale of enquiry, and in work leading to the definition of sediment delivery ratios for hierarchies of drainage basins (e.g. Trimble 1975, Schumm 1977). However, there seems to have been little attempt to take the spatial dimension into explicit account in studies of mass wasting.

In other areas of environmental science, more recognition has been given to the need to define the representativeness of spatially varying characteristics, whether static or dynamic. In soil science, the need to map 'homogeneous' units has led to enquiries which seek to define the amount of variability accommodated within such units (e.g. Beckett & Webster 1971). A similar concern in hydrology derives from the need to estimate catchment volumes (of precipitation, snow cover, soil moisture, etc.) from sets of point measurements (e.g. Miller 1977). Both of these fields are of direct interest to the study of mass wasting rates and processes, since both involve conditions which impinge upon the slope system. Both also suggest that the variability of conditions increases with the area of interest. This is the hypothesis which is tested here by reference to surficial movement rates observed during a period which might be described as one of steady state equilibrium (Schumm 1977).

This concern with the variability of geomorphic activity in space is complementary to that involving mass wasting controls (e.g. Carson 1967) and, like that, is important for three reasons. First, where the statistical distribution is known, variability as a measure of error is needed in experimental design and in any model testing, inference, prediction or retrodiction which derives from it. Second, it is important in extrapolating in a spatial sense, as well as the temporal one. Finally, a knowledge of variability, and its underlying distribution, is basic to the ergodic procedure which requires the assumption that statistical distributions in time and space are identical.

A MODEL FOR SPATIAL VARIABILITY

Two approaches to the estimation of spatial variability are evident in hydrology and pedology. Following the review of Beckett and Webster (1971), the variability of soil characteristics in space has been defined almost exclusively by the coefficient of variability (CV), estimated for pedons, soil series or some larger spatial unit (e.g. Bascomb & Jarvis 1976, Chittleborough 1978, Broadbent *et al.* 1980, Mausbach *et al.* 1980). This is not a good measure of variability, since it is scaled by the mean $(\bar{x})$ and so will reflect change in the position of a statistical distribution as well as its dispersion (Trudgill & Briggs 1978). If the form of the distribution is known to be similar for all the cases considered, the variance (s^2) or standard deviation (s) as a direct measure of scatter is preferable.

Alternatively, and more useful because it retains information that is explicitly spatial in nature, the spatial correlogram or variogram may be used (e.g. Davis 1973, p.384, Clarke 1976). These two are similar in form and inversely related. To my knowledge, they have not been applied to observations of mass wasting rates and only rarely to slope form data (e.g. Thornes 1972), probably because they require a systematic sampling design. However, Clarke (1976) used the correlogram in a spatial analysis of soil moisture volumes. A simpler procedure involves the plotting of interstation correlations (or variances) against the distance across which they are made (e.g. Hutchinson 1969, Greenland 1978, Caine 1979). This gives a curve which appears superficially similar to the correlogram (or variogram), to which it is not a good analog because it is based on a less complete spatial sampling.

If the dispersion of mass wasting measurements is estimated by the variance (s^2) and the distance separating measurement sites (D) is measured directly, the model of Figure 3.1 may be suggested as the expected form of the s^2–D relationship. (The equivalent plot of CV and D would have a similar form, *if the associated means are identical.*) This model for spatial variability is based on three assumptions:

(i)   where $D = 0.0$, $s^2 = 0.0$;
(ii)   where $D > 0.0$, $s^2 > 0.0$;
(iii)   beyond a critical distance (D_c), $s^2 = $ constant.

These combine to suggest a form in which s^2 reaches a limit at D_c, either as an asymptote or an inflexion. Further, in the context of mass wasting at least, we might expect D_c to be a vector variable which changes direction-ally: perhaps greater in the upslope–downslope direction (the direction of sediment flux) than in the cross-slope one.

This model implies that observations which are separated by less than D_c are not statistically independent. Therefore, the estimation of D_c for any set of conditions should be important in the design of experiments and in the evaluation of data derived from them. The objective of this paper is to test the model of Figure 3.1 with empirical data and, if it proves reasonable, to estimate D_c for conditions similar to those from which the data derived.

OBSERVATIONAL DATA

The data for which spatial variances have been estimated derive from three studies reported elsewhere (Caine 1963, 1968, 1976, 1981). A sum-mary of basic results is shown in Table 3.1. All three involve measured rates of surface soil movement based on marked particles and required approximately the same field procedure. These data provide relatively large samples (more than 15 observations) from which the variances as-

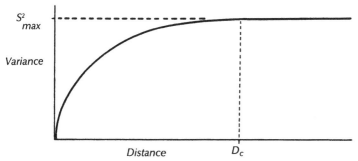

FIGURE 3.1
A model for the change in variance with distance. Beyond a critical distance (D_c), the variance is approximately constant.

sociated with a range of spatial scales may be estimated. Conditions vary greatly within and between the three field areas, but all measurements have been made on unvegetated gravel pavements undergoing quasicontinuous creep in alpine environments. On a single site, individual particles show the marked variability of movement evident in other tracer studies (e.g. Gardner 1979), though neighboring particles tend to move at similar rates (Fig. 3.2). None of the study sites is known to have been affected by catastrophic events during the periods of study and the presence of sorted stripes on some of them suggests that they are responding almost entirely to slow mass wasting processes, including gelifluction and frost creep. Therefore, the data they provide are assumed to be representative of landscape development in a period of steady state equilibrium, even though the mean rates include some high values (Table 3.1).

All data have been converted to annual rates and fitted to the log-normal model (Caine 1968, Carson & Kirkby 1972, p.202). Apart from the pragmatic reason that it appears to normalize the sample distributions, the log-normal may be justified by the 'semi-closed' nature of soil movement data and by the expectation that surface movement should reflect the product of a number of controls, rather than their sum (e.g. as in the **universal soil loss equation**). Clearly, the use of the log-normal model has an important influence on the sample variances which are defined here in $(\log_{10}$ units$)^2$. Perhaps the most important effect is that of making the means and variances independent.

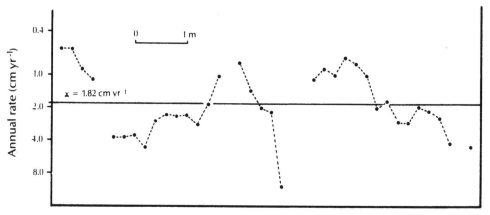

FIGURE 3.2
Mean annual movement rates at the ground surface, Middle Series, Hamilton Crags, Ben Lomond, Tasmania. *Rates are cm yr$_{-1}$ (log-scale) averaged over a 15.3-yr period (1963–79). The 35 particles shown were recovered from 40 originally inserted on this part of the site.*

For individual sites and combinations of them, sample variances have been estimated in the usual manner:

$$s^2 = \Sigma_i(x_i - \bar{x})^2/(N - 1) \qquad (3.1)$$

where the annual rates (x_i') are log-transformed as:

$$x_i = \log_{10}(x_i' + 0.1) \qquad (3.2)$$

and measured as a velocity (cm yr^{-1}). On the same spatial scale, the resulting sample variances are of similar magnitude (Table 3.2), despite the fact that the mean values vary across more than two orders of magnitude (Table 3.1). This indication of the independence of means and variances is supported by variance homogeneity testing, which shows fewer than 10% of the possible pairs of samples in the same area (and on the same scale) to differ ($p = 0.05$) with respect to their variances.

On a single site, i.e. on a smaller scale than that defined by between-site variances, the distance–variability relationship may be examined as a spatial series, since the marked particles were introduced systematically along a line. Unfortunately, testing by correlogram, variogram or power spectrum is not well suited to the problem considered here, since the empirical data provide only short sequences (less than 40 observations) in which missing data are quite common (due to burial or loss of the paint marking).

SPATIAL VARIABILITY

Variance estimates for spatial scales ranging from 1.0 m to 35 700 m are summarized in Table 3.3. They suggest that s^2 tends to increase with distance, more or less following the model of Figure 3.1 (Fig. 3.3). Not only is the pattern repeated in all three field areas, but the scales on which it is defined are surprisingly similar, indicating an unexpected measure of generality for the relationship.

As a simple test of the model, these data have been fitted to two regressions: one for values of D less than 15 m; the other for values of D greater than 3.5 m. In the first case ($D<15$ m), the power function ($y = ax^b$) is preferred because it provides the constraint of $s^2 = 0.0$ where $D = 0.0$. It also gives a slightly better statistical fit to the data, though obviously not a significant one (Table 3.4). For the three field areas, the results are sufficiently similar to suggest that they be pooled, which gives the general relationship (for $D<15$ m):

$$s^2 = 0.027\ D^{0.8} \qquad (3.3)$$

with $r = 0.897$ and $N = 16$. At distances greater than 3.5 m, two of the three correlations prove non-significant (Table 3.4) and the regression coefficients are all close to 0.0, which suggests that s^2 is approximately constant across this range of D. (It may be effectively invariant across a much greater range, since $s^2 = 0.348$ for all three field areas combined, i.e. with $D > 10^7$ m!) Therefore, s^2 is best estimated by its mean values in this range. In that case, the San Juan Mountain study area and the Ben Lomond area do not differ ($t = 0.524$ with 12 degrees of freedom) and may be combined to a single limiting estimate ($s^2 = 0.2$, with s.e. $= 0.02$). The data from the English Lake District suggest a lower limit (mean of $s^2 = 0.07$), perhaps reflecting the shorter period of measurement there: a single winter when the effects of bias (Caine 1981) and the dominance of frost creep may have reduced the measured variability.

The value of D_C (Fig. 3.1) may be estimated by solving the regression equations of Table 3.4 for D, given the limiting values of s^2 (i.e. the means

TABLE 3.1
Soil surface movement studies.

Field Area	Coordinates	Elevation (m)	Range of gradient	Observation period (yr)	Tracer caliber (cm)	Movement rates (cm/yr)	References
San Juan Mts, Colorado	37°47'N 107°33'W	4000	1°–18°	3	1	0.2–8.3	Caine 1976
Ben Lomond, Tasmania	41°35'S 147°40'E	1500	1°–10°	15	3	0.3–4.2	Caine 1968, 1981
Lake District, England	54°35'N 3°15'W	750	9°–15°	0.7	5	8.8–35.5	Caine 1963

TABLE 3.2
Surface movement rates (cm/yr): Eldorado Lake Site 5E.

Plot	N	Mean rate[a] $\log_{10}$ (cm/yr)	Variance (s^2) $\log_{10}$	$\bar{X} \pm 1s$[b] (cm/yr)
1N	15	−0.441 (0.36)	0.101	0.17–0.75
1S	14	0.245 (1.76)	0.186	0.65–4.75
2N	17	−0.201 (0.63)	0.055	0.37–1.08
2S	19	0.159 (1.44)	0.047	0.87–2.38
3N	18	−0.394 (0.40)	0.114	0.18–0.88
3S	20	−0.341 (0.46)	0.107	0.21–0.97

[a]Mean rate: mean of log-transformed annual rates, with geometric mean in parentheses.
[b]$\bar{X} \pm 1s$: range of geometric mean $\pm$ 1 standard deviation.

of s^2 at distances greater than 3.5 m). For the San Juan field area, this yields an estimate of 12.7 m; for the Ben Lomond data, one of 10.3 m; and for these two areas combined, an estimate of 12.3 m. Because of the low estimated limit of s^2, the data from the English Lake District give a much lower estimate of D_c (4.1 m).

On a smaller scale, individual sets of surface movement rates have been analyzed as Fourier series (although, as noted earlier, they are not well suited to this). The middle set of observations at Hamilton Crags, Ben Lomond (Fig. 3.2) is the best of those available for this because of its integrity. Part of this experiment originally involved 40 marked particles of which 35 were successfully identified after 15 years (in analysis, the missing data have been allocated the mean value of the available observations). This gives a series of 8.0 m length with a sampling interval of 20 cm, i.e. a short sequence for analysis of this sort.

TABLE 3.3
Sample variance (log₁₀) and separation distances (m).

Distance (m)	Eldorado Lake, San Juan Mts.	Ben Lomond, Tasmania	English Lake District
0.5	0.0164		
0.9		0.0192	
1.4			0.0284
1.6			0.0225
1.8		0.0839	0.0584
2.0	0.0495		
2.5		0.0730	0.0336
3.0		0.0996	
3.7			0.0732
5.0			0.0690
6.0		0.1398	
7.5		0.1070	
10	0.1759	0.1728	
15	0.1419		
18	0.1531		
20		0.1442	
30		0.1781	
40	0.2136		
120	0.2080		
200			0.0578
500	0.3188		
600			0.0609
1200			0.0618
1500		0.2901	
1600			0.0844
4500		0.2983	
36 700	0.2531		

The results of series analysis on these data are shown as a correlogram and a raw variance spectrum in Figure 3.4. The correlogram suggests a first-order autoregressive sequence, defined on the scale of about 20 cm (a shorter distance than that defined by Clarke (1976) for soil moisture amounts). This is supported by the variance spectrum which shows little contribution to the variance by high order components: the first 7 terms in the Fourier series account for almost 80% of the total variance in the set. Most of the variance is defined in the 3rd and 5th terms (more than 50%) which correspond to periodic distances between 1.75 m and 3.0 m, with little variability associated with distances of 1.0 m or less. This corresponds to the spacing of the zones of more and less rapid motion that are evident in Figure 3.2. Similar results have been obtained from other field sites but are less satisfactorily defined because they are based on shorter sequences, or ones with more missing data.

TEMPORAL VARIABILITY

The variability of surface movement rates with time is not the primary concern of this paper since it has been treated elsewhere (Caine 1981). However, it is important in considering space-for-time substitution. Other workers have noted that the variability of particle movement on debris slopes tends to decrease as the period of observation increases (e.g.

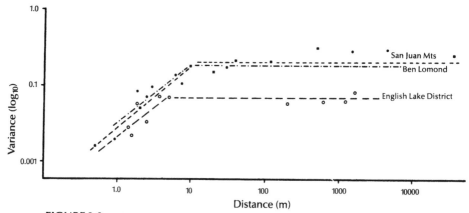

FIGURE 3.3
Movement rate variances and distance in three field areas. ◯, *English Lake District;* ●, *Ben Lomond, Tasmania;* ■, *San Juan Mountains, Colorado. Also shown are the power regressions for estimates at* D < 15 m *in each area and the mean variances (horizontal lines) for* D > 3.5 m. *Regressions are defined in Table 3.4.*

Gardner 1979) and this observation is supported by the data used here. For the Ben Lomond field area, log-variances defined in the time domain (in a manner equivalent to that used above for distance) are summarized in Table 1 of Caine (1981). These show a reduction, usually by a factor of two or more, during the year after insertion of the tracer particles with, predictably, convergence of the long-term variance estimates on those of

TABLE 3.4
Variance–distance regressions.

Field area	N	r	a	b	Range (m)	Model
Eldorado Lake	3	1.0	0.028	0.792	<15	power
Ben Lomond	7	0.884	0.034	0.738	<15	power
Lake District	6	0.767	0.022	0.763	<15	power
Eldorado Lake	3	0.999	0.012	0.016	<15	linear
Ben Lomond	7	0.878	0.041	0.013	<15	linear
Lake District	6	0.772	0.016	0.012	<15	linear
Eldorado Lake	7	0.326	0.202	1.4×10^{-6}	>3.5	linear
Ben Lomond	7	0.833	0.158	3.5×10^{-5}	>3.5	linear
Lake District	6	0.395	0.064	5.9×10^{-6}	>3.5	linear

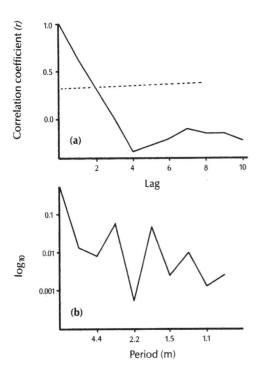

FIGURE 3.4
Series analysis of movement rates, Middle Series, Hamilton Crags, Ben Lomond, Tasmania.
(a) Correlogram, including the 0.05 probability limit on the correlations. (b) Raw variance spectrum. Both are based on $\log_{10}$ – tranformed annual rates with a sampling interval of 20 cm. The data are shown in Figure 3.2, except that the mean value has been inserted for missing data.

Table 3.3 corresponding to the distance across which observations were made. The decline in s^2 with time seems to reflect the experimental bias evident in mean movement rates (Caine 1981) and leads to a correlation between $\bar{x}$ and s^2 during a 'settling period' at the start of observations.

DISCUSSION

These results support the hypothesis that variability in surficial movement rates increases with the spatial scale considered. Although they do not define an unequivocal limiting value of s^2, they do suggest that its rate of increase with distance is reduced, effectively to zero, at about 12 m. Therefore, this length is defined as D_C for unvegetated gravel slopes in alpine environments. Further, the trend in s^2 and the value of D_C appear to be the same for a wide variety of site conditions (which induce marked contrasts in mean movement rates but not in the log-variances associated with them). This indication that the log-mean and log-variance are independent lends support to the use of a log-normal model for mass wasting data (Caine 1968, Owens 1969, Carson & Kirkby 1972, p.202).

The dependence of s^2 on D in measurements made within 12 m of each other suggests a lack of statistical independence which should be important in the design of experiments intended to define rates of surface erosion. First, it means that where observations are more closely spaced than 12 m, there will be redundant information in the results such that the effective sample size may be much smaller than the number of observations actually made. For maximum efficiency in a randomized or block design, the sampling interval and block dimension should probably be 10 m or more. The inverse of this is also true: areas of ground that are more than 12 m from a measurement site will not be represented directly in the sample and their characteristics (in terms of surface movement) will only be capable of definition if the experiment has been randomized. Extending this argument, we might suggest that D_C should define the maximum length of slope profile units used in synthetic modelling (Caine 1979).

This evidence has yet another use in experimental design. With s^2 independent of the mean movement rate and limited at about 0.2 ($\log_{10}$ units)^2, the probable range of observations may be defined and used to estimate sample sizes. (It should be recognized that $s^2 = 0.2$ is equivalent to a 95% confidence range which spans 1.8 orders of magnitude – a wide error envelope.)

The possibility that D_C may be directionally biased was mentioned earlier but not tested directly here. At Site 5E, Eldorado Lake, a set of 6 study plots arranged on a regular 2 x 3 grid with 15 m spacing allows a preliminary test of the cross-slope versus downslope influence (Table 3.2). This suggests that the variability is greater along the slope profiles (plots 1, 2 & 3) than along the contours (plots N & S) for equivalent distances. This is opposite to the result expected if the sediment flux were a homogenizing influence. However, since the grid spacing exceeds D_c, this is not a good test for directional bias. Such a test requires a grid experiment with more plots and spacing of less than 12 m so that D_c in both directions may be defined directly.

The similarity of sampling variances in the time and space domains is important to any procedure based on an ergodic assumption. This evaluation suggests that, for surface debris movement, time and space are not interchangeable on a small scale. The spatial variance increases with distance, up to 12 m; the temporal variance decreases with the period of observation, up to 1.0 yr to 1.5 yr. (Although, the variance in time, as in space, must be 0.0 when the time interval is 0.0!) This suggests that the ergodic assumption may not be justified in short-period, small-site experiments involving surface movement. Over periods longer than 2 yr and distances greater than 15 m, both temporal and spatial variances converge on the same constant estimate and so the assumption of ergodicity with respect to s^2 is reasonable.

The reduced variance in particle movement rates at short distances, and the spatial correlation which it indicates, suggests that the particles on these debris slopes do not move as individuals in response to random events. As on steeper debris slopes (Gardner 1979), particles respond to environmental events with a wider field of operation than that defined by the particle size (in the cases examined here, the field appears to be between 10 and 100 particle diameters in size). These events may reflect soil moisture, surface water flow or freezing conditions, all of which vary locally on about the right scale. Alternatively, the correlated movement pattern may arise from a linked instability such that the movement of one particle predisposes its neighbors to subsequent movement.

Finally, the model of increasing s^2 with D should fit other mass wasting processes, though the parameters defined for surface movement will not be directly applicable to them. Further, empirical estimation of D_c for, say, soil creep or gelifluction is likely to prove difficult because only small data sets may be available for those processes. Experiments to estimate D_c for other processes and conditions are needed. Summer (1980) describes one such experiment involving the distance–variability relationship in the context of a single geomorphic process. Her comparison of 700 paired observations shows a lack of correlation between the erodibility of vegetation-covered soil and distance, perhaps due to the confounding of extraneous effects.

CONCLUSION

This empirical work suggests a spatial correlation and a lack of statistical independence in estimates of one set of geomorphic processes measured over distances of less than 12 m. It also suggests a maximum variance for these data of about 0.2 ($\log_{10}$ units)^2 which should be useful in future experimental design. The apparent generality of these results is perhaps the most surprising outcome of the analysis: data from sites on three continents and with mean movement rates varying across three orders of magnitude all suggest the same conclusions.

REFERENCES

Bascomb, C. L. and M. G. Jarvis 1976. Variability in three areas of the Denchworth soil map unit. 1. Purity of the map unit and property variability within it. *J. Soil Sci.* **27**, 420–37.

Beckett, P. H. T. and R. Webster 1971. Soil variability: a review. *Soils & Fertilisers* **34**, 1–15.

Broadbent, F. E., R. S. Rauschkolb, K. A. Lewis and G. Y. Chang 1980. Spatial variability of nitrogen-15 and total nitrogen in some virgin and cultivated soils. *Soil Sci. Soc. Am. J.* **44**, 524–7.

Caine, N. 1963. Movement of low angle scree slopes in the Lake District, northern England. *Rev. Geomorph. Dyn.* **14**, 171–7.

Caine, N. 1968. The log-normal distribution and rates of soil movement: an example. *Rev. Geomorph. Dyn.* **18**, 1–7.

Caine, N. 1976. The influence of snow and increased snowfall on contemporary geomorphic processes in alpine areas. In *Ecological impacts of snowpack augmentation in the San Juan Mountains, Colorado*, H. W. Steinhoff & J. D. Ives, (eds), 145–200. Final report, San Juan Ecology Project, Colorado State University, Fort Collins.

Caine, N. 1979. The problem of spatial scale in the study of contemporary geomorphic activity on mountain slopes (with special reference to the San Juan Mountains). *Studia Geom. Carpatho-Balcanica* **13**, 5–22.

Caine, N. 1981. A source of bias in rates of surface soil movement as estimated from marked particles. *Earth Surf. Proc. and Landforms*, **6**, 69–76.

Carson, M. A. 1967. The magnitude of variability in samples of certain geomorphic characteristics drawn from valley-slopes. *J. Geol.* **75**, 93–100.

Carson, M. A. and M. J. Kirkby 1972. *Hillslope form and process*. Cambridge: Cambridge University Press.

Chittleborough, D. J. 1978. Soil variability within land systems and land units at Monarto, South Australia. *Aust. J. Soil Res.* **16**, 137–55.

Chorley, R. J. 1978. Bases for theory in geomorphology. In *Geomorphology: present problems and future prospects*, C. Embleton, D. Brunsden & D. K. C. Jones (eds), 1–13. Oxford: Oxford University Press.

Clarke, R. T. 1976. Statistical methods for the study of spatial variation in hydrological variables. In *Facets of hydrology*, J. C. Rodda (ed.), 299–314. Chichester: J. Wiley.

Davis, J. C. 1973. *Statistics and data analysis in geology*. New York: J. Wiley.

Derbyshire, E. 1976. Geomorphology and climate: background. In *Geomorphology and climate*, E. Derbyshire (ed.), 1–24. Chichester: J. Wiley.

Gardner, J. S. 1979. The movement of material on debris slopes in the Canadian Rocky Mountains. *Z. Geomorph.* **23**, 45-57.

Greenland, D. 1978. Spatial distribution of radiation in the Colorado Front Range. *Climat. Bull.* **24**, 1–14.

Haggett, P., R. J. Chorley and D. R. Stoddart 1965. Scale standards in geographical research. *Nature* **205**, 844–7.

Hutchinson, P. 1969. Estimation of rainfall in sparsely gauged areas. *Int. Assoc. Sci. Hyd. Bull.* **14**, 101–19

Mausbach, M. J., B. R. Brasher, R. D. Yeck and W. D. Nettleton 1980. Variability of measured properties in morphologically matched pedons. *Soil Sci. Soc. Am. J.* **44**, 358–63.

Miller, D. H. 1977. *Water at the surface of the Earth*. New York: Academic Press.

Owens, I. F. 1969. Causes and rates of soil creep in the Chilton Valley, Cass, New Zealand. *Arct. Alp. Res.* **1**, 213–20.

Schumm, S. A. 1977. *The fluvial system*. New York: J. Wiley.

Schumm, S. A. and R. W. Lichty 1965. Time, space and causality in geomorphology. *Am. J. Sci.* **263**, 110–19.

Summer, R. M. 1980. *Alpine soil erodibility on Trail Ridge, Rocky Mountain National Park, Colorado*. PhD thesis, University of Colorado, Boulder.

Thornes, J. 1972. Debris slopes as series. *Arct. Alp. Res.* **4**, 337–42.

Tricart, J. 1952. La geomorphologie et la notion d'echelle. *Rev. Geomorph. Dyn.* **3**, 213–18.

Trimble, S. W. 1975. Denudation studies: can we assume stream steady state? *Science* **188**, 1207–08.

Trudgill, S. T. and D. J. Briggs 1978. Soil and land potential. *Prog. Phys. Geog.* **2**, 321–32.

4

Variability in badlands erosion; problems of scale and threshold identification

Ian A. Campbell and John L. Honsaker

INTRODUCTION

In this paper we describe and analyze the effects of the spatial and temporal variability of geomorphic erosional and depositional processes over a ten-year period in an area of badlands in Dinosaur Provincial Park, Alberta. Our emphasis will be, first, to examine how a number of spatially different types of surface respond to externally induced processes by adjustments of form and surface character, especially elevation, through time; second, to discuss the problem of identification of thresholds and their existence, and thereby define response during subthreshold periods, in a landform type that is exceptionally sensitive to variations in geomorphic processes. To do this, and to place this particular study and its significance within the theme of this symposium, it is necessary to define the special role that the study of badlands has in geomorphology.

It is important to note that the focus of this symposium is on the spatial and temporal variability between thresholds (we use the term subthreshold), as opposed to across thresholds and, moreover, to consider this within the context of scale linkages – that is, to show how alterations in spatial and temporal scales present particular difficulties as one moves from the small, or local, situation to the general case. In these terms, the study of badlands is, perhaps, uniquely relevant since badlands are rapidly evolving landforms and may be regarded as small-scale analogs of many arid and semi-arid landscapes.

THE SPECIAL CASE OF BADLANDS

The barren, unprotected nature of the badland surfaces and the frequently non-coherent or poorly consolidated nature of many of the materials, together with their characteristically steep slopes, makes them ideal regions to study the operation and effects of geomorphic processes. Badlands, like all landscapes, develop forms that represent adjustments of their lithologic units and structures to external and internal processes. But, in the case of badlands, these adjustments occur with unbelievable rapidity in a normal geologic sense and there are great variations in the forms produced over very short distances. However, it is largely because of this tempo of response that badlands occupy a special case in the study of landform types, and the kind of information their study provides merits thoughtful consideration before it is incorporated into geomorphic theory.

It is not surprising that badlands have attracted detailed attention. They lie, moreover, within a group of landforms that fit the classification of 'transport-limited' types (Thornes 1976). That is, the processes of erosion are not limited by the availability of weathered material, which is readily available; rather, erosion is dependent upon the frequency and magnitude of sediment-transporting events above some critical limiting threshold. Because of the nature of some of the surfaces, these limiting threshold values tend to be very low and, in general, badlands are especially sensitive to relatively minor inputs of, for example, precipitation. Even small rainstorms of low intensity may produce significant effects, depending upon the character of the surface materials (Bryan & Campbell 1980), though this is not as simple a situation as the preceding statement may imply. In fact, it is the lack of a widespread homogeneity of response that gives badlands their unique topographic expression and quality – dense drainage networks and abrupt changes in slope forms.

The relevancy of these observations to the question of spatial and temporal variability between thresholds is readily apparent. How does one identify a threshold, and thereby define subthreshold responses, in a landscape which may have extremely low threshold values? It is often assumed that badlands represent ideal examples of naturally occurring accelerated erosion. If this is so, can an equilibrium situation exist, or be defined, for it is essentially departures from such equilibrium limits that one seeks as evidence of thresholds (Ritter 1978, p.12)? We will return to this during the discussion on the nature of thresholds and analysis of the badlands data.

There is another important reason, apart from their rapidity of evolution, why badlands act as useful data-collection and process-study landscapes and that is related to their scale. Many of the landforms present are miniature equivalents of much larger components in 'normal' landscapes, and the interrelationships between these small-scale forms and the geomorphic processes which operate on them are conveniently and uniquely expressed in badlands. It is, however, debatable as to how far one can extrapolate with validity from the small to the larger scale, particularly under conditions of accelerated erosion (Bloom 1978, p. 280). Hence, the second theme of the symposium – that is, scale-linkage in both a temporal and a spatial sense – may also be addressed in terms of its practice and appropriateness.

In summary, while the study of badlands possesses undoubted benefits for the geomorphologist, these very advantages (rapidity of response and compactness of scale) present special difficulties. Moreover, the extrapolation of the findings from badlands to other landforms requires caution.

THRESHOLDS AND BADLANDS

Having described some of the special geomorphic character of badlands, we can now define the particular relationships that thresholds and, by implication, subthresholds have to these landforms.

(a)Thresholds define the geomorphic limits between equilibrium and disequilibrium conditions in the landscape. Schumm (1973) regarded extrinsic thresholds as those where, as a result of an increase in external stress, there was a significant response in the geomorphic system. This may be contrasted with intrinsic thresholds (Schumm, 1973) in which thresholds inherent to the system may be exceeded as a result of the release of cumulative stresses which are not triggered by any particular externally derived event. Some mass movements, for example, lie within this category.

Bull (1979) has extended the concept of thresholds to fluvial systems by incorporation of the idea of critical power, i.e. the power needed to transport sediment load. In defining these terms, Bull (1979) compares the role of equilibrium and threshold concepts, noting that equilibrium occurs when a self-regulating feedback mechanism adjusts the system in such a way that changes do not occur with time.

An analogy may be made by considering a ball at rest on the top of a conical form. If the top of the cone has a concave depression in which the ball is free to move (rock), but is restrained from leaving the depression by the edges, then disturbances within certain finite limits will not dislodge the ball and the ball will return to its initial position once the distrubances stop. If the disturbing forces are large enough to keep the ball in motion, yet are never large enough to dislodge it, then the system is in a steady state of dynamic equilibrium. If at some time the disturbing forces are sufficiently large (threshold exceeded) to dislodge the ball entirely so that it seeks a lower level of rest (to a new depression?), then the system has acted as one in dynamic metastable equilibrium. If, however, the upper surface of the conical form has a flat top, then the ball rests in a state of unstable (static) equilibrium since any disturbance will destroy the state of balance. Thus, the balance threshold will be exceeded even by minute disturbances.

(b) The concept of a threshold is linked intimately to the concept of response to some driving force (Garner 1974, p. 26). Probably the most appropriate example is that of critical shear stress – the threshold value of a force which must be applied to move an object. In order to discuss thresholds (or critical shear stress) it is first necessary to have a good idea of the characteristics of that driving force. In particular, is the driving force steady or variable? If it is variable, as in the case of precipitation, what are the details of its intensity distribution or, better, the actual time series? In other words, what is the frequency of occurrence with which that critical threshold is applied? Moreover, are there possible modulating factors involved? This raises the question of complex response to a threshold limit and the role of positive- and negative-feedback mechanisms. Once the threshold is exceeded, is it amplified or dampened by subsequent events? In the case of a surface in the badlands, a storm of a particular intensity and duration may cause the detachment of certain sizes of particles and move them downslope. A subsequent storm of identical character, which is in itself a highly unlikely event, may cause more or less erosion, depending on the precise way in which that particular surface responds to the second storm because of the antecedent moisture conditions generated by the first storm and the time interval between the storms. The surface permeability may have changed in terms of runoff and infiltration so the effects of an identical rainstorm may be quite different. In this event, even very detailed knowledge of rainfall variability will be of little assistance in delimiting or identifying threshold conditions or subthreshold responses, unless the precise pattern of behavior of different surfaces is known over a very wide spectrum of precipitation and antecedent moisture conditions. Additionally, it is evident (Bryan et al. 1978) that dissimilar internal adjustments (intrinsic thresholds) occur in response to similar rainfall intensities and amounts within the same lithologic units, while other units behave quite differently.

(c) The concept of thresholds in geomorphic processes implies an inherently non-linear response of a landform to the process which is trying to alter it. Subthreshold processes may produce an amount of change which is small on some relative scale, but when the threshold or critical level of the process is exceeded, the rate of change suddenly becomes larger. How much change is sufficient to enable us to discern between these conditions? Subthreshold conditions may alternatively be described as the resiliency of the landform; at higher levels the resiliency is lost and the landform could be called 'brittle.' This question is important since it requires the ability to rigorously define the spatial and temporal scale constraints of the network within which the system operates. While an attempt has been made with some success to establish temporal boundaries (Schumm & Lichty 1965), the identification of spatial scale limits appears to be more elusive.

(d) Evidently, different landscapes possess varying degrees of resistance to change and this is expressed in terms of their capacity to absorb externally applied energy or to undergo internal adjustments without observable response. Nevertheless, it is true that change must be continual in all landscapes. Their dimensions and volumes alter with time and at different rates. However, the perception of these changes in form and their variations are seen in the context of the operational scale, both spatial and temporal, at which the landscape is viewed, i.e. from the molecular and intergranular adjustments within and between soil particles to the evolution of valley-side slopes and larger forms.

(e) Seen from this perspective, there probably are not discernible thresholds that are capable of precise definition; rather, there exists a graded continuum in which the cumulative effects of thresholds at smaller scales (spatial and temporal) make their presence felt at the larger. There will, therefore, be a tendency, as the scale increases, for the consequences of cumulative stress to become dramatically more apparent, so that the effects of exceeding certain stress limits or thresholds become more evident as the size of the responding landscape unit increases, or as the time scale at which the landscape is viewed is extended.

For example, Patton and Schumm (1975) found in an area of northwestern Colorado that in drainage basins larger than about 10 km^2, critical threshold valley slopes existed above which gullying was initiated in the alluviated valley floors. Schumm (1977) refers to this as an example of a geomorphic threshold, a special type of intrinsic threshold. The analysis by Patton and Schumm (1975) required the separation of basins larger than about 10 km^2 in area from those which were smaller. This implies that critical threshold values were a function of the size of the landform unit, since below the 10 km^2 limiting value there are almost equal numbers of gullied and ungullied basins and the ungullied ones span a wider range of slope values, including steeper slopes, than do the gullied basins.

(f) If this observation on a graded continuum of thresholds is correct, then it is clearly related to the concept of magnitude and frequency of geomorphic processes as expressed by Wolman and Miller (1960). Here again there exists a continuum from high frequency–low magnitude events to low frequency–high magnitude events.

(g) Together, the two concepts of thresholds and magnitude and frequency of geomorphic processes mean that as the spatial and temporal scale of the landscape increases so the magnitude of threshold which must be attained to produce discernible change must also rise. In effect, the magnitude of thresholds and intervals between related responses increases with the spatial scale of the landscape unit in the same manner in which the concepts of steady, graded and cyclic time (Schumm & Lichty 1965) require a temporal perspective.

(h) In the case of badlands, the incorporation of the above concepts into the nature of their geomorphic evolution patterns is of particular concern. Badlands, or large areas of them, almost by definition must contain many units and slopes which are continually at, or very close to, threshold condition. As 'transport limited' features poised in this position, they are susceptible to transporting events of very low magnitude and, therefore, of comparatively high frequency (Bryan *et al.* 1978, Bryan & Campbell 1980).

(i) Badlands, thus, are seen as potentially disequilibrium landforms since they are characterized by accelerated erosion and are perpetually on the very margins of threshold conditions. Clearly, however, the threshold limit is not constantly exceeded, for even in badlands comparatively lengthy periods of time (days or even weeks) pass during which patterns of visible or measurable change are minimal. But the spacing of the thresholds – that is, the subthreshold period – must be greatly telescoped in comparison with more benign or resistant landscapes.

(j) In badlands, as in all landscapes, there is a nested or multiple series of thresholds that affects different units of the landscape in various ways. The unique quality that badlands possess is in their selective sensitivity and rapidity of response to externally derived stresses. In the badland slope profile, adjacent units respond to similar levels of extrinsic thresholds quite differently, reflecting the durability and intrinsic qualities of any particular lithology. Therefore, their intrinsic thresholds will also vary. There are, then, various thresholds even at the same spatial and temporal scale and this is the reason for the marked variations in erosional rates that are typical of badlands environments.

THE STUDY AREA

The area of badlands that features in this study is in Dinosaur Provincial Park, Alberta (Fig. 4.1). They form part of a 300 km stretch of badlands that fringe the Red Deer River in central and southern Alberta. The badlands in Dinosaur Provincial Park are cut into the Upper Cretaceous Oldman

Formation, a series of channel, levée and floodplain deposits laid down in a vast alluvial and deltaic plain. The badlands form a broad, fretted escarpment along both sides of the Red Deer and have a total relief of about 100 m. They have been cut entirely since the retreat of the Wisconsin ice sheet from the region, about 17 000 years ago, and represent a remarkable example of rapid geomorphic development.

The stratigraphic sequence consists of an almost horizontal series of interbedded sandstones and shales of varying degrees of compaction. Changes in lithology in both vertical and horizontal directions are frequent and abrupt, for there was much fluvial reworking of the sediments during their deposition. Hard, well-indurated and deeply rilled sandstones with typically impermeable surfaces alternate with friable shales that characteristically display a dense network of deep dessication cracks. Slope angles vary from horizontal to near vertical over short distances and high drainage densities dominate many of the outcroppings. The end result is an assemblage of forms that typifies badlands development as it is expressed in a variety of locations throughout the world.

Climate

The climate of this portion of Alberta is semi-arid, with the short warm summers and cold winters that characterize much of the Canadian prairies. In the well-known Köppen system of climatic classificiation, this region of

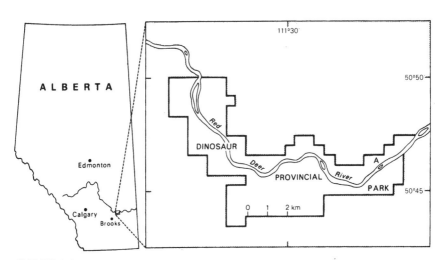

FIGURE 4.1
Map showing location of Dinosaur Provincial Park and study area drainage basin 'A'.

southern Alberta is type BSk (cold winter steppe). Much of the precipitation is concentrated in the summer period (Table 4.1) when, particularly in the June–August period, it falls from locally generated convectional thunderstorms. The rainstorms are of high intensity and great spatial and temporal variability (Longley 1972).

In the case of Brooks (Table 4.1), which has one of the longest and most reliable records near Dinosaur Provincial Park, the values of the probability of a wet day (0.2 mm or more) following a wet day vary from 0.36 (April) to 0.52 (June) over the period April–September; comparable probability values for a dry day following a dry day range from 0.76 (June) to 0.85 (April) over the same April–September period (Longley 1972). These values reflect the fact that April is generally dry and June relatively wet.

It was also determined (Longley 1972) that correlation coefficients of monthly amounts of precipitation, between 24 regional stations, show these are seldom above 0.40 over distances of more than about 300 km and that their correlations in July are only significant for distances of less than 150 km. The variation is a reflection of the localized convective nature of the rainstorms, especially in July.

Field measurement techniques

The measurement techniques and procedures on which this study is based have been previously described (Campbell 1970, 1974); they are reviewed here to make the following statistical analysis fully comprehensible.

A rigid, light-weight, movable frame which rests on permanently installed posts is used as the reference plane from which, in a five by five grid pattern over an area of 1 m², 25 sliding rods touch the ground surface

TABLE 4.1
Climatic data for Brooks, Alberta.

	Jan.	Feb.	Mar.	Apr.	May	June	July	Aug.	Sept.	Oct.	Nov.	Dec.	Year
Mean daily temperature (°C)	−13.1	−10.4	−4.7	5.2	11.6	15.3	19.4	17.5	11.9	6.0	−3.1	−8.8	3.9
Mean rainfall (mm)	0.2	0.5	1.3	10.6	38.8	37.4	37.8	52.1	31.0	8.6	1.5	0.7	240.1
Mean snowfall (mm)	167.0	167.0	195.5	100.9	20.3	—	—	—	17.7	91.4	121.9	142.0	1023.7
Mean precipitation (mm)	17.0	17.3	20.8	21.6	40.8	37.4	37.8	52.1	32.7	17.8	13.7	15.5	344.0

(Fig. 4.2). Measurements are taken to the nearest 0.5 mm from the top of the frame to the top of the rods. It is therefore easy to measure and record changes in the surface height and form of the ground beneath the reference plane.

Nine such plots were installed in 1969 (Fig. 4.3) on a variety of units and slopes (Table 4.2) and twice each year (spring and fall), over the decade 1969–79, the 25 measurements at each plot were recorded. Thus, 225 measurements were taken each time, 450 each year and 4500 over the total experimental period.

Such measurements allow a variety of statistical and graphical techniques, including computer-produced contour diagrams, to be used in interpreting the variations, rates and effects of geomorphic processes.

STATISTICAL COMPARISONS IN SPACE AND TIME

The badlands system would seem to represent a simplified version of the variable list of Schumm and Lichty (1965). The surface position measurements are the dependent variable 'hillslope morphology', and also

FIGURE 4.2
The field measurement frame shown here in position over plot 8. The fragile shale crust is not disturbed by the light-weight rods.

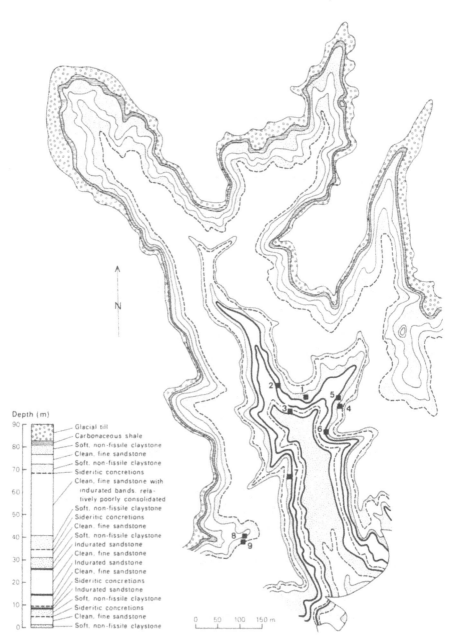

Depth (m)

Glacial till
Carbonaceous shale
Soft, non-fissile claystone
Clean, fine sandstone
Soft, non-fissile claystone
Sideritic concretions
Clean, fine sandstone with
    indurated bands, rela-
    tively poorly consolidated
Soft, non-fissile claystone
Sideritic concretions
Clean, fine sandstone
Soft, non-fissile claystone
Indurated sandstone
Clean, fine sandstone
Indurated sandstone
Clean, fine sandstone
Sideritic concretions
Indurated sandstone
Soft, non-fissile claystone
Sideritic concretions
Clean, fine sandstone
Soft, non-fissile claystone

FIGURE 4.3
*Geology of the study area basin showing the nine plot locations
(source: Faulkner 1970).*

include smaller-scale elements of the drainage morphology. In the list mentioned, this variable is dependent on others only in the dynamic equilibrium case, when any internal or external threshold of the independent variables is being exceeded. In the badlands, the independent variables which act are rainfall (climate), runoff, relief and lithology. The vegetation factor is practically absent.

The point erosion rates

If erosion over all the surfaces was constant and uniform, the successive differences in readings of each of the 225 points would have the same value, assuming equal times elapsed between them. An analysis of variance on these differences enables testing for the effects of time and place on the mean rate of erosion. To clarify the meaning of this analysis, we first examine the components of these two factors. *Time* includes the external variables of precipitation (rainfall and snowmelt runoff) which drive the erosion, and also any internal geomorphic non-linear responses or delays which may be describable as intrinsic thresholds. *Place* includes all differences between the test plots, i.e. relief (elevation and aspect), type and composition of material, and mean slope angle. Rather than the word 'space', we use *place* to better indicate the constraint to locations on a surface. The extrinsic and intrinsic components of the variation of erosion in *time* cannot be distinguished without additional data, e.g. precipitation. The components of *place* can only be distinguished by comparing subsets of the plots, which have some of the components in common.

TABLE 4.2
Plot characteristics.

Plot	Bedrock and surficial materials
1	consolidated floodplain clays and shales with a disaggregated and fractured upper layer 5–10 cm deep
2	alluviated pediment bench covered with laminar deposits of fine sand and silt
3	alluviated pediment bench covered with laminar deposits of fine sand and silt
4	dense compacted channel sandstone and levée deposits
5	well-indurated shale base armored with a thin, scattered cover of small clay-ironstone fragments
6	highly dessicated sandy-shale with a sparse cover of clay-ironstone fragments
7	friable poorly compacted shale with a scattering of clay-ironstone fragments
8	floodplain clays and shales with a fractured upper layer 5–10 cm deep
9	floodplain clays and shales with a fractured upper layer 5–10 cm deep

The results of the analysis of variance on the uncorrected erosion data (directed perpendicularly to the measuring frame, thus roughly normal to the surface) may be stated as follows (Table 4.3). The mean erosion rate is affected by both *place* and *time* (significance at the 5% level). We note that the estimated variance due to *time* (mean sum of squares) is much larger than that due to *place*. The mean rate is also affected by the interaction between *time* and *place* factors, i.e. by the particular erosional history of each place. The estimated variance is similar to (but must be distinguished from) that due to the *place* factor alone. The same conclusions hold when the two alluvial plots (2 and 3) are excluded (their mean erosion is opposite in sign [aggradation] to the others), as well as plot 1 which correlates poorly with the others. The above results lead to these conclusions.

(a) The time series behavior of erosion is not equivalent between places.

(b) The erosion is significantly dissimilar between different places at any time.

(c) The erosion is significantly affected by the previous history of erosion on the surface involved. In other words, the ravages of time occur in different ways at different places. To glean any more information from the data, we must examine the response of each *place* in detail.

Correlations between plots

Further analysis may be made by considering the specific spatial arrangements of the 25 point measurements in each plot.

(a) The overall erosion rate at each plot may be defined as the mean of

TABLE 4.3
Analysis of variance: differences between successive readings; all nine plots included.

	Degrees of freedom	Sums of squares	Mean sums of squares	F statistic
time	19	1420	74.7	178
place	8	95	11.9	28
interaction	152	1028	6.7	16
error	4320	1812	0.4	
total	4499	4356		
	excluding plots 1, 2 and 3			
time	19	1199	63.1	139
place	5	34	6.9	15
interaction	95	349	3.6	8
error	2880	1304	0.4	
total	2999	2888		

all 25 successive differences, scaled to units of volume per unit time per unit level area by dividing by the true difference and by the cosine of the frame slope angle (Fig. 4.4). The regression lines shown on the figure indicate no significant tendency of any of the erosion rates to change over the decade of the measurements.

(b) Roughness is defined as the variance of the residual distances of the 25 points from a planar trend surface through them. It is a measure of the deviation of the surface from a smooth, inclined plane (Fig. 4.5).

(c) A '**clinomorphic ratio**' or **normal/slope** relief is the relief ratio (maximum:mean) taken in a direction normal to the trend plane. It can be considered a measure of shape: near 1 for a strongly rilled or cracked surface, 0.5 for a smoothly undulating slope, and smaller for rocky or clumped surfaces. The applicability is most directly related to the historical-development picture of erosion, where a trend from 1 or 0 toward 0.5 is expected. It is also a point-sample analog of the hypsometric integral (Strahler 1952). The values of this ratio for the plots all lie between 0.35 and 0.75, and are clustered about the expected median of 0.5. There are no evident correlations between plots, nor any apparent systematic changes. We have concluded that this development-oriented variable is less informative on the present time scale than the above two.

(d) The actual slope angle of the trend plane measured from the horizontal. The trends in slope angle are small – only a few hundredths of a degree per year. Again, no consistent pattern appears; five of the plots decreased in slope (1, 3, 6, 7 and 9), numbers 4 and 8 became slightly steeper, and 2 and 5 did not change significantly.

Each of these derived variables was calculated as a time series for each of the plots. The correlation between erosion rates (Table 4.4) is significant between the set of plots numbered 4 through 9, but numbers 1, 2 and 3 are more irregularly correlated. Among the latter, the two alluvial plots (2 and 3) correlate most significantly. If we take the mean erosion rate of

TABLE 4.4
Badlands effective erosion rates: correlation matrix between nine measured plots.

		Plot number								
		1	*2*	*3*	*4*	*5*	*6*	*7*	*8*	*9*
Plot number	*1*	1.0000								
	2	−0.2349	1.0000							
	3	0.0132	0.8406	1.0000						
	4	0.2744	0.4845	0.6694	1.0000					
	5	0.2419	0.5675	0.7397	0.9813	1.0000				
	6	0.3439	0.3543	0.5580	0.8476	0.8415	1.0000			
	7	0.1251	0.5209	0.6271	0.9291	0.9306	0.8462	1.0000		
	8	0.7837	−0.1272	0.1694	0.6083	0.5762	0.6794	0.4940	1.0000	
	9	0.5122	0.0087	0.2724	0.7852	0.7064	0.7354	0.6978	0.7084	1.0000

20 measurements; 18 degrees of freedom.
5% significant, $R = 0.4438$.

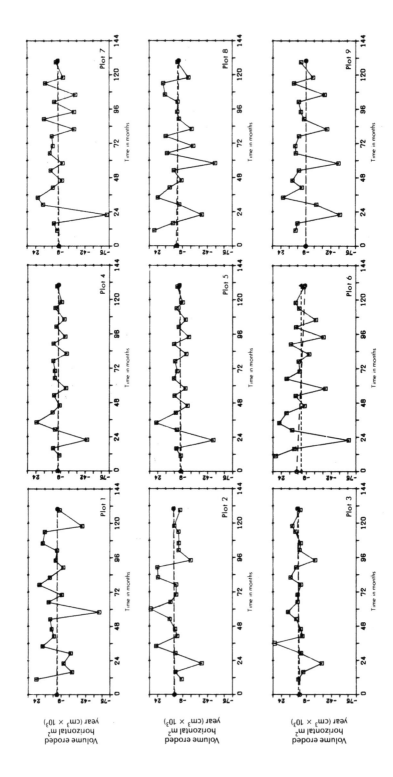

FIGURE 4.4
Overall volumetric erosion rates (surface changes) 1969–79.

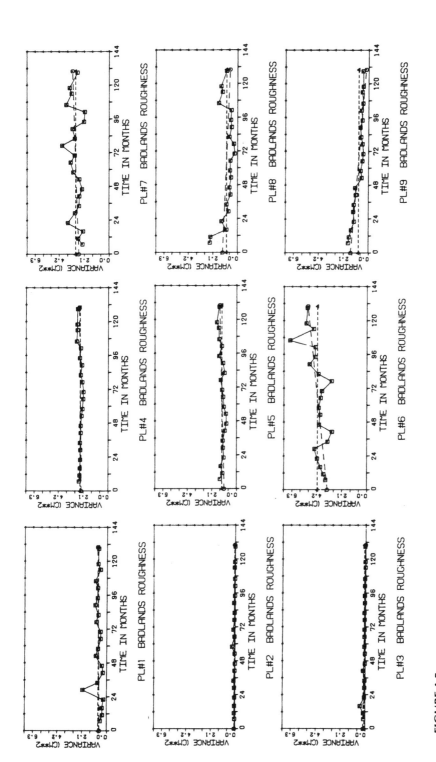

FIGURE 4.5
Roughness variations of the plot surfaces.

plots 4 though 9 as a best estimate for this region of badlands, we get a lowering of 5.34 mm per year. At this rate, it would take nearly 20 000 years to erode down from prairie level to the river, about 100 m below.

RAINFALL AND ITS EFFECTS ON THE BADLANDS

The only sources of erosive shear stress on the badlands formations are neatly divided by season. The most important of these is the rain actually falling on the surface during the summer, together with simultaneous runoff from the surfaces above. Secondarily, the spring runoff from the melting of accumulated winter snow may contribute to a small amount of erosion. One might also mention wind erosion, but this would appear to be unimportant by itself; however, the wind associated with summer thunderstorms, which are the main source of precipitation in the area, may modify the shear stress of the rainwater slightly.

Thus, one is dealing with sporadic events of erosion, occurring on a few days during any year. Our erosion data are cumulative over each season, so we can compare them with seasonally accumulated rainfall. Monthly climatological records are published for six stations in the area, at distances ranging from 25 to 35 km from the study site (Atmospheric Environment Service 1969–76). Monthly rainfall correlates well between all stations, so we are justified in taking a mean value of the six as a good representation of rain on the badlands. There are four of these stations in the vicinity of Brooks, and also one each at the towns of Duchess and Pollockville. The problem of occasionally missing reports disappears when all six are combined.

Longley (1975) has described a correction due to river valley air circulation, which reduces rain in the valley by about 15% below that measured at stations on the prairie surface where the above stations are located. Since this correction is probably independent of season, it is only necessary to note that the actual rainfall in any of the seasons on the site may be reduced by this amount. We are concerned here mainly with relative changes in rainfall, which should not be affected by this factor.

In particular, we are interested in the non-linear response of erosion to the rainfall. Bryan and Campbell (1980) have attempted to assess this by observing sediment flow during certain storms. There is evidence for material transport when the rate of rainfall exceeds threshold values which depend somewhat on the surface variables. The most fruitful analysis of the erosion data comes to light when the events of the rainfall record (Fig. 4.6) are compared to erosion events at corresponding times.

The most concentrated rain in the period 1968–76 occurred in June 1970, when four storms within two weeks deposited 166 mm of rain averaged over five stations. The heaviest of these was on June 16, when 56 mm fell on Brooks-One Tree. The average accumulated rain for the summer of 1970 was 272 mm (to study month 22), at which time all of the plots except number 1 exhibited a strong downward erosion rate. Plots 6 and 7 showed increased roughness at this time, and plot 1 became rougher slightly later. The clinomorphic ratios of plots 1 and 6 decreased and that of 2 increased.

There were two more episodes of heavy rain in the period, occurring in the summers of 1973 and 1975. Plot 1, which was inactive in 1970, responded very strongly to the first of these. The 1975 event was less concentrated than the others, as 282 mm of rain was spread out over three months in this case.

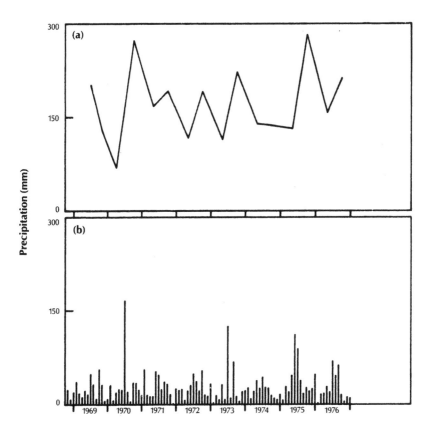

FIGURE 4.6
Rainfall record, 1969–76. *Graph (a) shows the seasonal patterns of precipitation (summer and winter) between the two measurement times; graph (b) shows the mean monthly amounts.*

Differences in erosion rate responses to the heavy rain events

(a) Plot 1 had lowered just prior to the 1970 event, and also responded more strongly than the others to the 1973 event. There was no response in 1975. This highly modified or delayed response may possibly imply an intrinsic threshold mechanism associated with the character of this plot. As the relief and lithologic variables of plots 1, 8 and 9 are similar, it is not clear what factor causes the distinctive response of plot 1.

(b) The alluvial plots, numbers 2 and 3, might be expected to respond oppositely to the higher surfaces which are sources of their material. The upward responses in 1973 and 1975 indeed behaved in this way, but the heavy rains of 1970 eroded them like the other plots. This is a highly non-linear response, which cannot be described by a simple threshold model. One could postulate two thresholds; a lower one applied to aggradation from above, and a higher one for degradation under very large flow rates or shear stress. Alternatively, a cubic polynomial response function could be devised to fit these responses – but this seems somewhat artificial, and requires three parameters plus a threshold!

(c) The best-behaving plots are numbers 8 and 9, which are adjacent to one another and responded to all three events (1970, 1973 and 1975). Their extrinsic rainfall threshold thus seems to be fairly low, and there is probably no intrinsic threshold.

(d) Plots 4 and 5 did not respond in 1973 or 1975. This implies that their extrinsic threshold is higher than those mentioned above. This statement may be qualified, since no inference can be made about intrinsic thresholds.

(e) Plots 6 and 7 responded to two of the three events. This suggests a possible interaction between intrinsic and extrinsic thresholds as in plot 1, or else an extrinsic threshold very near the rainfall intensities which occurred in 1973 and 1975.

The qualitative results of these responses are summarized in Table 4.5.

When the responses to the three events are disregarded, the remaining erosion rates all have similar variance within each plot. This could be taken

TABLE 4.5
Qualitative threshold levels.

Plot	Extrinsic (rain)	Intrinsic
1	high?	present, high?
2 and 3	double, low/high	absent
4 and 5	high	unknown
6 and 7	moderate	possibly present
8 and 9	low	absent

as representative of the subthreshold activity at the scale we are now considering. The residual rates are not strictly random, but show seasonal periodicity for plots 4, 5, 6 and 7. This does not contradict the high thresholds deduced above, for we may assume that we see here the detection of effects from the next-smaller scale. A much smaller threshold associated with this scale is exceeded virtually every year, and produces a low-amplitude erosion.

We are now led to ask whether there are any true subthreshold periods at all. Strictly, these must be limited to a few days or weeks at a time when no rain falls and erosion is suspended, as it also is during the frozen period in winter; however, in the latter case, this is more appropriately potential erosion awaiting moisture release upon meltout. Amounts reported as a 'trace' of rain are probably not sufficient to cause a measurable change on these surfaces. Aside from these conditions, nearly every other rainfall or runoff will have some effect, if viewed on a small enough scale.

CONCLUSIONS AND DISCUSSION

Our 9 plots exhibit a broad range of response characteristics. We cannot claim that this sample exhausts all of the possible responses. Indeed, the catastrophic events such as collapse of pipes or undermined slopes which are evident throughout the area are not represented at all. What we can say is that no single response function linking erosion to rainfall is adequate to represent all of our study plots, thus no single threshold of rainfall can be established for this area. The indirect evidence for intrinsic thresholds is more uncertain, but it seems that spatial irregularity is also characteristic of this kind of landform response, if its existence can be shown.

It is interesting to consider the scales of landforms both in space and time, going from the very small to the very large. On the smallest scale, let us begin just above the molecular processes of erosion of soluble materials.

To move even the smallest grain, the shear stress of water (or wind) must exceed a certain minute value. The timescale associated with such granular motion would reflect discrete jumps, such as saltation, below the region of sediment flow. As the spatial scale is increased, the shear stress threshold would probably be attributed to cohesion of the material, and the slightly slower motion of clumps of matter would set the timescale.

When viewed from the perspective of larger time and space scales, these microthresholds appear so small as to be negligible, and the associated processes are treated as being continuous, or 'steady state.' Here, the observable events are of the kind described in this paper, and time scales range from days to seasons and years. There are, of course, the larger scales of valley systems and mountain ranges above this scale, with their long timescales for development and degradation. Much of the discussion of other authors mentioned above is concerned with thresholds (intrinsic and extrinsic) limiting the evolution of these larger-scale landforms.

Our 1 m^2 plots are midway on the grand logarithmic scale (Table 4.6). This table includes the overall range within which the questions of self-similarity in both space and time may be explored. Of course, there may be some arbitrariness in the matching of space and time units, but with the logarithmic base of 1000, a re-adjustment factor of 10 or so hardly matters!

The question which arises from these musings is whether we have a continuum of scales which appear similar to one another if each one is viewed at the appropriate resolution. Mandelbrot (1977) has described the shapes of rivers and coastlines in these terms. It is conceivable that these so-called self-similar forms can also be found in the space and time dimension together, and that the concepts of responses to driving forces exceeding thresholds extend all the way to 'cyclic' large-scale changes over geologic time, where the threshold stress is so large that it is exceeded only by collisions between continents. If one is careful to observe according to the proper time–space relationship, dynamic metastable equilibrium may be found to describe processes occurring on virtually every scale.

TABLE 4.6
Possible relationships between erosion scales.

Element	Size (m) or time (yr)	Causal event
dust	10^{-6}	shower
particle	10^{-3}	thunderstorm system
slope segment	10^{0}	seasonal cycle
valley side	10^{3}	climatic change
mountain range	10^{6}	plate tectonics

ACKNOWLEDGEMENTS

The authors thank the Natural Sciences and Engineering Research Council of Canada for grant support and Alberta Recreation and Parks for their co-operation.

REFERENCES

Atmospheric Environment Service 1969–76. *Monthly record; meteorologic observations in Canada.* Downsview, Ontario.

Bloom, A. L. 1978. *Geomorphology: a systematic analysis of late Cenozoic landforms.* Englewood Cliffs, NJ: Prentice-Hall.

Bryan, R. B. and I. A. Campbell 1980. Sediment entrainment and transport during local rainstorms in the Steveville badlands, Alberta. *Catena* 7, 51–65.

Bryan, R. B., A. Yair and W. K. Hodges 1978. Factors controlling the initiation of runoff and piping in Dinosaur Provincial Park, Alberta, Canada. *Z. Geomorph.* **29**, 151–68.

Bull, W. B. 1979. Threshold of critical power in streams. *Geol. Soc. Am. Bull.* **90**, 453–64.

Campbell, I. A. 1970. Micro-relief measurements on unvegetated shale slopes. *Prof. Geog.* **22**, 215–20.

Campbell, I. A. 1974. Measurements of erosion rates on badlands surfaces. *Z. Geomorph.* **21**, 127–37.

Faulkner, P. H. 1970. *Aspects of channel and basin morphology in the Steveville badlands, Alberta.* Unpubl. M.Sc. thesis, Univ. Alberta.

Garner, H. F. 1974. *The origin of landscapes: a synthesis of geomorphology.* New York: Oxford University Press.

Longley, R. W. 1972. *Precipitation in the Canadian prairies.* Ottawa: Environment Canada.

Longley, R. W. 1975. Precipitation in the valleys. *Weather* **30**, 294–300.

Mandelbrot, B. B. 1977. *Fractals: form, chance and dimension.* San Francisco: W. H. Freeman.

Patton, P. C. and S. A. Schumm 1975. Gully erosion, northwestern Colorado: a threshold phenomenon. *Geology* 3, 88–90.

Ritter, D. F. 1978. *Process geomorphology.* Dubuque: Brown.

Schumm, S. A. 1973. Geomorphic thresholds and complex response of drainage systems. In *Fluvial geomorphology*, M. Morisawa (ed.), 299–310. London: George Allen & Unwin.

Schumm, S. A. 1977. *The fluvial system.* New York: Wiley.

Schumm, S. A. and R. W. Lichty 1965. Time, space and causality in geomorphology. *Am. J. Sci.* **263**, 110–19.

Strahler, A. N. 1952. Hypsometric (area-altitude) analysis of erosional topography. *Geol. Soc. Am. Bull.* **63**, 975–1007.

Thornes, J. B. 1976. *Semi-arid erosional systems.* Geog. Papers no. 7. London School of Economics and Political Science.

Wolman, M. G. and J. P. Miller 1960. Magnitude and frequency of forces in geomorphic processes. *J. Geol.* **68**, 57–74.

5

The ergodic principle in erosional models*

Richard G. Craig

INTRODUCTION

For at least 20 years geomorphologists have developed successively more complex models of slope erosion. To date, at least two dozen differential equations have been suggested to describe slope behavior. These models are almost exclusively phrased in two-dimensional form since higher-dimensional formulations defy analytical solution (Smith & Bretherton 1972).

Young (1972) has recognized four classes of slope evolution models: (1) based upon direct recession, (2) based upon creep and flow mechanisms, (3) involving form interaction, and (4) based on process measurements. Members of the third set of models are rarely couched as differential equations, instead they are implemented directly in discrete form on a computer. The last type, typically, is based on empirically derived equations describing the observations of process.

Among the first two sets of models, there appears to be a convergence of approach yielding two basic equations:

$$\frac{\partial y}{\partial t} = -b\frac{\partial y}{\partial x} \qquad (5.1)$$

where y = elevation, t = time, x = position, expressed as distance from divide, b = a positive constant.

This appears to describe the situation in which the slope suffers parallel retreat through time, and so b has been variously termed the 'recessional coefficient' (Hirano 1975) and the 'coefficient of retreat' (Nash 1981). The equation is applicable under any conditions in which the rate of denudation is proportional to slope. Adjustments can be made to reflect a situation in which weathering acts normal to the slope (Scheidegger 1961).

Where denudation is proportional to the convex curvature, whether the process is creep (Hirano 1975), plastic flow (Souchez 1966) or any subduing mechanism, the appropriate equation is evidently:

$$\frac{\partial y}{\partial t} = a\frac{\partial^2 y}{\partial x^2}, \tag{5.2}$$

and *a* is a positive constant, termed the 'coefficient of debris diffusion' (Nash 1981), the 'subduing coefficient' (Hirano 1975), the 'coefficient of diffusion' (Culling 1965) or describes the ratio of specific gravity of the regolith to its viscosity (Souchez 1966). A related equation is capable of describing the form of accumulation slopes such as alluvial fans (Scheidegger 1961).

These two equations have been brought together by Hirano (1975) in a form which describes the combined effects of weathering, creep (or any diffusive process) and surface wash (any process of direct removal, including solution). The equation given is:

$$\frac{\partial y}{\partial t} = a\frac{\partial^2 y}{\partial x^2} + b\frac{\partial y}{\partial x} \tag{5.3}$$

where *a* and *b* retain their earlier meaning.

These equations implicitly assume the ergodic principle, namely that one can substitute space for time. For example, Equation 5.3 says that the change in elevation through time at a given point is proportional to: (i) the slope at that point, and (ii) the instantaneous change in slope at that point. It is generally assumed that *a* and *b* are positive constants so that 'steeper slopes change faster' and 'more pronounced breaks in slopes (e.g. cliffs) are more subject to wear.' A large variety of solutions is available depending upon the choice of *a* and *b* each is applicable under certain restricted conditions.

There is considerable difficulty in determining which equation should be applied to a particular problem. Only a few areas have been subjected to the kind of intensive field work required to establish which processes are active in an area, and to assess the relative contribution of each process. In addition, it is difficult to establish that an equation such as (5.2) actually does reflect the changes in a slope subject to creep. Could we not envisage other equations that perhaps more adequately describe the response to creep?

Thus, within the context of the ergodic formalism, we are faced with two difficult problems:

(a) How do we learn what processes must be modelled at a given site?
(b) How do we assure ourselves that the right equation is chosen to represent that process?

Most of us will accept that the first question can (at least theoretically) be studied through appropriate field procedures. Caine (1976) has shown how to estimate the energetics of each process active in a landscape and Ritter (1978) has summarized much of the available literature. Few of us would be sure how to proceed against the second problem. Some studies have suggested that it too can be investigated in the natural laboratory. For example, Schumm (1967) reports slope values and relates them to rates of erosion. However, the data do not appear to distinguish whether

$$\text{erosion} \propto \sin (\text{slope angle})$$

or

$$\text{erosion} \propto \tan (\text{slope angle}).$$

Thus, we are evidently constrained in our attempts to formulate adequate models of slope behavior, not to mention the more difficult problem of *landform* behavior where a large number of slopes must be studied. We are stifled because we cannot extract information relevant to the ergodic formalism.

This paper shows a method of extracting such information *from the landform* in a fashion that lends itself readily to construction of useful models of landform behavior. To do this we must first take a closer look at precisely what the ergodic principle says about landforms. Stated bluntly, the principle is: 'time can be equated with space.' More exactly, the statement should read: 'time averages can be replaced by space averages.' Placed within the geomorphic context (Melton 1958, p. 49), this has traditionally been interpreted to concern substitution of 'laws of *changes* in space for laws of *changes* in time.' Thus, the geomorphologic use is only an analogy to the physicists application which was formulated with a very different object in mind. The difficulty encountered by the physicists was that the phenomena being investigated (molecular behavior) occurred extremely rapidly compared to the duration of observation. Thus a set of observations *over time* was used to infer its instantaneous character (Khinchin 1949). Of course, the changes in time we are talking about are decreases in elevation at a point (i.e. erosion). If there is some regularity (pattern, structure) to the changes in elevation over time (which assumption is at the very heart of all attempts to understand the dynamics of landforms), then there must be a corresponding regularity to the changes in slope over space. And if we find changes in slope over space that follow a regular pattern and are capable of yielding a reasonable dynamic for erosion, then we are much more adequately justified in accepting the ergodic principle. Thus, a critical examination of the ergodic principle rests upon an examination of the spatial patterns of slopes.

We might anticipate that the sequential behavior of a landform at a point (the dynamics) will be related to the behavior of sequences of points. If we are to reconstruct a contiguous sequence of times in the history of the point, it is reasonable to search for the sequential history of contiguous points. That is, we must examine a traverse of the landform for possible pattern. If the temporal (sequential) behavior of a point is describable by a simple model, then it is parsimonious to expect the behavior of a sequence of points will yield, if properly coaxed, a model of equivalent simplicity.

GEDANKEN EXPERIMENT ON SLOPES

Let us imagine perhaps the simplest assumption we can about the behavior of points through time (t); the point elevation (E) is constant (k), neither suffering erosion nor uplift.

$$E(t) = k, \forall\, t \tag{5.4}$$

If this were so, we might perhaps expect a sequence of elevations $[E(X)]$ to be constant (the plateau, the peneplain).

$$E(X) = k, \forall\, X \tag{5.5}$$

But in fact, we do not expect this to occur and must abandon such a model until such time as an immortal landform is documented. Perhaps the next simplest assumption, and one which is markedly easier to accept, is that the change in elevation is constant:

$$E(t+1) - E(t) = k, \forall\, t \tag{5.6}$$

Translated into terms of space, such an assumption also says changes in elevation (slopes) are constant:

$$E(X+1) - E(X) = k, \forall\, X \tag{5.7}$$

Such a model (and the differential equations that are constructed on this basis) lacks the quality of reality in that it implies that the landform consists of a single slope that has a constant angle. It does not stop going up at the one end and does not stop going down at the other! Of course, we can circumvent this embarrassment by restricting our attention to a single slope and letting X index the distance from the 'divide', i.e. the highest part of the slope. But such myopic models are inherently unsuitable for description of an ensemble of slopes–the landform. Few would accept the temporal implication of the model either. Rather, we anticipate that a point will decrease in elevation not at a constant rate but that (in the absence of uplift) the initial erosion will be rapid and the rate of erosion will decrease through time (Schumm 1963). This might occur if the change in elevation were proportional to the last change in elevation:

$$E(t+1) - E(t) = k[E(t) - E(t-1)] \qquad (5.8)$$

or

$$E^1(t) - kE^1(t-1) = c = 0 \qquad (5.9)$$

Such a model would correspond closely to the concept promulgated by W. M. Davis nearly 100 years ago (Davis 1899), and supported by a long line of followers. Translated into spatial terms we have:

$$S(X) = kS(X-1) \qquad (5.10)$$

where $S(X)$ is the slope from point $X + 1$ to point X. This is an improvement in the appearance of a slope in that it will generate a form such as in Figure 5.1.

But we must restrict ourselves to a certain spatial range; looking too far to the left to examine the form of the slope would reveal the unlimited

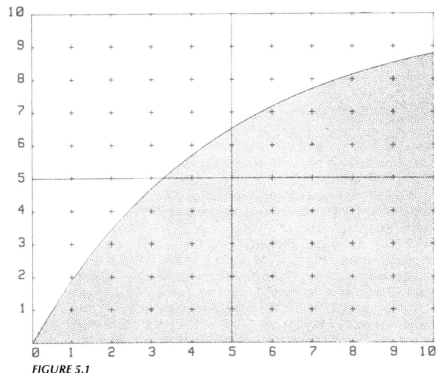

FIGURE 5.1
Realization of a slope created by the model of Equation 5.8. Value of k used is 0.90.

range given to that slope. We never know just where the slope begins. Similarly, the slope becomes increasingly more level as we go further to the right. This is also true in the temporal version where we cannot pick a starting point and it cannot be predicted from the model. So we could call this the 'Golden Age of Greece Model'. Things began well in some unknown and unknowable past and have steadily worsened. We can reject the model on the basis of our observations. Landforms survive in great variety despite the antiquity of the Earth. Thus, any model that yields a monotonic decay must be rejected.

However, we do not know the precise times when decay will be reversed, any more than we are able to predict the next earthquake or the next volcanic eruption. Thus, we might expect a reasonable model of landform change (note the shift from pure erosional concerns) will include a random component in which cumulative probabilities of positive and negative changes are equal. We must accept this in the absence of any indication that the continents systematically have become higher or lower through time. There may be long spells of lows and highs (Gilluly 1949, de Sitter 1956), but it seems somewhat optimistic to believe that nature is so good at arithmetic that Equation 5.8 always holds exactly (that is $c = 0$ with no residual). If we adjust this to a statistical regularity, we obtain a much more reasonable model. Thus to allow positive and negative changes on an irregular basis, it is reasonable to extend the model by adding a random variable (e) in place of the constant equal to zero.

$$E^1(t) - kE^1(t-1) = e(t) \qquad (5.11)$$

Because we would expect that positive and negative changes are equally likely and because we are dealing with continuous (not discrete) changes, it is not unreasonable and is certainly parsimonious to assume that e follows the normal distribution.

In addition to describing the behavior of the landform over time, the model describes an interesting landform over space (for instance, Fig. 5.2) which is precisely the ARIMA (1,1,0) model of elevations (Box & Jenkins 1976). Such reasoning can continue but, no doubt as you may have recognized already, the greater the departure from the initial assumption (Eq. 5.4), the more varied are the paths. I have explored only one of them here and no doubt other reasonable alternatives could be suggested.

The entire exercise has little meaning unless we can relate the implied spatial behavior with some quantifiable features of the landform. The next section will describe a method which will allow us to identify appropriate spatial and temporal models from the landform and, perhaps even more interestingly, will allow us to make very precise estimates of such parameters as *a*, *b* and the distribution of e.

BOX AND JENKINS MODELLING: THE STAGES

If a sequence of variables is described by a model such as Equation 5.11, it is reasonable to expect adjacent values to be correlated with one another (autocorrelated). Indeed, that is precisely what Equation 5.11 indicates. Conversely, it has been shown (Box & Jenkins 1976) that with

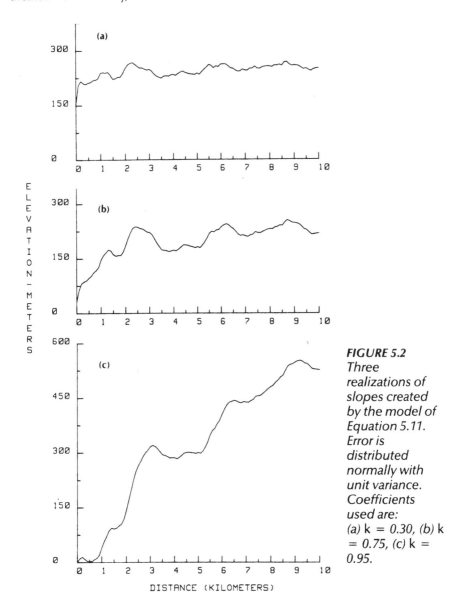

FIGURE 5.2
Three realizations of slopes created by the model of Equation 5.11. Error is distributed normally with unit variance. Coefficients used are: (a) k = 0.30, *(b)* k = 0.75, *(c)* k = 0.95.

sufficient information on the autocorrelation properties of a series, the particular form of model (chosen from a class of discrete models) which that series follows can be identified. There are two diagnostic tools required to identify models, they are the **autocorrelation function** (ACF) and the **partial autocorrelation function** (PACF). Properties of both have been rigorously described by Box and Jenkins (1976) and have been illustrated in a geomorphic context by Thornes (1972, 1973). It is useful to recall that the autocorrelation equation is exactly analogous to the traditional correlation coefficient and it can be computed for various lags in the data.

where:
$$r_k = C_k/C_0, \quad k=1 \ldots N \tag{5.12}$$

$$C_k = \frac{1}{N} \sum_{i=1}^{N-k} (x_i - \bar{x})(x_{i+k} - \bar{x}) \tag{5.13}$$

r_k is the autocorrelation coefficient at lag k and C_k is the autovariance at lag k.

The ACF is a plot of r_k versus k, an example is given in Figure 5.3. Because the autocorrelation has a transitive nature, it is not surprising to find that a series which follows Equation 5.11 will have, in addition to a large value in the ACF at lag one (governed by the magnitude of k in Equation 5.11), a large value at lags 2, 3, 4 and so on. The magnitude decreases only slowly. This is in spite of the fact that nothing in Equation 5.11 explicitly states that $E(1)$ is related to $E(3)$ or $E(3)$ is related to $E(5)$, etc. The reason for the high value of r_2 is that $E(1)$ is highly related to $E(3)$ because of the explicit strong relation of $E(1)$ with $E(2)$ and of $E(2)$ with $E(3)$; thus the value of $E(3)$ is influenced by $E(1)$ through the intermediate step at $E(2)$. Statisticians summarize this effect by saying that r_2 is correlated with r_1. For a model such as Equation 5.11, the degree of relation decreases slowly, so that r_2 will be only slightly lower than r_1, and so on, until finally the relation becomes essentially zero. That is why the ACF in Figure 5.3 takes on the particular form that it does.

Unfortunately, it is not possible to go directly from an ACF having the form shown in Figure 5.3 and infer that the series must be governed by an equation having the form of 5.11. This is because there is at least one other equation that will yield the same form as Figure 5.3 displays. The second equation is of the form:

$$S(X) = \phi S(X-1) + e(X) - \theta e(X-1) \tag{5.14}$$

It is possible to distinguish between these two because the models can be fitted to the data just as in the case of the classical regression analysis. When the coefficient ϕ is estimated for a series that follows the form of Equation 5.11, it will be found to be zero, or not significantly different from zero. But this is somewhat inefficient; the model can be diagnosed more completely prior to the fitting stage by making use of the second tool of time series analysis–namely the PACF. This is totally analogous to the partial correlation coefficient of multiple regression. In this case, we would like

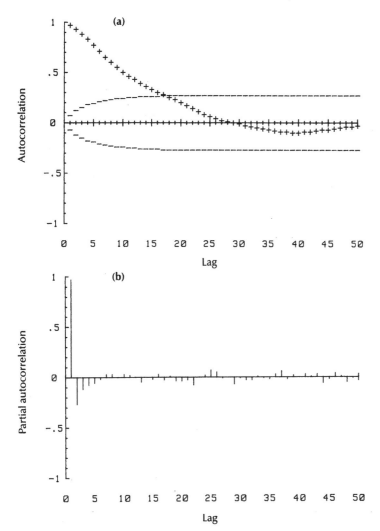

FIGURE 5.3
(a) ACF and (b) PACF of a non-stationary series following an ARIMA (1, 1, 0) model. Functions were computed from a traverse of 200 observations at a spacing of 48 m in the Snow Shoe SE Pa 7½' quadrangle. Note extremely slow decay of ACF and large value at lag one of PACF.

to show the value of the interrelation between $E(1)$ and $E(3)$ when the carryover effect through E(2) is (statistically) removed. Computation of the partial correlation coefficient (ϕ_{ii}) is done using the following equations:

$$\phi_{11} = r_1 \tag{5.15}$$

$$\phi_{22} = (r_2 - r_1^2)/(1 - r_1^2) \tag{5.16}$$

One can generally distinguish one model from another by examining both the ACF and PACF of the series and of a sufficient number of derivative series (i.e. the series of first differences, or the series of second differences, etc.). Thus, we may not be able to distinguish one model from another by examining the ACF and PACF of an elevation series alone. However, examination of those functions together with the ACF and PACF of the sequence of slopes (first differences of elevations) may be, and in general is, sufficient to distinguish the models.

Using the appropriate number of ACFs and PACFs we can actually identify the model which may describe the surface processes which are shaping the landform. In general, we find that the models will be members of the general class of **autoregressive-integrated-moving-average,** ARIMA (p,d,q), models having the form:

$$E^d(i) = \phi_1 E^d(i-1) + \ldots + \phi_p E^d(i-p) + e(i)$$
$$- \Theta_1 e(i-1) - \ldots - \Theta_q e(i-q) \tag{5.17}$$

where $E^d(i)$ is the ith value of the dth difference of the elevation series. That is, E^0 is the original elevations, E^1 is the slope series, E^2 is the series of changes in slope, so that:

$$E^1(i) = E^0(i) - E^0(i-1) \tag{5.18}$$

and in general

$$E^d(i) = E^{d-1}(i) - E^{d-1}(i-1) \tag{5.19}$$

The values of p, d and q are to be determined by use of the diagnostic tools, the ACF and PACF described above. A few examples of the diagnostic process will be discussed. The first step is to determine the order of d. If the chosen d is too small, the ACF will decay extremely slowly, starting from a value of $r_1 \approx 1.0$. If such behavior is found in either the ACF or the PACF, attention should be shifted to the difference of that series, until slow decay is replaced by rapid decay or by isolated 'spikes' in the plot. Figure 5.3 shows the ACF and PACF of a series requiring differencing, and Figure 5.4 shows the ACF and PACF of the differenced series. The exponential decay of the ACF together with a single spike at lag one of the PACF (Fig.

5.4) indicates that the series has $p=1$ and $q=0$, as we have seen already that $d=1$, the model is an ARIMA (1,1,0) or:

$$E^1(i) = \phi_1 E^1(i-1) + e(i) \qquad (5.20)$$

Interestingly enough, this corresponds to Equation 5.11 presented earlier. Let us examine another ARIMA model that arises in the study of slopes.

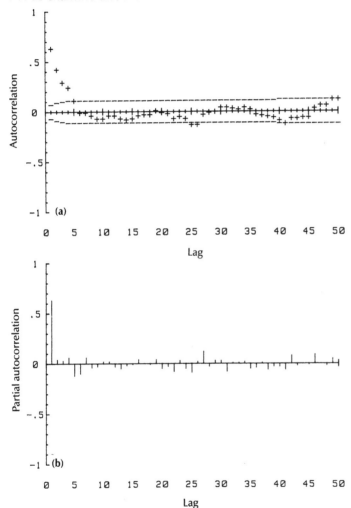

FIGURE 5.4
(a) ACF and (b) PACF of the first difference of the series of Figure 5.3. Note the rapid decay of the ACF and the single 'spike' at lag one of the PACF. This value (0.63) at lag one is a good estimate of ϕ_1 for the ARIMA (1, 1, 0) model.

THE ARIMA (2,1,0)

The form of the model is:

$$E^1(i) = \phi_1 E^1(i-1) + \phi_2 E^1(i-2) + e(i) \tag{5.21}$$

In this case, four basic forms are possible in the ACF and PACF (Fig. 5.5). This model should be of extreme interest to geomorphologists because Equation 5.3, governing slope erosion, can be expressed in the form:

Erosion = a(change in previous slope) + b(previous slope) (5.22)

If phrased in the notation introduced previously, we have:

$$E^1(t) = aE^2(X-1) + bE^1(X-1) \tag{5.23}$$

and under the ergodic assumption

$$E^1(t) = E^1(X) \tag{5.24}$$

so that

$$
\begin{aligned}
E^1(X) &= aE^2(X-1) + bE^1(X-1) \\
&= a\{[E(X-1)-E(X-2)] - [E(X-2) - E(X-3)]\} + b[E(X-1) - E(X-2)] \\
&= (b+a)[E(X-1) - E(X-2)] - a[E(X-2) - E(X-3)] \\
&= (b+a)E^1(X-1) - aE^1(X-2)
\end{aligned}
\tag{5.25}
$$

and if we add a term, e, expressing our uncertainty

$$= (b+a)E^1(X-1) - aE^1(X-2) + e(X) \tag{5.26}$$

which is equivalent to Equation 5.21.

Thus, a stochastic partial differential equation describing slope processes suggests that a landform whose temporal changes are controlled by a combination of slope wash and creep (recession and smoothing) will display a particular spatial structure that yields an ACF and PACF as in Figure 5.5. Conversely, a landform displaying such a structure can be described by a differential equation such as 5.3. If we can use the data from the traverse to estimate ϕ_1 and ϕ_2, then it is possible to estimate a and b, and so to know the relative contributions of the two processes shaping the landform in question. In particular, we have:

$$
\begin{aligned}
\phi_1 &= b + a \\
\phi_2 &= -a
\end{aligned}
\tag{5.27}
$$

Thus, if $\phi_2 = 0$, there is no creep and the surface is totally shaped by surface wash.* The surface in that case is totally controlled by creep. A pure creep process will have the form:

$$E^1(X) = aE^1(X-1) - aE^1(X-2) + e(X) \tag{5.28}$$

*Using the terms surface wash and creep *sensu lato* to include any processes which behave according to Equations 5.1 and 5.2 respectively (Culling 1965).

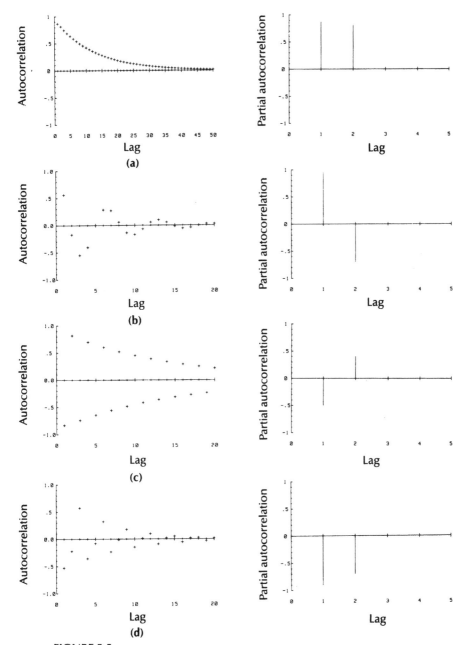

FIGURE 5.5
The four theoretical forms of the ACF and PACF of an ARIMA (2, 0, 0) model. (a) Both coefficients positive, (b) $\phi_1 > 0$, $\phi_2 < 0$, (c) $\phi_1 < 0$, $\phi_2 > 0$, (d) both coefficients negative.

while a pure surface wash model will have the form:

$$E^1(X) = bE^1(X-1) + e(X) \tag{5.29}$$

which is exactly an ARIMA (1,1,0). We can use the equations to generate landforms with the appropriate autocorrelation structure and so learn what forms each process is capable of generating. This has been done and a few examples are shown in Figure 5.6.

It is useful to know that a mixed creep and surface wash model will never be confused with a pure creep model, in spite of the fact that both are, technically, ARIMA (2,1,0) models. This could only happen when

$$\phi_1 = -\phi_2 \tag{5.30}$$

which implies

$$a + b = a \tag{5.30}$$

so that $b = 0$, which is a pure creep model. There are, from statistical considerations (Box & Jenkins 1976), three constraints which an ARIMA (2,1,0) model must satisfy in order to be invertible, these are:

$$\phi_1 + \phi_2 < 1$$
$$\phi_2 - \phi_1 < 1 \tag{5.31}$$
$$|\phi_2| < 1$$

Together, these define a permissible region for parameters of the ARIMA (2,1,0), as shown in Figure 5.7. Also shown are the regions attributable to each of the pure processes, surface wash and creep. We can summarize the results concerning these processes:
  (i)   pure slope wash: $\phi_1 = b$ and $\phi_2 = 0$;
  (ii) pure creep: $\phi_1 = a$ and $\phi_2 = -a$;
  (iii) mixed process: $\phi_1 = a + b$ and $\phi_2 = -a$.

Although Equations 5.15 and 5.16 are satisfactory for estimating the PACF to identify the appropriate model, they are not ideal for actually estimating the values of the coefficient ϕ_1 and ϕ_2 because they do not make full use of the data. To do this, a maximum-likelihood procedure is available that requires non-linear methods and is best handled by a commercially available computer program. In addition to ϕ_1 and ϕ_2, these programs provide improved estimates of the mean and residual variance of the series as well as useful validation tools that ensure that the series was properly identified in the first place. Uncertainty bounds around the estimates of ϕ_1 and ϕ_2 show if either parameter is essentially zero. Note that if the model being considered here is correct, the value of ϕ_1 is zero only if $b = -a$. It is interesting to speculate what this last condition might indicate. If the values of a and b are taken as a measure of the relative energy expended

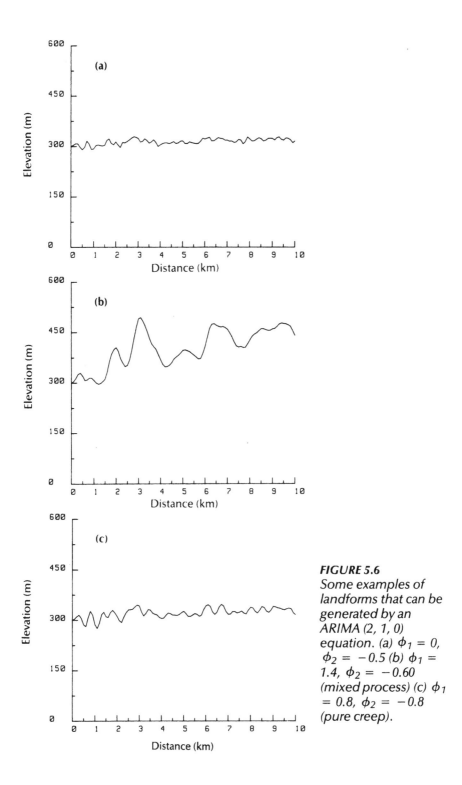

FIGURE 5.6
Some examples of landforms that can be generated by an ARIMA (2, 1, 0) equation. (a) $\phi_1 = 0$, $\phi_2 = -0.5$ (b) $\phi_1 = 1.4$, $\phi_2 = -0.60$ (mixed process) (c) $\phi_1 = 0.8$, $\phi_2 = -0.8$ (pure creep).

in each process, it would seem that creep must be a more efficient process since equal energy distribution (as measured by the magnitude of *a* and *b*) results in a form dominated entirely by creep. This may explain the predominance of convex or concavo-convex slopes to the near exclusion of purely concave forms, as would be produced by surface wash (Carson & Kirkby 1972).

Interpreting μ and σ

It is of interest to examine the meaning of the remaining term in the equation, namely the e's. As has been mentioned, the program which estimates ϕ_1 and ϕ_2 also provides an estimate of the variance of e. It is usually assumed that the e's are random variables which are distributed normally (and independently), with mean zero. Thus the variance of e conveys all of the remaining information in the landform.

If the values of ϕ_1 and ϕ_2 measure the relative contribution of creep and surface wash, then the e's must measure the degree to which the form is *not* determined by these processes alone. It is not surprising that the actual form departs somewhat from the ideal that these processes would generate. There are a number of non-random features in the landscape that also affect the form. For instance, it is commonly recognized that the action of slope processes will produce different forms depending upon the lithologies present. Additionally, structure – the spatial arrangement of lithologies – may exert an effect upon the final form. Another factor always requiring assessment is the degree to which the landform is a 'relic' form inherited from a time when the process mix was different (Twidale 1976).

Just as the form of the land is capable of yielding information on the processes responsible for its shape, so too can we extract information about the components which contribute to the variance of e. We shall first

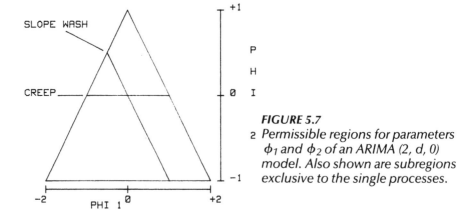

FIGURE 5.7

2 *Permissible regions for parameters ϕ_1 and ϕ_2 of an ARIMA (2, d, 0) model. Also shown are subregions exclusive to the single processes.*

address ourselves to the lithologic component. If the departures from the pure process–form model are in fact related to lithologic conditions, it is possible to measure the extent of the lithologic contribution by recording the lithologies underlying each point along traverses of the landform grid. In practice, we record map units rather than lithologies because: (i) it is more convenient, and (ii) it gives more detail and usually can be converted to lithologic groupings if required. Because we are dealing, in Equation 5.21, with slopes, the values of e*(i)* do not describe variation in *elevation* in response to additional factors such as lithology; rather, they represent variations in *slope*. Thus we would like, first of all, to quantify the effect of lithology on slope. In this case, of course, slope is defined as differences in elevation rather than by some angular measure. We can readily convert to angular measure if the horizontal distance between points is kept fixed. Since horizontal spacing is constant, difference in elevation between adjacent points increases as the tangent of the slope angle. Since the identification and estimation procedures described earlier also require fixed horizontal spacing, we adopt that convention. The spacing between points is important; but there is a way to compute the best spacing (Craig 1981a), at least in the typical cases where ϕ_1 is considerably larger than ϕ_2. Thornes (1973) has described the scale problem that exists in this methodology. However, it has been shown that resampling an ARIMA $(l,d,0)$ process yields the same model with systematically changed parameters (Craig 1981b). It would be useful for purposes of characterization to choose a spacing so that, on the average, each point will fall in the outcrop belt of the map unit immediately adjacent to the last point. It is especially undesirable that map units be 'skipped over'; but the grid cannot be made arbitrarily small or the E's will be autocorrelated as suggested above.

Our method has given us a means of identifying the processes responsible for the landform and has yielded a means of quantifying that form, namely a spacing at which observations of slope recorded on a grid pattern can be expected to reflect the influence of additional sources of variation on the results of the process. Using this spacing, elevation and lithology have been recorded for the example area (see Craig 1981a, Figs 1 and 10). The slopes observed have been classified according to the lithologies that occur at the end of that slope. Note that this incorporates the information about lithology *and* structure. These data may be summarized in the form of frequency histograms, as in Figure 5.8. The form displayed is typical of that observed in a large number of distributions collected in the Appalachians of central Pennsylvania (Craig 1979, 1981c). Because the model assumes the e's are distributed normally, it is of considerable value to test these distributions for normality. This was done using three measures: skewness (Fisher's g_1), kurtosis (Fisher's g_2) and the goodness-of-fit (chi-squared). The results are presented in Table 5.1.

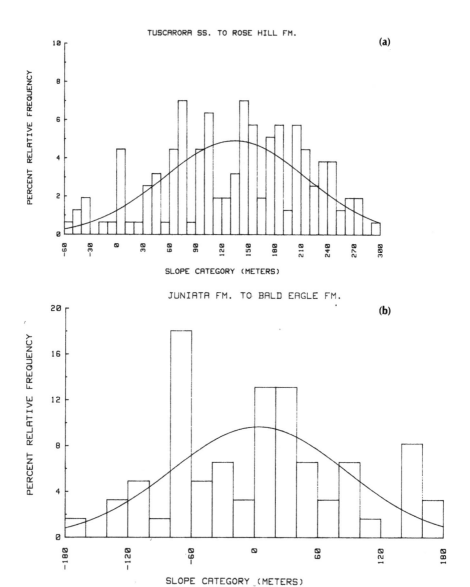

Tests of normality provide adequate reason to suspect that when re-corded in this fashion we obtain random samples from population(s) of slopes that is (are) distributed normally. Under the assumption of normal-ity, the distributions are completely characterized by two parameters, the mean, μ, and the variance, σ^2. Estimates of these parameters for 68 lith-ology pairs are also presented in Table 5.1. Since we are dealing with a

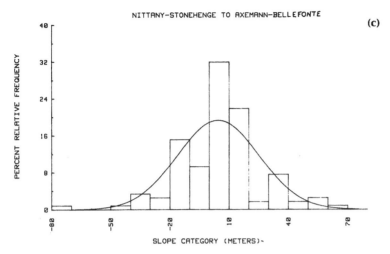

(c)

FIGURE 5.8
*Frequency distributions of three examples of
lithology pairs: (a) first lithology (sandstone) is much
more resistant than second lithology shale; (b) both
lithologies are resistant; (c) both lithologies are weak.*

distribution of values which is apparently a function of the two lithologies at the ends of the slope, it is useful to index the parameters (and their estimators) by subscripts indicating which lithologies are involved, for example, μ_{AB} and σ^2_{AB} represent the mean and variance respectively of the distribution of slopes which occurs whenever lithology A is at some point X and lithology B occurs 1 km away in the direction of the traverse. Because a slope from A to B is also a slope from B to A, slopes are an antisymmetric relation, that is $S(A,B) = -S(B,A)$ and so:

$$\mu_{AB} = -\mu_{BA} \tag{5.32}$$

it is also obvious that:

$$\sigma^2_{AB} = \sigma^2_{BA} \tag{5.33}$$

Of course, the mere collection and classification of slope data as described above do not guarantee the exercise has meaning, even if the population(s) are distributed normally. Such an exercise has the greatest value if it can be demonstrated that the lithologies are related to significant differences between parameters of the slope distributions. Therefore, before any attempt is made to provide an interpretation of these parameters it would be wise to investigate whether it is reasonable to believe that the parameters are indeed a function of the indexing lithologies. In this case, the function to be considered is quite simple, we merely mean that there exist some lithologies *I, J, K* and *L,* which are elements of L, the set of all lithologies in the region of interest, such that:

$$\mu_{IJ} \neq \mu_{KL} \tag{5.34}$$

TABLE 5.1
Statistics of slope distributions of 68 pairs of lithologies from the Appalachians of central Pennsylvania. Asterisks indicate values that would be exceeded with probability less than 0.05 if the population is actually normally distributed. Lithology numbers are keyed to the stratigraphic sequence (1 = youngest).

Lithology first	second	Sample size	Mean	Sigma	G(1)	G(2)	Probability chi-square
3	3	530	12	95	4.46*	32.54*	0.02
12	12	506	−14	148	−5.11*	11.98*	0.00
5	5	494	−17	150	−0.72	6.49*	0.12
4	4	345	−19	104	1.08	1.45	0.75
21	21	330	15	112	−1.07	1.16	0.99
19	19	280	5	64	−0.02	13.84*	0.75
7	7	264	49	192	0.49	−1.31	0.27
18	18	238	14	77	−2.14	4.63*	0.01
6	6	235	−26	250	−2.59	1.31	0.19
4	5	228	101	155	0.60	−0.01	0.12
3	5	218	156	133	3.15	−0.75	0.00
5	3	218	−148	148	−5.36*	3.26	0.00
13	13	205	26	206	−1.73	−0.06	0.59
5	6	193	232	227	−0.87	1.36	0.90
5	4	186	−127	132	0.24	2.52	0.61
19	18	183	1	110	6.66*	13.13*	0.70
20	20	176	−4	73	0.41	2.34	0.90
4	3	119	−60	92	0.81	1.26	0.31
20	19	119	15	68	−0.23	3.43	0.87
15	14	117	−165	305	2.66	0.38	0.13
7	8	116	66	193	−1.38	−0.75	0.57
12	13	115	−212	228	0.40	−2.00	0.05
6	7	113	437	208	−0.13	−0.61	0.43
15	17	108	247	217	−1.84	−0.20	0.17
13	14	106	−353	269	−1.31	−0.48	0.93
20	21	95	−61	83	0.36	−0.31	0.92
12	11	91	−128	249	7.55*	12.56*	0.09
17	15	89	−374	229	1.46	−1.27	0.39
15	15	86	−18	221	1.03	−0.63	0.58
18	17	160	−138	173	−0.62	−2.27	0.00
11	11	158	57	154	0.40	0.12	0.58
14	13	157	442	267	−1.39	−1.69	0.05
14	14	155	88	381	1.41	1.86	0.39
3	2	150	−81	85	1.27	5.41*	0.09

TABLE 5.1 (continued)

Lithology first	second	Sample size	Mean	Sigma	G(1)	G(2)	Probability chi-square
2	3	148	64	82	1.15	2.29	0.61
3	4	145	50	112	2.06	0.15	0.19
17	18	143	138	164	1.92	0.70	0.13
6	5	141	−317	228	−3.37	−0.03	0.07
13	12	140	138	149	0.76	−0.24	0.99
18	19	137	41	88	1.44	0.86	0.69
21	20	131	60	101	0.49	6.10*	0.93
17	17	121	20	203	0.36	0.98	0.08
11	12	85	67	193	−0.16	−0.27	0.70
16	15	80	−54	221	−1.76	−0.39	0.33
17	16	76	−361	197	2.56	1.40	0.27
19	20	67	−18	73	−0.08	0.35	0.82
14	15	63	152	261	1.88	1.02	0.54
15	16	61	12	271	0.53	−1.11	0.11
16	17	54	378	192	−0.77	−0.40	0.30
15	13	52	278	209	−1.16	−0.30	0.87
16	18	52	475	169	1.20	−0.89	0.82
18	16	52	−464	175	2.09	0.72	0.53
2	5	51	241	114	2.44	0.12	0.02
5	2	46	−267	144	−1.56	−0.17	0.64
16	14	42	−331	235	1.94	3.80*	0.69
3	6	41	365	177	1.61	1.20	0.53
17	19	40	137	154	2.96	4.33*	0.48
12	14	39	−519	256	1.96	0.17	0.31
15	18	39	341	147	0.80	−0.67	0.35
16	16	39	−16	237	−3.13	3.51	0.25
14	12	37	568	237	−1.50	0.08	0.58
2	4	31	80	94	−0.47	−0.08	0.19
11	10	30	117	156	−0.22	−0.13	0.45
14	16	30	131	340	0.44	−0.92	0.04
12	15	28	−421	252	0.15	−0.61	0.23
4	6	27	382	399	−3.00	1.82	0.18
6	3	26	−358	179	−0.55	−0.16	0.92
7	6	15	−513	152	0.59	0.00	0.56

For example, we may find that the average slope when going from I = shale to J = limestone is different from the average slope when going from K = sandstone to L = limestone. If such is the case, we must not have:

$$H_0: \mu_{11} = \mu_{12} = \ldots = \mu_{1N} \ldots = \mu_{NN} \tag{5.35}$$

where there are N lithologies under scrutiny.

Such a null hypothesis may be investigated by means of a one-way analysis-of-variance (Table 5.2). For the region of central Pennsylvania, the results show that, with little likelihood of being in error, we may reject the null hypothesis in favor of the alternative that slopes do indeed differ as a function of lithology. It is equally important to note that the variability of slopes differs significantly as a function of lithology.

Roughly 67% of the variance of the e's, which itself is the residual variance not explained by the basic erosion Equation 5.11, is explained by the lithologies. This, in conjunction with the 76% of the variance of the landform explained by Equation 5.11, gives a total of 92% of the variability of the landform that is explained by the equation–lithology combination. One must begin to question the actual importance or necessity of invoking the previous history of the landform (e.g. peneplanation or climatic change) in order to understand the current landform in this region.

Since the means and variances of the individual lithology pairs evidently do contain useful information, it is of some value to interpret what these parameters tell us about the lithologies. Figure 5.9 shows no reason to suggest any general relation between the degree of slope (mean) and the variability (variance). This lack of relation between μ_{AB} and σ^2_{AB} is important, for it shows that two independent parameters are required to characterize the landform. These, of course, are in addition to the previous parameters identified, namely ϕ_1 and ϕ_2. It has been suggested that these latter two relate to the dominant erosional processes; it remains to be considered what μ_{AB} and σ^2_{AB} represent.

It is suggestive to substitute the statistician's terms, we may then call μ_{AB} the *expected* slope between two points (subject to the distance constraints imposed earlier) underlain by these lithologies and σ^2_{AB} the *uncertainty* in that slope. That is, if we choose at random on the spatial grid an *A–B* slope, we expect that it will have the value μ_{AB}, although we know from σ^2_{AB} that it will rarely assume that exact value. But if the ergodic hypothesis is correct, $E^t(t) = E^t(X)$, and if we now consider what slopes will

TABLE 5.2
Analysis of variance of the slopes for lithology pairs in which two distinct lithologies occur. The null hypothesis is that slopes are the same for each pair. This is rejected in favor of the alternative that significant differences occur between pairs.

Source of variation	Degrees of freedom	Sum of squares	Mean square	F ratio
between pairs	29	189 587 160	6 537 488	189.4
within pairs	2688	92 782 377	34 517	
total	2717	282 369 537		

occur at one point at any given time, we expect that the value of the slope will be μ_{AB}. That slope will, over time, vary from μ_{AB} according to the magnitude of σ^2_{AB}, but will tend to take on values close to μ_{AB} and, over sufficient periods of time, such values will be observed most frequently. Thus, under the ergodic hypothesis, the slopes observed at one point, through time, can be considered to be 'draws' from this (normal) distribution of 'typical' slopes. That the most frequently observed slopes will be close to μ_{AB} and that they will never tend to remain long either larger or smaller than μ_{AB} would suggest that a natural and reasonable interpretation of μ_{AB} is that it represents the stable slope for that lithology pair. Indeed, because it is so readily obtained, in an objective manner it may prove to be a useful definition of stable slope. It appears to satisfy our intuitive notion of what it means to be stable in a dynamic environment. If μ_{AB} represents the stable slope between those lithologies, then the particular value of μ_{AB} must convey information on their relative erodibility. For example, A stands 100 m above B because A is more resistant to erosion, the additional potential energy represented by that 100 m is required to supplement the energy of the erosional process sufficiently to overcome the greater strength of A, and results in equal rates of denudation on both units (thus maintaining μ_{AB} constant). Interpretation of the remaining parameter, σ^2_{AB}, is a more difficult task.

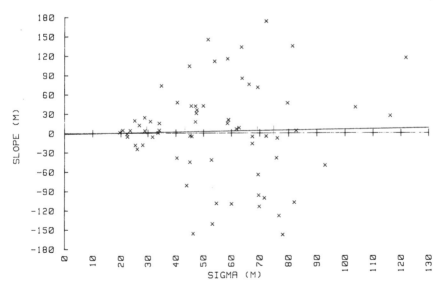

FIGURE 5.9
Relations between mean slope and standard deviation of slope for 68 lithology pairs.

It is perhaps wise to begin by stating the obvious. If a pair ($A-B$) shows large variance in slopes, it means we will frequently encounter a slope which is quite steep compared to the stable slope and, just as frequently, we will observe comparatively quite gentle slopes (Fig. 5.10). The exact value of the stable slope will be subject to some uncertainty. We would expect this sort of behavior when at least one of the lithologies in question is relatively resistant to erosion; so that the slopes are capable of sustaining virtually any inclination that chance events, such as stream corrasion at the base of the slope, might impose. But we already believe that μ_{AB} measures the relative erodibility of the lithologies. Why then do we not see a relation between μ_{AB} and σ^2_{AB} (Fig. 5.9)?

In fact, we do! If we examine the question more carefully, we note that if lithology A is much more resistant than lithology B, then $\mu_{AB} >> 0$. However, $\mu_{BA} << 0$, since $\mu_{AB} = -\mu_{BA}$, so that because the two values are both plotted on Figure 5.9, the true relation is obscured. If, instead, we treat slope as a directed quantity and consider only those slopes with a positive sense, which is justified from the antisymmetry, we obtain a significant positive relation between $|\mu_{AB}|$ and σ^2_{AB} (Fig. 5.11), although the relation is far from perfect due to the fact that two resistant lithologies that are close in erodibility will also have high variance. Whatever σ^2_{AB} measures about $\{A,B\}$ it is, at least in this area, related to the erodibility of $\{A,B\}$.

Because of the relation of high values of σ^2_{AB} with lithologies of nearly equal erodibility, it might be tempting to say that these slopes are more stable and hence think of higher values of σ^2_{AB} as indicating greater stability. Precisely the opposite is true. If a large range of slopes occurs over space, we expect from the ergodic hypothesis that a large range of slopes will

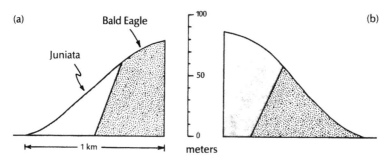

FIGURE 5.10
Examples of slopes can be drawn from a normal distribution with high variance ($\sigma = 83$ m). This is the same lithology pair depicted in Figure 5.8b. (a) A steeper (uphill) slope will occur about 17 times in 100. (b) A steeper (downhill) slope will occur with the same frequency. About 66% of the slopes between these two units will be more gentle than the slopes depicted here.

occur at one point over time. Since adjacent slopes are unrelated (this condition was imposed by the sample spacing), we similarly expect time sequences of slopes to be unrelated. (Elevation changes will be Markovian, and can be modelled as a random walk with two reflecting barriers (Kemeny & Snell 1960).) Thus, any slope at that point will be followed by a random draw from a distribution with mean μ_{AB} and variance σ^2_{AB}. If σ^2_{AB} is large, we expect large differences between successive realizations taken at unit increments. Therefore, the rate of change of slope at that point is larger than at some point where the lithologies are $\{C,D\}$ and $\sigma^2_{AB} >> \sigma^2_{CD}$. Slopes subject to high rate of change per unit time are legitimately considered unstable. Conversely, where the variance is low (Fig. 5.12), we expect successive realizations of slope at one point to be very similar; the rate of change of slope will be small and such slopes may be termed appropriately *stable* slopes (Young 1970).

Thus, μ_{AB} is the stable slope from A to B, its magnitude represents the relative erodibility of slopes underlain by $\{A,B\}$, which incidently is expected reasonably to be identical to $-\mu_{BA}$. Stability of slopes decreases with increasing values of σ^2_{AB}. We can now see that it makes sense for high values of $|\mu_{AB}|$ to show some relation to high values of σ^2_{AB}, because the more resistant lithologies can be expected to be subject to more intense forms of erosion and mass wasting if denudation is to continue at the same rate as that of less resistant units.

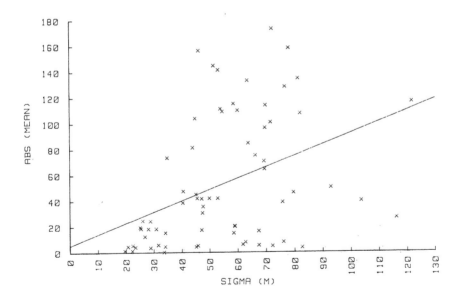

FIGURE 5.11
Relation between the absolute value of mean slope and standard deviation of slope for 68 lithology pairs.

Synthesis

The relations derived above have been programmed for a computer so that the erosional history of an entire area can be traced (Craig 1979, 1980). A prerequisite to proper application of the program is a digital representation of the landform at a spacing that reflects the value of Phi (1) (Craig 1981a, 1981b). This ensures that the effects to be modelled can be described by the parameters of lithology discussed above; so that erosion at each point is considered an independent realization from the known distribution of slope values.

The program operates in the following manner. It is assumed that erosion can occur at any point on the landform. A point is chosen at random and one of the neighboring points is assumed to have been eroded last. A new slope is drawn from the distribution of slopes appropriate to that pair. If the change to this slope does not result in a decrease in elevation, the slope is left unchanged. Erosion at that point is thus more likely when the 'original' slope is steeper than the mean than if it is less than the mean. Of course, erosion at the other point is more likely in the latter case, so that the combined effects of erosion at the two points make the mean the most likely result. If erosion occurs at this point its neighbors are considered, and so on, until no points remain to be considered. Because the changes tend to spread out from the original point, the entire process is called a **diffusion cycle.** After each of these diffusion cycles, a new point is chosen and the entire process repeated. Conceivably, the entire grid could be affected by a single diffusion cycle. In practice, only a limited area was so affected and the area affected showed remarkable consistency in size, an example is given in Figure 5.13. Variations in area were small in proportion to grid size and the area affected showed no tendency to either increase or decrease during the course of the simula-

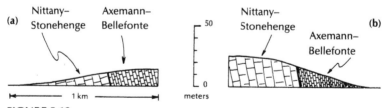

FIGURE 5.12
Examples of slopes that can be drawn from a normal distribution of slopes with low variance ($\sigma = 20\,m$). Same lithology pair as in Figure 5.8c. (a) A steeper (uphill) slope will occur 17 times in 100. (b) A steeper (downhill) slope will occur 17 times in 100. About 66% of the slopes will be more gentle than the slopes depicted here.

tion. Figure 5.14 shows the sizes of areas involved in each step of a single 786-step simulation. The least-squares fit of the line shown is not significantly different from one with a slope of zero. This pattern is typical of that observed in a large number of simulations. The average (mean) size of area affected during each of 36 simulations is shown in Table 5.3. The consistency of the results is quite remarkable and leads to the suspicion that this may be an inherent characteristic of landform change in this area.

It is not surprising that the area involved in a change should show a great deal of regularity. This is to be expected from the nature of the model

TABLE 5.3
Mean number of points affected by erosion in each cyle of a simulation. Each point represents an area of 1 km.²

Simulation number	Mean area eroded	Mean area eroded given erosion occurs	Probability of erosion occurring
1	138	29	0.21
2	284	158	0.56
3	143	23	0.16
4	139	32	0.23
5	102	24	0.24
6	74	13	0.17
7	191	55	0.29
8	139	26	0.19
9	482	222	0.46
10	103	22	0.21
11	53	9	0.17
12	4	1	0.15
13	138	22	0.16
14	160	43	0.27
15	102	21	0.20
16	156	32	0.21
17	193	110	0.57
18	190	42	0.22
19	209	36	0.17
20	170	82	0.48
21	391	263	0.67
22	205	42	0.21
23	117	16	0.14
24	213	41	0.19
25	133	25	0.19
26	105	63	0.60
27	203	43	0.21
28	159	35	0.22
29	104	16	0.15
30	120	19	0.16
31	145	36	0.25
Mean	163	61	0.27

derived for the area. Equation 5.11 shows that the change at one point is a function of change at adjacent points. Thus, in a probabilistic sense, whether a point (P) changes or not, $\overline{P}$ depends upon the adjacent point $P\text{-}1$:

$$
\begin{aligned}
p(P|P\text{-}1) &= p_1 \\
p(\overline{P}|P\text{-}1) &= 1 - p_1 \\
p(P|\overline{P\text{-}1}) &= p_2 = 0 \\
p(\overline{P}|\overline{P\text{-}1}) &= 1 - p_2 = 1
\end{aligned}
\tag{5.36}
$$

If the first point to be considered (P') does not change, no other points will change on that diffusion cycle. To have a total of N points change requires (for $N > 0$) that the first point changes, $N\text{-}1$ more points change *in a row* and then one point does not change.

As would be expected, approximately one-half of the steps do not result in erosion. This is because the first step of the diffusion cycle has probability 0.5 of resulting in a change or not, since the expected value of slope at that point is exactly μ. Thus, the expected area to be changed is

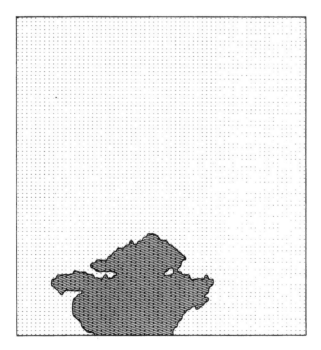

FIGURE 5.13
Map showing the area eroded (ruled) during
one step of the simulation.

conditional upon what happens on the first step. Given that a change occurs, the expected area of change is 22 km².

Besides the area affected, the mean depth of erosion during each cycle (given that any occurs) does not show a tendency to increase or decrease during the course of a simulation (Fig. 5.14). This appears to be in consonance with the suggestions of Hack (1960), but differs markedly from the predictions of the evolutionary model (Davis 1899) in which the volume of erosion would systematically decrease with time.

Neither theory would predict the pattern of *variability* in erosion which is actually observed. Figure 5.14 includes a plot of the variance in erosion for each cycle, as computed from the set of all points actually eroded in that cycle. As can be seen, there are large fluctuations in the variability which, together with the approximately constant mean, suggest that on some diffusion cycles all points are eroded about the same amount, while on others some points are eroded deeply and others are barely eroded at all. From the number of diffusion cycles, the mean size affected and the total grid size it is clear that each point is being eroded, on the average, three times. If the system acts to 'remain stable' or if not stable to move in the direction of increasing stability, we would expect the variability of erosion to remain constant or decrease. That we do not observe this is a discrepancy that must be investigated.

We are accustomed to the notion that a slope may be stable. Indeed, it has been suggested here that the variance of the distribution of slopes on a given lithology pair may be taken as a measure of that stability. If decreasing variance signals greater stability, the ultimate in stability would occur when $\sigma^2 = 0$, in which case the slopes would always (spatially and temporally) equal the mean. Thus, not only does σ^2 seem quite suitable as a measure of stability but we are apparently justified in defining μ as *the stable slope* for that pair. This corresponds closely with Young's (1970) definition of stable slope. Can we extend the notion of stability to the landform as a whole?

A natural extension would be to define a *stable landform* as one in which all slopes are stable. This appears to agree with the inexact uses of the term suggested by numerous authors (Gilbert 1909, Hack 1960). It also suggests a natural measure of the stability of the landform, namely the (standardized) difference between observed, $S(i)$, and expected (stable) slope, $E[S(i)]$, summed over all slopes of the landform. A slight refinement makes the measure considerably more useful. If we square each of these standardized differences, elementary probability theory (Lancaster 1969) shows that the resultant sum follows an important and well-known statistical distribution, the chi-squared distribution:

$$\sum_{i=1}^{N} \{S(i) - E[S(i)]\}^2 \qquad (5.37)$$

This fact allows us to make statements about the probability of observing that the particular landform under the null hypothesis is in fact stable. (Put more precisely, under H_0 the observed landform is a random sample of the set of possible landforms that are stable.) Low probabilities thus provide an *objective* signal that the landform is not in equilibrium, hence may be 'relic'.

This measure has been applied to the simulation with interesting results. It has been found that the degree of fit to the hypothesized model, as measured by this method, displays non-random quasiperiodic fluctuations which suggest that the landform is oscillating about some preferred value, but never reaches it. Such behavior is enigmatic in light of current theories of landform change, and demands explanation.

Is there a stable landform?

In an attempt to explicate this phenomenon, consider the following simplified landform (Fig. 5.15). It is trivially true that any closed traverse (loop) which begins and ends at the same point on the grid must show a

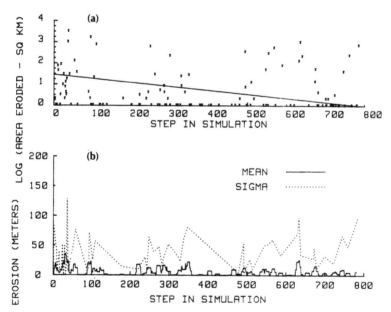

FIGURE 5.14
Plot showing trends in: (a) mean size of area eroded on each step of a 786-step simulation; regression line shown is not significant, data have been plotted on a log scale and zeros are excluded for clarity; (b) mean and standard deviation of depth of material eroded in each step.

net difference in elevation of zero. The simplest loop is just a circuit around one square of the grid and so consists of exactly four slopes. If we adopt the convention that going uphill is a positive change and downhill a negative change, then the algebraic sum of changes in any loop must be zero. Denoting slopes as pairs of letters which indicate the lithologies at each end of the slope (e.g. *AB*) we have for the small loop in the upper left of Figure 5.15a.

$$AB + BC + CD + DA = 0 \qquad (5.38)$$

But substition of the *stable* slope value in each case gives:

$$3 - 4 + 3 - 1 \neq 0 \qquad (5.39)$$

In other words, a stable landform *cannot exist* on the lithology grid depicted!

This situation is probably not unusual. Indeed, since failure at a single loop is sufficient to make the landform unstable, this situation is probably typical of all but the most trivial lithology grids (i.e. those with only one or two rock types). In the area of the Appalachians investigated (Craig 1981a, Figure 10), 967 unstable sites exist. This is roughly 18% of the total area. The result of this intrinsic lack of stability is that each time a surface process acts to bring one slope closer to its stable value, the same process typically drives some adjacent slope further from overall stability. In its simplest form, this would produce a cycling effect in which each slope is brought in turn to stability, but when the loop is closed the first slope is now not at the stable value and so change begins anew.

In a more complicated lithology grid, the presence of a single unstable loop will influence adjacent slopes, even if they are inherently stable. For example, in the simple case of Figure 5.16, the central unstable loop shares

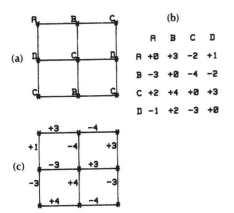

(b)

	A	B	C	D
A	+0	+3	-2	+1
B	-3	+0	-4	-2
C	+2	+4	+0	+3
D	-1	+2	-3	+0

FIGURE 5.15
A small segement of the landform lithology grid sampled from the Appalachians of central Pennsylvania. (a) Lithologies at each of nine points on the grid. (b) Stable slope values between each lithology pair. (c) Grid with stable values substituted.

a slope with each adjacent loop; if that slope is oscillating in value, it will change the stability relations of the peripheral loops. Such changes can be propagated through the system to eventually affect slopes far removed from the instability. The overall result is that the entire grid can display oscillations in its stability.

Such oscillations are a drastic departure from the behavior predicted by current geomorphic theory. For instance, the evolutionary concepts of Davis (1899) imply that the landform changes (after initial uplift) will always move towards a more stable condition. Hack (1960) describes landform behavior as:

'It is assumed that within a single erosional system all elements of the topography are mutually adjusted so that they are downwasting at the same rate. The forms and processes are in a steady state of balance and may be considered as time independent. Differences and characteristics of form are therefore explainable in terms of spatial relations in which geologic patterns are the primary consideration...' (p. 85).

But, as we have seen, it is precisely those geologic patterns and spatial relations that lead to an inherent instability in the landform. Neither theory leads to a consistent system when applied to a whole landform.

The situation is an example of the behavior of complex systems (Beer 1959, Ashby 1964) in which the whole is not equal to the sum of its parts. Such behavior cannot be understood if we consider single slopes in isolation. An analogous pattern of interactions has been described recently by Schumm (1977) for the fluvial system. Such notions are gaining currency in other fields, such as ecology (Sprugel & Bormann 1981). Diagrammatically, we could summmarize the notion as in Figure 5.17. Earlier theories considered landform behavior as an (more or less rapid and possibly asymptotic) approach to the central point of stability (Howard 1965). The

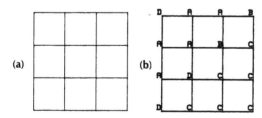

FIGURE 5.16
Larger lithology grid showing a central loop that is unstable and the surrounding loops, all of which are (in themselves) stable. Instabilities in the central loop will drive changes in the otherwise stable peripheral loops.

idea proposed here is that this is not the only form of 'stable behavior'. Orbits of the central point are also 'stable' patterns. If elliptical, they produce measurable variations in the stability without disrupting the overall pattern of behavior. There is no systematic tendency to produce a unique final state. Curiously, these are precisely the diagrams used by mathematicians to describe the qualitative behavior of systems of partial differential equations too complex to be solved analytically (Hirsch & Smale 1974).

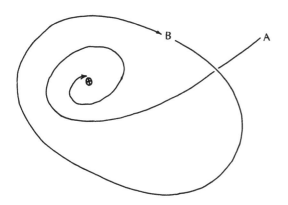

FIGURE 5.17
Conceptual models of landform change. (a) Traditional conception of continually increasing stability. (b) Pattern of change in which there is no monotonic increase in stability. Periods of close approach to stability are inevitably followed by larger divergences.

Conclusions

We can define the landform as a set of observed elevations arrayed according to the parameters ϕ_1 and ϕ_2 that are related to the intrinsic size of lithologic elements. The absolute values of the parameters are determined by that size, and their relative values vary in response to the intensity of competing erosional processes. The size of the lithologic units moderates the process system. Differences in elevation between adjacent points in the landform are related to the nature of the lithologies at those points. Slope distributions over space provide a clue as to their behavior over time; they tend towards stable values. The 'erodibility' of a lithology is related to the mean of that distribution. The variance is a measure of the stability of the individual slopes. These four parameters $\{\phi_1, \phi_2, \mu, \sigma^2\}$ constitute a compact description of the landform. Taken together they provide a method of quantifying the landform which yields an objective statistic measuring its stability. It is unlikely that stable landforms occur over any appreciable area. This is because the landform itself is 'self-actuating' due to the complex interactions of the competing slope units.

The procedures outlined here neither predict nor require that all land-forms will show the same kind of lithologic control as the Appalachians. Indeed, some areas may show no lithologic control at all. However, it does provide a *test* for the existence of such control.

In addition, the model does not predict or require *temporal* stability of the parameters μ and σ^2. In fact, changes in these parameters can be expected if the climate changes. However, we may make good estimates of the new values of μ and σ^2 under temporally altered climatic conditions if we use new parameters estimated from areas climatically similar to the altered state (Hoyt 1981). Thus, the ergodic hypothesis allows useful models of landform behavior.

Acknowledgements

Ms Dorothy Thompson lent her expertise in computer text-editing to help in the preparation of the text. Barry Miller provided valuable criticisms of earlier drafts. Responsibility for any errors that remain rests with the author.

References

Ashby, W. R. 1964. *An introduction to cybernetics.* London: Methuen.
Beer, S. 1959. *Cybernetics and management.* New York: Wiley.
Box, G. E. P. and G. M. Jenkins 1976. *Time series analysis: forecasting and control,* 2nd edn. New York: Holden-Day.
Caine, N. 1976. A uniform measure of subaerial erosion. *Geol. Soc. Am. Bull.* **87**, 137–40.
Carson, M. A. and M. J. Kirkby 1972. *Hillslope form and process.* Cambridge: Cambridge University Press.
Craig, R. G. 1979. *A simulation model of landform erosion.* Unpubl. PhD disserta-tion, The Pennsylvania State University.
Craig, R. G. 1980. A computer program for the simulation of landform erosion. *Comput. Geosci.* **6**, 111–42.
Craig, R. G. 1981a. Criteria for constructing optimal digital terrain models. In *Applied geomorphology,* R. G. Craig & J. C. Craft (eds). London: George Allen & Unwin.
Craig, R. G. 1981b. Sampling an autocorrelated process: the AR(1). *J. Int. Assn Math. Geol.* (in press).
Craig, R. G. 1981c. Quantification of slope-lithology relations: an example from Pennsylvania. *Geol. Soc. Am., Abstracts with Programs* **13**(3), 127.
Culling, W. E. H. 1965. Theory of erosion on soil-covered slopes. *J. Geol.* **73**, 230–54.
Davis, W. M. 1899. The geographical cycle. *Geog. J.* **14**, 481–504.
Gilbert, G. K. 1909. The convexity of hilltops. *J. Geol.* **17**, 344–50.
Gilluly, J. 1949. Distribution of mountain building in geologic time. *Geol. Soc. Am. Bull.* **60**, 561–90.

Hack, J. T. 1960. Interpretation of erosional topography in humid temperate regions. *Am. J. Sci.* **258A**, 80–97.

Hirano, M. 1975. Simulation of developmental process of interfluvial slopes with reference to graded form. *J. Geol.* **83**, 113–23.

Hirsch, M. W. and S. Smale 1974. *Differential equations, dynamical systems and linear algebra*. New York: Academic Press.

Howard, A. D. 1965. Geomorphological systems–equilibrium and dynamics. *Am. J. Sci.* **263**, 302–12.

Hoyt, B. R. 1981. *Climatic effects on slopes on sedimentary rocks*. Unpub. MS thesis, Kent State University.

Kemeny, J. G. and J. L. Snell 1960. *Finite markov chains*. New York: Van Nostrand.

Khinchin, A. I. 1949. *Mathematical foundations of statistical mechanics*. New York: Dover.

Lancaster, H. O. 1969. *The chi-squared distribution*. New York: Wiley.

Melton, M. A. 1958. Geometric properties of mature drainage systems and their representation in an E4 phase space. *J. Geol.* **66**, 35–54.

Nash, D. 1981. FAULT: a FORTRAN program for modelling the degradation of normal fault scarps, *Comput. Geosci.* (in press).

Ritter, D. F. 1978. *Process geomorphology*. DuBuque: Brown.

Scheidegger, A. E. 1961. *Theoretical geomorphology*. Englewood Cliffs, NJ: Prentice-Hall.

Schumm, S. A. 1963. *The disparity between present rates of denudation and orogeny*. U.S. Geol. Survey Prof. Paper 454-H.

Schumm, S. A. 1967. Rates of surficial rock creep on hillslopes in western Colorado. *Science* **155**, 560–61.

Schumm, S. A. 1977. *The fluvial system*. New York: Wiley.

Sitter, L. U. de 1956. *Structural geology*. New York: McGraw-Hill.

Smith, T. R. and F. P. Bretherton 1972. Stability and the conservation of mass in drainage basin evolution. *Water Resources Res.* **8**, 1506–29.

Souchez, R. 1966. Slow mass-movement and slope evolution in coherent and homogeneous rocks. *Soc. Belg. Geol. Bull.* **74**, 189–213.

Sprugel, D. G. and F. H. Bormann 1981. Natural disturbance and the steady state in high-altitude balsam fir forests. *Science* **211**, 390–93.

Thornes, J. B. 1972. Debris slopes as series. *Arct. Alp. Res.* **4**, 337–42.

Thornes, J. B. 1973. Markov chains and slope series: the scale problem. *Geog. Anal.* **5**, 322–8.

Twidale, C. R. 1976. On the survival of paleoforms. *Am. J. Sci.* **276**, 77–95.

Young, A. 1970. Concepts of equilibrium, grade and uniformity as applied to slopes. *Geog. J.* **136**, 585–92.

Young, A. 1972. *Slopes*. New York: Longman.

6

The geomorphology of the Sangamon surface: its spatial and temporal attributes

Leon R. Follmer

INTRODUCTION

In the part of the world affected by continental glaciation, well-developed soils formed between the major glacial episodes. In the midwestern United States, the soil found beneath the deposits of the last glaciation (Wisconsinan) and developed upon the deposits of the previous glaciation (Illinoian) is known as the Sangamon Soil. Much interest has been focused on regional correlations of stratigraphic features associated with the Sangamon Soil and in interpretations of environmental conditions during and following the formation of the soil. However, the geomorphic aspects of the Sangamon Soil have received relatively little attention in comparison, although many generalized interpretations have been made as to the character of the Sangamon surface.

The Sangamon Soil, or at least diagnostic parts of its profile, is easily recognized in outcrops and cores. As a consequence of its notable character, it is often described by name or by its features in reports, notes or logs by geoscientists, engineers, water-well drillers, etc. In spite of the general knowledge of the Sangamon Soil, little is known about the details of the spatial characteristics of its geomorphic surface. Three principal reasons explain this shortcoming: (i) no study has been directed toward this objective, (ii) the Sangamon is buried and the character of its surface must be interpolated from observation points in outcrops and drill sites which are, for the most part, spaced too far apart to resolve any detail of its surface, and (iii) the Sangamon in most places contains a complex sequence of overlapping soil profiles, and its top is difficult to determine. Therefore, the objectives of this study are to study the nature of the spatial characteristics and the major temporal aspects of the Sangamon surface in Illinois. Temporal attributes in terms of soil stratigraphy and chronostratigraphy have been controversial problems among Quaternary stratigraphers in the Midwest. Central to most of the problems is the basis on which the top of the Sangamon Soil is defined. An associated and interdependent problem is its age. This stems from the lack of absolute age determinations and also because its age is dependent upon what is taken to be the top of the Sangamon.

Previous studies of the Sangamon soil

The Sangamon Soil had an important place in the conceptual development of both Quaternary stratigraphy and buried soils in the Midwest, a role which I have reviewed comprehensively elsewhere (Follmer 1978). The most important early work was by Leverett (1899), but although his interpretations of the glacial geology of Illinois were remarkably good, his understanding of soil was limited and his description of the Sangamon geomorphic surface perfunctory. Between 1916 and 1930, the genetic concept of soil formation developed rapidly, incorporating studies of the Sangamon Soil (Follmer 1978). These and subsequent studies provided a better understanding of the Sangamon but also led to a number of stratigraphic and pedologic problems. The problems stem from what was thought to be the Sangamon Soil stratigraphic unit but contained (i) several parent materials, (ii) several pedogenic profiles, and (iii) a complex of pedogenic features (Follmer 1978, Follmer *et al.* 1979). A critical part of the interrelated problems is the genesis of the upper horizons of the profile. Three interpretations have been proposed: (i) intense silicate weathering (the gumbotil of Leighton & MacClintock 1962), (ii) accumulation of water-transported sediments derived from the surrounding slopes (the accretion-gley of Willman *et al.* 1966), and (iii) pedogenic incorporation of a thin loess deposit into the underlying soil (Thorp *et al.* 1951, Simonson 1954).

Current concepts of the Sangamon Soil evolved mainly from the work of Thorp *et al.* (1951), Simonson (1954), Frye *et al.*(1960a, 1960b), Leighton and MacClintock (1962), Willman *et al.* (1966), Willman and Frye (1970), Johnson *et al.* (1972), Ruhe (1974), Follmer (1978), Follmer *et al.* (1979). The following features of the Sangamon Soil can be derived from the discussion by Willman and Frye (1970).

(a) It underlies Wisconsinan-age deposits and had developed on Sangamonian, Illinoian or older deposits.

(b) It is composed of two genetic types:

(i) accretion-gley – a gray to blue-gray massive clay, with some pebbles and sand, formed by accretion under poor drainage and gleying conditions; may have sharp contact with the till below, gradational upper boundary;

(ii) *in situ* – a red-brown to sometimes dark gray profile developed in place under moderately good drainage to sometimes poor drainage on till or other glacial deposits; has a distinctive clayey B2-zone, typically has Mn–Fe pellets and staining; gradational lower boundary, sharp upper boundary.

(c) It is leached of carbonates to about 1.8–3 m; depth of leaching slightly diminishes northward; leached below the till contact in some accretion-gley profiles.

(d) It is strongly developed:

(i) *in-situ* profiles are more strongly developed in the south than in the north;

(ii) accretion-gley profiles are similar throughout Illinois.

(e) It is not restricted to the Sangamonian time interval:

(i) soil formation started in some localities during the Illinoian stage;

(ii) early Wisconsinan sediments were incorporated into the gley profiles to some extent.

Evident from the description is a clear emphasis on the B horizon. This is for good reason as the B horizon is the most diagnostic feature in strongly developed profiles. In contrast, the organic-rich A or O horizon is most readily recognized in the less-developed or cold-climate type soils.

BOUNDARIES, GEOMORPHIC SURFACES AND EVENTS

The upper and lower boundaries of a buried soil profile are frequently gradational and difficult, if not impossible, to determine precisely. This has led to different interpretations of how and where the top of the Sangamon is recognized and defined (Johnson et al. 1972, Frye et al. 1974, Follmer 1978, Follmer et al. 1979, Ruhe & Olson 1980). At issue is the A horizon or the uppermost horizon of the Sangamon soil.

The A or O horizon must be recognized or accounted for because its presence in a normal form indicates a stable geomorphic surface at the site of observation. The absence of an A horizon above a B horizon indicates an unstable geomorphic surface which, without exception, must be an erosion surface. In a geomorphic analysis of the Sangamon Soil, or any soil surface, the areas affected by erosion are distinguished from the areas receiving the sediments generated by the erosion and from the areas unaffected by either process – the stable areas. The land surface of these three areas forms a geomorphic surface at one moment in time, but the land surface of each area has definite spatial and temporal relationships to the others.

In practice, the concept of a geomorphic surface has been used in two ways: (i) a portion of the land surface that is specifically defined in space and time (Ruhe 1969) and is composed of an erosion surface and the contemporaneous, adjacent, depositional surface, and (ii) a portion of a land surface that is underlain by a laterally continuous, well-expressed soil and is dependent upon the continuous nature of the soil to provide the general unifying character for its recognition and mapping, such as the Yarmouth-Sangamon surface of Iowa (Ruhe 1956, 1969). The two concepts differ in the principal criteria but agree on the separation of an older, stable, geomorphic surface from younger erosional and depositional geomorphic surfaces. The difference between the two concepts is temporal, since the stable surface was a former surface of geologic construction or erosion before soil formation. Therefore, in a relative sense, a geomorphic surface underlain by a soil is considered to be a stable geomorphic surface.

The stable and eroded Sangamon Soil surfaces in Illinois are not difficult to determine where the soil horizons appear normal and the contacts with the overlying deposits are distinct. However, where the Sangamon Soil was slowly buried by thin increments of sediment, the soil surface is masked and the horizons have anomalous thicknesses and characters. In this setting, three general conditions can be recognized. (a) If the rate of burial was slow enough, the A and B horizons grew upward and produced an overthickened B horizon. (b) If the rate of burial was somewhat faster and/or the soil-forming processes were slowed for some reason, the added material developed A horizon characteristics, i.e. accumulation of organic matter, porosity, granularity, etc. In this case, the A horizon thickens upwards leaving the B horizon more or less unchanged. (c) If the burial rate was too fast, then the incorporation ability was surpassed and a distinct boundary can be found between the soil and the new deposits.

The fact that a soil can grow upwards in the right environment means that the top of a soil is time-transgressive wherever sediment accumulation is occurring. Therefore, in the strict sense, the age of a geomorphic surface is not the same from place to place. For example, where a stable surface lies adjacent to a buried surface (depositional area), the geomorphic surface at the boundary between the two areas splits (Fig. 6.1). The original or older surface (1) passes under the younger deposits. Consequently, the new surface (2) on the younger deposits merges at its border with the surface of the stable area. In erosional areas, the original surface (1) is destroyed and the age of the erosion surface is the same as the age of the surface in the depositional area. However, this assumes that the area of erosion is the source for the sediments in the depositional area. If this is not the case, then the erosional and depositional events are separated in time, which results in an age difference between the depositional area surface (2) and the erosional area surface (2').

The age of a geomorphic surface, in the strict sense, is determined by the age of the deposit or by the age of the last erosional event affecting the deposit (Daniels *et al.* 1971). Originally, the stable surface (1) was a former depositional or erosional surface, which establishes the age of surface 1. However, in the stable area, surface 1 is coincident with or can be considered a logical extension of surface 2. In concept, the spatial and temporal features of each surface can be easily differentiated, but in practice it is often difficult because the level of discrimination must be considered. Where the distinction can be made, the concept of a 'ground surface' can be applied, which is defined as 'all those erosional and depositional surfaces and layers which have developed in a landscape during one interval of time and upon which a unit mantle of soils has developed' (Butler 1959). In this concept, a unit of time is required for the devlopment of a soil, or at least some soil or physical characteristics which allow one 'ground-surface' unit to be distinguished from another. Therefore, a new geomorphic surface is made up of the areas affected by erosion and deposition over some period of time. The age of the new surface is assigned to the time when the surface gained stability or the time when soil formation began. Following this concept, the stable surfaces (the merged surfaces of 1 and 2 of Fig. 6.1) are distinguished from the new surfaces (2 and 2'). If the age of surface 2 is the same as surface 2', as can be determined at this level of discrimination by a soil continuum from the depositional to the erosional area, then surfaces 2 and 2' are linked together as one geomorphic surface.

The level of discrimination of geomorphic surfaces can be applied at any scale from watershed studies to Quaternary glaciation. However, the smaller the scale of the spatial and temporal attributes of a geomorphic surface, the less significant they are to landscape studies. Events of erosion, deposition and soil formation require a minimum magnitude before they

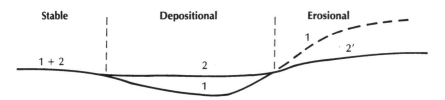

FIGURE 6.1
Principal components of geomorphic surfaces.

can be recognized and mapped across a landscape with some certainty. Therefore, a geomorphic event can be considered as an event that is characterized by a particular geomorphic condition (stable or unstable) that was subjected to a certain geomorphic process(es) (none, slopewash, solifluction, etc.) over some period of time. An event of this nature has a unity or a continuity that is important to a soil–geomorphic concept of a landscape. This concept divides time into intervals or events of geomorphic activity and nonactivity. Only during these events of nonactivity or stability can soil formation occur on the previously unstable geomorphic surfaces.

For the purposes of a soil–geomorphic analysis, geomorphic surfaces can be ranked into two categories. The higher rank is given to the geomorphic surface that is underlain by a soil profile. The lower rank is given to a geomorphic surface that is not underlain by a soil profile. For example, in the Midwest during the Pleistocene, sequential advances of continental glaciers deposited large amounts of drift and caused signficant, localized erosion. Between the events of glacial activity, soils formed on the stable surfaces. Then it follows that the new surfaces on which soils (modern and paleosols) formed are more important to chronological analysis and are given a higher rank than the surfaces that do not relate to soil surfaces. The lower rank surfaces are constructional (glacial, fluvial, eolian, etc.) and erosional surfaces that were soon buried. The geomorphic surfaces that are coincident with soil surfaces at this scale form the basis for the glacial and soil stratigraphy of the Quaternary.

TRUNCATED SOILS

Buried soils, which have been partially truncated, present a special problem in the interpretation of geomorphic surfaces. If the lower part of a profile is spared by an erosional event, the soil-forming processes during the following interval of stability will form a new soil through the old profile. However, in studies on the Sangamon Soil it was found that a new sediment is laid down upon most of the erosion surface as it develops on gentle slopes. This sediment has its main source from the erosion of the old soil surface (Ruhe 1969) or may include significant amounts of eolian additions (Frye *et al.* 1974). The sediment is mainly a slopewash deposit but has been called 'pedisediment' or colluvium.

In concept, the slope sediment generated during partial truncation of the Sangamon Soil may cover a large portion of the area subjected to erosion. Considering the process, the sediment would be expected to accumulate below shoulder slope positions, mainly on back slopes, foot slopes and alluvial positions. No sediment of this origin is expected on the highest parts of the landscape. In the lower landscape positions, the slope sediments merge into the alluvial deposits. During the following soil-forming event, a new soil is formed in the two materials – the new sediment and the remainder of the old soil. The new soil inherits some of the horizons of the old soil and, when the soil-forming processes proceed far enough, the features of the old soil are masked or obliterated by the process of soil welding (Ruhe & Olson 1980). The new soil profile can appear to be unaffected by the buried erosion surface. The old surface can be deduced from lithologic discontinuities in the profile (Ruhe 1956, Follmer 1970, Johnson *et al.* 1972, Frye *et al.* 1974, Follmer *et al.* 1979). Where the old soil contains sand and gravel, the erosion surface is generally indicated by a stone line – a zone or layer enriched in sand and gravel produced by the erosional event (Ruhe 1959).

A problem that comes out of this discussion is the fact that two geomorphic surfaces are identified – an erosion surface on the older truncated profile and a depositional surface on the slope sediment. The two features are closely related in time because the erosion was the cause of the slope sediment. Ruhe (1969) interpreted an erosional event within Sangamonian time in Iowa with these relationships. However, he named the event, the slope sediment and the (younger) soil 'late Sangamon'. More importantly, he did not differentiate the upper and lower boundaries of the late Sangamon sediment as separate geomorphic surfaces, but treated the whole sediment as a geomorphic feature.

Buried soil profiles that contain a partially truncated soil profile are variable because they contain varying portions of the old profile. Most of the problems of partially truncated buried soils stem from this inherent variability and the lack of agreement on criteria for recognition and definition of buried soils and surfaces. Early workers in Quaternary stratigraphy in the Midwest demanded an organic-rich horizon before a soil was recognized (Follmer 1978). After the genetic concept of soil formation was accepted, the B horizon became the key horizon, particularly in the profiles that did not have an organic-rich A or O horizon. Light-colored A horizons were often interpreted as younger material with a younger soil developed in them. The different types of buried soils, from a material point of view, were a major issue between two schools of thought advocated by Willman *et al.* (1966) and Leighton and MacClintock (1962). Their arguments were specifically focused on the Sangamon and Yarmouth Soils and brought much attention to the compositional and stratigraphic significance of the Sangamon in Illinois, but not to its geomorphic attributes.

In many of the studies that identified the Sangamon surface, the Sangamon Soil was interpreted to have been truncated and then covered by a new material that was subsequently weathered. The weathering formed a new soil profile in the new material and the underlying truncated Sangamon Soil or older materials, depending on the area and local circumstances. In Iowa, Ruhe (1969) interpreted this series of events to have occurred within Sangamonian time, therefore naming the new soil the late Sangamon. In Illinois, Frye *et al.* (1974) interpreted the erosional event to have occurred at the end of the Sangamonian. The latter interpretation is largely based on evidence for loess in the new material they called colluvium, in which they found fresh, weatherable minerals and a clay mineral assemblage that indicated a genetic association with the overlying early Wisconsinan loess. Therefore, Frye *et al.* interpreted the new material to be the first deposit of Wisconsinan age and named the new soil the Chapin Soil. The concepts presented in both studies demonstrate that two prime relationships must be considered before the top of the Sangamon Soil can be identified – the relation of erosion events and deposition–burial events to soil horizons. In general, they are both age determinants.

LOESSAL ADDITIONS

The Sangamon Soil at all locations studied in Illinois is, without exception, immediately overlain by loess or organic-rich deposits in stable or depositional environments (Follmer 1978). Where uneroded by the Wisconsinan glaciers in northeastern Illinois, the same relationships hold (Horberg 1953). In the early work, the 'key beds' of organic material and weathered Illinoian till were correlated as the Sangamon. This level of discrimination was satisfactory for the stratigraphic purposes of that time, which used the Sangamon to differentiate between Illinoian and Wisconsinan time and deposits. Later studies showed that the correlation between the organic soil (mid-Wisconsinan) and weathered till (Sangamonian) was inaccurate. The error was caused by the miscorrelation of the early Wisconsinan loess (Roxana Silt) which ranges from about 0.3 m (Follmer 1970, Johnson, *et al.* 1972) to 10 m (McKay 1979). Where the Roxana is about 1 m or more thick, it has been recognized as a silt deposit overlying the Sangamon Soil; but where less than about 1 m thick, it has been included in the upper part of the Sangamon Soil (Thorp *et al.* 1951, Follmer 1970). Where thin, the Roxana is leached, weathered and expresses A and B horizon characteristics. Also, the lower 1–2 m in thicker sections have these characteristics.

In an evaluation of all of the previous studies, several conclusions can be drawn. Many site investigations have been made that identify the surface of the Sangamon and reveal its spatial characteristics over short distances. These studies have been sufficient to establish that the character of the Sangamon surface appears to parallel the character of the modern surface. Previous studies have indicated that a temporal focus is needed, i.e. one geomorphic surface must be identified as the Sangamon surface. The standard practice used in Illinois and in this report for the temporal definition of the Sangamonian/Wisconsinan boundary, is based on the presence of Roxana Silt (or other early Wisconsinan glacial deposits in other areas) upon the uppermost Sangamon Soil horizon present (Willman & Frye 1970). This provides a unique feature that can be identified with careful study and mapped as a unique geomorphic surface. Although previous studies have shown that in many places soil formation continues from Sangamonian into Wisconsinan time, the deposition of new material upon the Sangamon Soil represents a change of conditions. Then the soil formed in the new material represents a younger soil-forming event.

STUDY AREA

An area of about 12 km^2 south of Casey, Illinois, was selected for detailed study (Fig. 6.2). The study area lies on a flat drainage divide more or less surrounded by the headwaters of streams. Leighton *et al.* (1948) included the Casey area in the Springfield Plain, a physiographic subdivision of the Illinoian till plain. Earlier, MacClintock (1929) designated the extensive, nearly level area covering much of the Springfield Plain, including the Casey area, as flat drift-plain. MacClintock described the topography of the area as controlled by non-morainic drift with steep-sided valleys in a dendritic pattern separating areas of flat drainage divides. The upland in the Casey area has a slight regional slope (0.13%) to the south and a maximum local relief of 16 to 25 m. The upland has numerous shallow depressions, many of which are indistinct but can be interpreted from soil maps published by the Soil Conservation Service.

This area was selected because of the clear pattern of modern soils which reflect the underlying Sangamon surface, the typical character of the flat upland divides, and the typical characteristics of the parent materials as determined from 5-m test cores within the flat drift-plain region. Most of the eastern part of the flat drift-plain region is covered by <1.5 m of Peoria Loess and <1.0 m of Roxana Silt. Both thicken to the east and west (Fehrenbacher *et al.* 1965, McKay 1979). In the Casey area and to the south, the presence of the Roxana has been questioned and the Sangamon Soil has been considered as 'normal', overthickened or truncated by previous workers.

A Giddings soil sampling machine was used to collect 59 continuous soil cores through the soil profiles to calcareous Illinoian till. A sufficient number of sites was studied using a grid pattern so that the number of

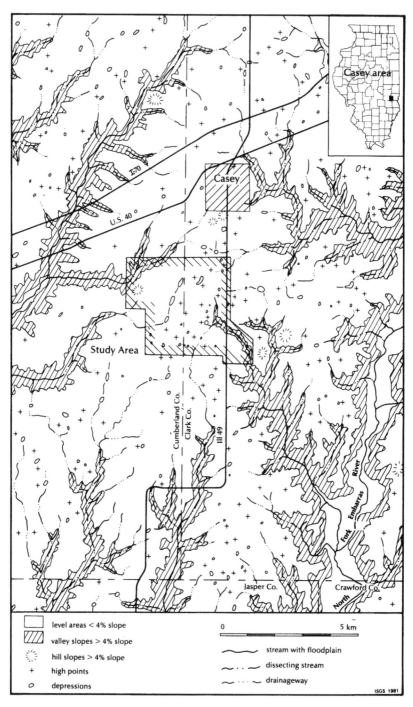

FIGURE 6.2
Physiography of the Casey area.

each modern soil type observed was approximately in proportion to the areal extent of each type (Fig. 6.3). Additional sites were cored for added control on the more prominent topographical features missed by the grid, such as centers of depressions and crests of swells. Several catenas, toposequences from high points to the centers of depressions, were selected for critical study. All cores from all sites were described in pedologic detail (Follmer 1970).

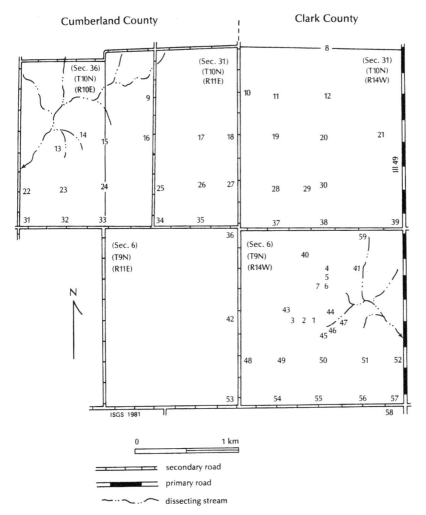

FIGURE 6.3
Site locations in the study area.

Landscape and modern soils

The study area is typical of the flat drift-plain region of the Illinoian till plain – peculiarly flat. Little attention has been given to why the region is so nearly level, but speculations have been made that this character reflects the relatively flat, underlying Pennsylvanian bedrock and the relatively fast advance and retreat of large-scale Illinoian glaciers. The incised streams of the regions form a dendritic pattern which produces somewhat rectilinear tracts of land 2 to 5 km wide and 15 km or more long, with very irregular boundaries caused by the headward advance of lower-order streams. A

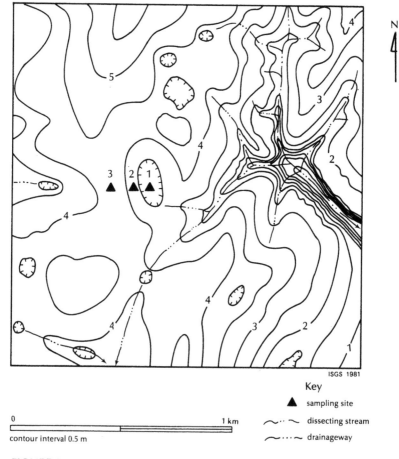

ISGS 1981

Key

▲ sampling site

⌒·· ⌒ dissecting stream

⌒···⌒ drainageway

0   1 km

contour interval 0.5 m

FIGURE 6.4
Topography of Section 6, T. 9 N., R. 14 W., Clark County.

lack of integrated drainageways occurs on the flat upland divides away from the actively eroding streams. Very shallow depressions are common-place across these flat areas, and are occasionally in a series. Man-made drainage improvements have eliminated many of the smaller depressions which make them difficult to recognize, except for the drainage charac-teristics of the modern soils that indicate their former topographic condition.

A very subtle swell and swale topography characterizes most of the flat-divide surface, with many of the swales forming closed depressions. Es-

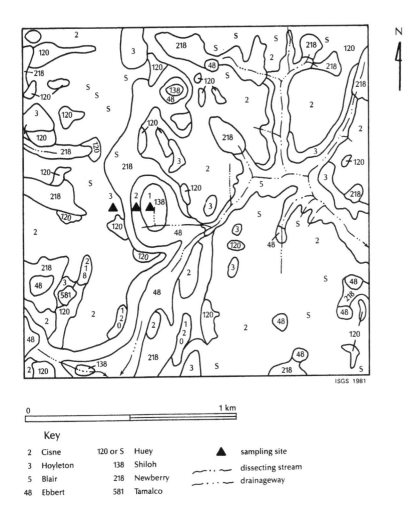

0   1 km

Key

2	Cisne	120 or S	Huey	▲	sampling site
3	Hoyleton	138	Shiloh		dissecting stream
5	Blair	218	Newberry		drainageway
48	Ebbert	581	Tamalco		

FIGURE 6.5
Soil map of Section 6, T. 9 N., R. 14 W., Clark County.

sentially, all of the local slope segments have less than 2% slope, except for the few kame-like features which may have side slopes that range up to 7% in the Casey area. The swell and swale topography forms a cyclic continuum across the upland areas. The cycles are rarely smooth but have ripples of local highs and isolated lows or depressions. For better control, a standard topographic survey of Section 6, T. 9 N., R. 14 W., was performed to determine the elevation of all microtopographic features (Fig. 6.4). The major cycles with more than about 0.6 m relief between high points extend over a distance of about 500 m in the Casey area, and may range up to 5 km on rare occasions in other parts of the region. The contour interval (0.5 m) was not sufficient to reveal most of the shallow depressions that were observed. Therefore, the positions of most of the depressions were interpolated.

The distribution of modern soil types on the upland surface closely follows the microtopographic features. Very poorly drained soils occur in the depressions and drainageways. Poorly drained soils are on the nearly level portions with less than about 2% slope. Somewhat poorly drained soils commonly occur on distinct convex slopes that are in the range of 1% to 5% but are usually too short to show on the map of Section 6 (Fig. 6.5). Better-drained soils are rare and only occur on the summits of distinct swells and hills. None of the better drained types of soils is large enough to be mapped at a scale of 1 : 15 840.

The modern soil Cisne dominates the landscape that has less than about 2% slope and consequently forms the matrix in which all other soils on the flat drift-plain surface occur (Fig. 6.5). Some soils, such as Blair, are exclusively border related where the entrenchment of headward advancing tributaries have generated better local drainage. Blair is a variable soil and occurs on slopes of about 2% or more where the loess cover thins and the Sangamon Soil occasionally outcrops. Hoyleton is a somewhat poorly drained soil that is commonly found around the heads of incised drainageways and on the summits of low swells where the internal and external drainage is better than Cisne. Tamalco is similar in drainage to Hoyleton but is alkaline because of a high exchangeable sodium content and is restricted to shoulder and summit positions. The poorly to very poorly drained Ebbert is found on level to depressional areas. The accretionary, very poorly drained Shiloh soil occurs in the lowest part of the larger or more distinct depressions. Newberry usually lies between Cisne and Ebbert and thereby, in many respects, is transitional between the two.

Another sodium-affected soil is Huey. This unusual soil is an exceptionally poor, agricultural soil and its occurrence is unpredictable from a geomorphic point of view. Small but significant occurrences of Huey are shown with an 'S' on the soil map. Frazee *et al.* (1967) suggested that the underlying till may be more permeable under the Huey than elsewhere. In most areas, Huey is found in an intricate pattern with Cisne and appears on airphotos as light-colored, indistinct mottles.

Field interpretations of the parent materials

The Peoria Loess in Illinois has been studied in great detail because of its continuous cover of nearly all of the present upland surfaces in the state. In spite of the many observations of a weathered silt or loess underlying the Peoria in the state beyond the Wisconsinan border, little attention has been given to the correlation of the Roxana Silt to the regions where it is thin. Because weathering and soil formation have modified the Roxana in these areas, its identity is obscured and questioned. A careful analysis of the principal components, those features most readily recognized in a profile, must be used to construct a model from which the less apparent features can be interpreted. In the Casey area the Peoria Loess, the Sangamon B horizon and calcareous Illinoian till are the principal components most easily identified. Within the undissected part of the area, the Peoria Loess thickness ranges from 1.00 to 1.45 m, averaging 1.25 m on the basis of 59 observations. Underlying the Peoria is weathered material that ranges from 1.8 to 2.7 m in thickness and averages 2.35 m. This weathered zone has been generally interpreted as the Sangamon Soil developed in Illinoian-age materials, but is now known to represent a merging of the Sangamon Soil with the Farmdale Soil (Follmer 1978, Frye et al. 1974, Follmer et al. 1979).

Commonly, the pedologic horizons are poorly expressed in many profiles of the Sangamon Soil, particularly the poorly drained types that tend to be macromassive. The massiveness is due in part to changes caused by burial. Most notable of these changes are the loss of the organic carbon content, an increase in bulk density, and loss of soil structure. These changes constitute retrogressive development. Better-drained types are the most resistant to these changes but lose nearly all of their organic carbon. Other significant changes are in the soil chemistry. Concretions and other precipitates may form as a consequence of post-burial conditions. Base saturation of most buried soils is nearly 100%, which indicates resaturation from a base-containing leachate from overlying materials. Therefore, much care must be taken to distinguish the genuine soil characteristics from those acquired after burial.

In the thin loess region of south-central Illinois, the most diagnostic buried soil horizon is the B2 horizon of the Sangamon Soil. Usually, it is recognized by its blocky soil structure and as a zone of maximum clay accumulation. Some profiles of the Sangamon Soil below a water table are gleys, a zone of reduction (interpreted from the gray, green or bluish colors) described as the B (Bg). These gley soils are poorly drained (containing yellowish-brown to red mottles in the B) to very poorly drained (essentially free of mottles) and are found on level to depressional areas on the Sangamon surface. Normally, the more poorly drained the Sangamon Soil, the more poorly expressed are its horizons.

There are two significantly different types of gleyed Sangamon Soils.

One developed on slowly accumulating materials in very poorly drained positions and is known as an accretion-gley. The other developed on poorly drained, stable upland flats in Illinoian or older materials, it was not subjected to truncation or accretionary processes and is known as an *in-situ* profile. The accretion-gley profile usually has no argillic horizon and usually has a constant to decreasing clay content with depth. The accretion-gley profile is difficult to divide into horizons because of its gradual color and textural changes and its macromassive characteristics. Some accretion-gley soils are quite thick and exhibit vertical continuity in their profiles, which indicate that as accretion occurred the accumulated material was blended with the underlying material and the soil horizons developed upwards, producing overthickened A and Bg horizons. The *in-situ* profile frequently contains an argillic B2 horizon that usually has a distinct blocky structure. If these features are absent, the *in-situ* profile can easily be confused with the accretion-gley. Upper horizons can be observed that commonly express both A and B horizon characteristics, which confounds interpretations. The two types of gleyed profiles can be differentiated on the basis of their B horizons. The very poorly drained accretion-gley usually has a thick B2g horizon and may have some evidence of stratification; whereas, the *in-situ* profile has a B2g with a normal thickness of 0.5 to 0.8 m. On the Illinoian till plain, the *in-situ* profile is most often developed in till or ice-contact deposits that can range from fine to coarse material.

In the thin loess areas of south-central Illinois, two reasonably distinct near-surface planes can be observed – the base of the Peoria Loess and the top of the B2 horizon of the Sangamon Soil. These planes are nearly parallel, giving rise to a zone 0.6 to 1.5 m in thickness that is continuous across the landscape. For the purpose of identifying parent materials, the till-derived portions of the Sangamon Soil are here assigned to zone III, or the third parent material with reference to the surficial Peoria Loess as zone I (Follmer 1970). Therefore, the zone that underlies the Peoria and overlies till-derived soil horizons and contains Roxana Silt and/or accretionary sediments is here described as zone II.

The pedological characteristics of zone II are poorly expressed or confounded. This zone is usually macromassive with tendencies for weak platy or weak granular structure in the upper 'A horizon' position. The lower part of zone II is often macromassive, but with subtle features of both A and B horizon characteristics, such as (i) silt coatings (silans); (ii) a very weak horizontal 'cleavage' or coarse platy characteristics; (iii) random, discontinuous clay coatings (cutans); and (iv) a weak, medium to coarse, angular, blocky structure with coatings (only apparent upon drying).

Historically, zone II has been called fossil soil, gumbotil, or accretion-gley by the geologists, or a buried A (IIAb) horizon, a buried B (IIBb) horizon or a IIC horizon by the pedologist. However, when the most pronounced features of the profile, the base of the Peoria Loess and the Sangamon IIIB2 horizon, are considered, zone II appears to be an A horizon because of its morphological characteristics and its stratigraphic position. In the Casey study area, zone II is a loam to clay loam in texture, ranging in thickness from 0.4 to 0.6 m on the drainage divides and is a silty clay loam to clay loam ranging up to 1.6 m thick in the depressions. The average thickness for zone II is 1.01 m, which is considerably overthickened for an A horizon that has developed under normal processes.

Compositional characteristics of Cisne and Shiloh profiles

Many parameters of the Cisne and Shiloh profiles were investigated, but few revealed any discontinuity between zones II and III (Follmer 1970). Heavy minerals of the 0.062–0.0250 mm fraction were studied and only one suite of minerals was found which is the same as the local calcareous till. Weatherable minerals are depleted, but in an irregular manner. In the three profiles studied, much mixing is indicated from the many fresh, unaltered grains that were found in all weathered samples alongside many partly altered and extremely altered grains. Spectrographic analysis for Ti and Zr revealed an enrichment of about two- or threefold in the medium and coarse silt fraction of the surface horizon of the Modern soil, which decreases with a relatively smooth trend downward to the calcareous till (Tables 6.1 and 6.2). The ratio of $TiO_2 : ZrO_2$ in the medium silt fraction (16–31μm) indicates a discontinuity at the base of the Peoria Loess in the accretionary Shiloh profile, but between zones II and III from the *in-situ* Cisne profile on the drainage divide. The ratio in both profiles reveals an increase at the base of the weathered Sangamon profile, suggesting a discontinuity in parent materials or a depletion of Ti-minerals due to weathering.

Clay mineral distribution in the Shiloh profile (Table 6.1), taken near the center of a depression, is dominated by expanding clay minerals from the Modern soil surface to the underlying till-derived horizons of the Sangamon Soil. At this point in the profile, the content of expandables decreases from about 80% to 17% in the calcareous till, IIIC. The illite content follows an almost inverse trend to that of the expandable and shows a minimum value in the Peoria Loess. Through zone II, the illite content is relatively constant and then, below the base, the illite rises from less than 37% to 70% in the calcareous till. The chlorite plus kaolinite content remains low through the entire weathered profile and shows a slight increase to about 13% in the IIIC horizon. The clay mineral trends in this profile indicate only one very thick solum.

The clay mineral distribution in the Cisne profile (Table 6.2) from the drainage divide is significantly different from the accretion-gley profile.

The contrast illustrates the importance of the local environmental conditions and the effect of these conditions on the clay mineral distribution. The expandable clay mineral content is lower and discontinuities are revealed at the base of the Peoria Loess and the top of the *in-situ* Sangamon Soil, thus delineating zone II. Surficial horizons of soils contain hydroxides, organic matter, mineral degradation products and interstratified clay

TABLE 6.1
Physical and mineralogical parameters of Shiloh, profile 1.

Parent material	Depth (m)	Horizon	Gravel >2 mm	Sand >2 μm	Silt 62-2 μm	Clay <2 μm	16-31 μm / 31-62 μm	TiO_2/ZrO_2 (16-31 μm)	Exp (%)	Il (%)	Ch + K (%)	DI	HSI	Bulk density (g/cc)
Peoria Loess	0–0.15	Ap	t	4	65	31	3.5	2.4	70	21	9	1.5	7	–
	0.15–0.28	A1	t	4	64	32	3.5	2.7	72	17	11	1.0	10	1.43
	0.28–0.38	A3	t	5	62	33	3.2	2.5	74	16	10	1.2	13	–
	0.38–0.51	A3	t	4	60	36	3.6	1.8	79	13	8	1.1	16	–
	0.51–0.64	A3	t	4	58	38	3.3	2.3	81	12	7	1.0	20	1.44
	0.64–0.76	A3	t	4	57	39	3.6	3.1	82	11	7	1.0	20	–
	0.76–0.91	B1	t	3	58	39	4.0	2.4	84	10	6	1.0	25	1.48
	0.91–1.07	B2	t	3	57	40	4.0	3.2	86	9	5	1.0	25	–
	1.07–1.22	B2	t	4	60	36	3.8	3.2	82	11	7	1.0	26	–
	1.22–1.32	B3	t	4	64	32	3.4	3.9	81	12	7	1.1	26	1.42
	1.32–1.42	B3	t	6	64	30	3.0	3.4	80	12	8	1.1	20	–
					Farmdale Soil surface									
Roxana Silt Sandy Silt Facies	1.42–1.50	IIAb	t	14	52	34	2.2	4.1	76	14	10	1.0	11	–
	1.50–1.57	IIAb	t	16	50	34	2.0	3.2	75	14	11	0.9	7	–
	1.57–1.65	IIAb	t	15	50	35	–	–	72	18	10	1.0	8	1.43
	1.65–1.73	IIAb	t	14	51	35	2.3	2.6	73	16	11	1.0	9	–
	1.73–1.80	IIB1b	t	19	46	35	1.9	2.4	78	13	9	1.0	11	–
	1.80–1.87	IIB1b	t	18	46	36	–	–	77	13	9	1.0	10	–
	1.87–1.95	IIB1b	t	17	47	36	1.8	3.0	78	13	9	1.0	10	1.57
	1.95–1.98	IIB1b	t	16	48	36	–	–	78	13	9	1.0	10	–
	1.98–2.08	IIB1b	t	18	45	37	1.9	3.3	80	12	8	1.0	11	–
	2.08–2.14	IIB1b	t	21	43	36	1.9	2.8	80	12	8	1.0	12	–
	2.14–2.21	IIB2b	t	20	43	37	2.0	3.2	82	11	7	1.1	17	1.51
					Sangamon Soil surface									
Berry Clay	2.21–2.29	IIB2b	t	20	41	39	1.4	3.4	82	12	6	1.5	17	–
	2.29–2.43	IIB2b	t	24	38	38	1.5	3.4	82	12	6	1.4	24	–
	2.43–2.54	IIB3b	0.9	32	35	33	1.3	3.3	82	12	6	1.6	27	–
	2.54–2.64	IIB3b	0.9	31	36	33	–	–	82	13	5	1.6	28	–
	2.64–2.74	IIB3b	1.0	29	–	–	–	–	82	13	5	1.6	27	–
	2.74–2.84	IIB3b	1.4	27	40	33	1.4	3.5	82	13	5	1.5	29	1.63
	2.84–2.95	IIB3b	1.2	29	39	32	1.4	3.1	78	17	5	2.7	29	–
					Illinoian till surface									
Illinoian Till	2.95–3.05	IIIB3b	4.6	35	34	31	–	–	55	37	8	3.2	11	–
	3.05–3.15	IIIB3b	2.7	35	38	27	–	–	51	41	8	3.6	14	–
	3.15–3.25	IIIB3b	5.3	39	37	23	1.2	4.6	36	55	9	4.2	6	–
	3.25–3.40	IIIC	7.4	43	39	18	1.1	5.8	17	70	13	3.6	–	2.04

Gravel–weight per cent of bulk sample.
Sand, silt, clay–per cent of <2 mm.
Exp.–expandable clay minerals; Il–illite; Ch + K–chlorite plus kaolinite.
DI–X-ray counts per second 10A peak/7A peak.
HSI–height of 1.7 mm peak in mm.

minerals. Any of these materials can cause erroneous results in the calculation of the clay mineral percentages from peak heights. When the diffraction intensities of the x-ray peaks are low or the background values are unusually high, the clay mineral percentages are not calculated. This phenomenon, in addition to other parameters, serves to identify oxidized, buried A horizons. The Sangamon Soil in Profile 3 has an expandable clay mineral content that decreases from a maximum of 48% in the upper IIIBb to 13% in the calcareous IIIC horizon. The illite content increases with depth from 33% to 75% in the IIIC, which amounts to a 42% depletion of illite in the upper part of the Sangamon Soil. This significant illite depletion has been reported in previous studies on the Sangamon Soil (Willman *et al.* 1966).

The particle-size analysis of the weathered profile in the thin loess area of south-central Illinois is perhaps the most useful in distinguishing the different types of materials in the profiles. Trends in the particle-size distribution reveal zones with uniformity or a curvilinear character. Departures in the trends between zones with relative uniformity are discontinuities. Where a particle-size discontinuity is caused by a change in parent materials, it is a lithologic discontinuity. Where a discontinuity is caused by horizonation processes, it is a pedologic discontinuity. The character of the trends and discontinuities helps to distinguish between pedogenic effects and inherited characteristics. Parent materials such as loess and till are relatively uniform within each unit and most often contrast with each other in some parameter. Subsequent weathering and soil formation change the parent material with an initial uniformity to a nonuniform material which is expressed in smooth, often curved trends. However, some parent materials are known to be non-uniform and must be dealt with individually. Pedologic discontinuities such as shown between the A2 and B2 in the Cisne profile are caused by eluviation–illuviation processes.

Depth to calcareous till IIIC ranges from 3.0 to 4.4 m across the nearly level divide in the study area (Follmer 1970). The calcareous till is characteristically dense and has a gravel content that usually ranges from 5% to 7%. In the Shiloh profile, the gravel content trend changes at 2.95 m, which suggests a boundary with a fining upward sequence of accretionary material. This interpretation is supported by the trends in the other particle-size parameters (Table 6.1). The gravel trend in the Cisne profile continues upward, with a slight decrease through the Sangamon Soil horizons formed in the till, until it drops to less than 2% in a horizon with A2 horizon characteristics (Table 6.2). Above this horizon, the gravel content ranges from 1% to 2% in a 0.4 m layer interpreted to be zone II that expresses a mix of A and B horizon characteristics. The Peoria Loess in both profiles has only a trace of gravel as well as zone II in accretionary environments.

The sand content in both profiles decreases upward through the Sangamon Soil and zone II. The relatively smooth trending decrease can be explained by weathering, but note that the silt content continues to increase upwards across the Sangamon boundary regardless of the type of soil horizon. Clay content trends are clearly affected by the pedogenesis causing the lower values in the Sangamon A2 horizon. The relatively smooth trends of the sand, silt and clay parameters that pass through the

TABLE 6.2
Physical and mineralogical parameters of Cisne, profile 3.

Parent material	Depth (m)	Horizon	Particle size parameters					$\frac{16-31\,\mu m}{31-62\,\mu m}$	$\frac{TiO_2}{ZrO_2}$ (16–31 µm)	Clay mineral parameters					Bulk density (g/cc)
			Gravel >2 mm	Sand 62-2 µm	Silt 62-2 µm	Clay <2 µm			Exp (%)	Il (%)	Ch+K (%)	DI	HSI	
	0–0.20	Ap	t	16	67	17	3.3	3.0	–	–	–	–	–	–
	0.20–0.36	A21	t	15	70	15	3.4	2.6	–	–	–	–	–	1.56
	0.36–0.43	A21	t	13	70	17	3.5	2.9	–	–	–	–	–	1.51
	0.43–0.48	A2B2	t	6	61	33	3.5	2.8	–	–	–	–	–	1.57
Peoria Loess	0.48–0.56	B2	t	6	55	39	3.1	2.8	60	24	16	1.0	7	1.57
	0.56–0.64	B2	t	7	54	39	3.5	–	66	20	14	0.9	8	1.48
	0.64–0.71	B2	t	7	53	40	3.4	3.1	69	19	12	1.0	8	1.51
	0.71–0.81	B2	t	7	56	37	3.5	2.7	71	19	10	1.2	9	1.57
	0.81–0.91	B3	t	8	59	33	3.3	2.6	70	21	9	1.5	9	1.64
	0.91–1.02	B3	t	8	59	33	–	–	71	20	9	1.5	10	–
	1.02–1.12	B3	t	9	60	31	3.5	2.7	72	19	9	1.6	11	–
	1.12–1.17	B3	t	10	61	29	–	–	73	19	8	1.7	11	1.64
					Farmdale Soil surface										
	1.17–1.27	IIB/A	1.0	19	52	29	2.4	3.1	64	24	12	1.3	6	1.79
Roxana Silt Sandy silt facies	1.27–1.37	IIB/A	1.1	22	50	28	–	–	–	–	–	–	–	1.78
	1.37–1.47	IIB/A	1.9	21	51	28	2.3	3.4	66	22	12	1.3	7	–
	1.47–1.57	IIB/A	1.6	25	48	27	–	–	69	19	12	1.1	8	–
					Sangamon Soil surface										
	1.57–1.68	IIIA2b	2.5	33	44	23	1.7	3.8	–	–	–	1.0	–	1.83
	1.68–1.78	IIIA2b	2.8	33	44	23	2.1	3.3	–	–	–	1.0	–	–
	1.78–1.88	IIIB1b	2.3	31	40	29	–	–	40	40	20	1.2	–	–
	1.88–1.98	IIIB1b	4.2	32	43	25	–	–	48	33	19	1.2	–	–
	1.98–2.08	IIIB1b	3.4	35	38	27	1.7	2.9	32	47	21	1.4	–	–
	2.08–2.18	IIIB1b	2.1	33	40	27	1.5	2.9	41	41	18	1.5	–	–
Illinoian till Glasford Formation	2.18–2.31	IIIB1b	3.4	31	38	31	–	–	39	42	19	1.4	–	–
	2.31–2.44	IIIB2b	3.6	33	34	33	1.5	2.6	40	42	18	1.6	–	–
	2.44–2.54	IIIB2b	6.3	34	32	34	1.4	2.7	43	39	18	1.6	–	–
	2.54–2.67	IIIB2b	3.9	33	33	34	–	–	36	45	19	1.6	–	–
	2.67–2.79	IIIB2b	4.5	33	34	33	–	–	32	52	16	2.1	–	1.79
	2.79–2.92	IIIB3b	3.5	40	30	30	1.2	3.0	33	53	14	2.5	–	–
	2.92–3.05	IIIB3b	3.8	43	28	29	1.1	3.9	26	63	10	4.0	–	–
	3.05–3.15	IIIB3b	2.3	55	24	21	–	–	20	69	10	4.3	–	1.81
	3.15–3.25	IIIB3b	4.7	53	24	26	21	–	–	–	–	–	–	–
	3.25–3.35	IIIB3b	6.0	48	29	23	0.9	4.6	21	70	9	5.0	–	–
	3.35–3.50	IIIB3b	4.2	48	30	22	1.0	4.4	12	78	10	5.1	–	–
	3.50–3.60	IIIC	5.8	46	28	22	1.1	4.7	13	75	12	4.1	–	2.12

Gravel–weight per cent of bulk sample.
Sand, silt, clay–per cent of <2 mm.
Exp.–expandable clay minerals; Il–illite; Ch + K–chlorite plus kaolinite.
DI–X-ray counts per second 10A peak/7A peak.
HSI–height of 1.7 mm peak in mm.

Sangamon surface indicate that soil formation was not interrupted but continued as new material was added to the developing profiles. The top of zone II is clearly marked by the decrease in sand and increase in silt at the contact with the Peoria Loess. This boundary is the most prominent and consistent morphological characteristic observed in the study area and is a characteristic found in most profiles in the loess-covered Illinoian till plain in Illinois.

Silt fractions were studied in order to determine more precisely the nature of the upward increasing silt content through zone II. Loess near its source is coarse, having a slightly larger 31–62 μm fraction than the 16–31 μm fraction. Together, these fractions make up about 60% to 65% of the total loess (Follmer *et al.* 1979). Within a short distance from the source, the 16–31 μm fraction becomes dominant and the coarser fractions decrease to low values (Follmer 1970). In the study area, the Peoria has a distinct mode in the 16–31 μm interval, referred to as the medium silt fraction, which ranges from 19.0% to 26.4% of the total <2 mm fractions (Follmer 1970). The 31–62 μm interval, coarse silt, in the Peoria ranges from 5.5% to 8.5% and yields medium to coarse silt ratios of 3 : 4 (Tables 6.1 and 6.2). Glacial till in the study area is characterized by nearly equal-size fractions with very small, inconsistent modes in the silt and fine sand fractions with values of 6.0% to 12.0% (Follmer 1970). The lowermost till-derived horizons have about equal amounts of medium and coarse silt, which range from 6.0% to 10.3% and yield ratios of 0.9 : 1.2 (Tables 6.1 and 6.2). With the underlying till for reference, zone II is definitely enriched with silt which contains 15.1% to 19.3% medium silt (Follmer 1970). The coarse silt content in all profiles remains relatively constant – 7.5 ± 1.5% – through all three zones, while the medium silt increases upwards reaching its maximum values in the Peoria Loess. Therefore, as the silt ratio increases upwards it is due to the increase of the medium silt content.

The medium: coarse silt ratio excludes most of the pedogenic changes that would affect each fraction individually during weathering. The silt fractions coarser than about 20 μm are relatively immobile compared to clay. Therefore, the trend of the silt ratio through a soil will reflect the nature of the parent material plus the effects of additions and removals of mobile constituents. Because both the 16–31 μm and 31–62 μm fractions are subjected to nearly the same concentration and dilution effects, a ratio of these fractions would remove much of the pedological effects and would reveal a close approximation of the initial silt-size distribution characteristics.

A soil that is developed in a single material should have a constant silt ratio trend through the profile. Indeed the Cisne profile, on a stable drainage divide, does have a nearly constant trend through the Peoria Loess. However, Shiloh has a curved trend with ratios up to 4.0, which suggests that the finer fractions were enriched during the accretionary origin of the parent material. A distinct discontinuity is evident when zone II is encountered in both profiles where the ratios drop to 2.2 and 2.4, respectively. But from that point downward there is no discontinuity through the remainder of the Cisne profile. Also, the ratio is not constant but decreases with depth. The values greater than about 1.0 indicate that zone II is enriched with medium silt and the gradually decreasing ratio suggests that a silty material is blended with the till with decreasing effects downward. The silt ratio in the Shiloh decreases downward but appears to be incremental, as might be expected from an accretionary soil in a depression. A small discontinuity is present in the middle of zone II at 2.21 m in the Shiloh profile, where the ratio decreases frm 2.0 to 1.4. The lower portion is more like till and is interpreted to be slopewash (Berry Clay) derived from weathered Illinoian drift during Sangamon Soil formation. The upper portion is not like till because it contains nearly a twofold enrichment of medium silt and essentially no particles >2 mm. This portion is interpreted to be derived from slopewash that had been enriched with a medium silt-sized material.

The results of the laboratory studies, combined with the field observations, support the interpretation that zone II is a silt-rich zone that is continuous across the Illinoian till plain. The only mechanism that can explain an enrichment of the medium silt over the coarse silt in a deposit that is continuous across an undulating landscape is an eolian depositional process. Weathering and sedimentological processes are incapable of generating an enrichment of medium silt over coarse silt on the divides of an undulating topography. Therefore, zone II is correlated to the sandy silt facies of the Roxana Silt, which was informally named by Johnson *et al.* (1972), partly based on the work by Follmer (1970). Follmer calculated the amount of Roxana of loessal origin that would be necessary to explain the medium silt enrichment and found that it is equivalent to a layer of loess 0.20 – 0.35 m thick. This accounts for 50% to 87% of the total thickness of zone II in the Cisne profile. Also, the heavy mineral analysis of the sand fraction showed that it came from the underlying horizons, which accounts for another 19% to 25% of the total material in zone II of the Cisne profile.

Sangamon surface criteria

Where the Roxana is thick, it is the sole parent material for the Farmdale Soil (Willman & Frye 1970). But where the Roxana is thinner than the Farmdale solum, the Farmdale Soil inherits the underlying material, which is usually the Sangamon Soil developed in older materials, such as in the Casey study area. However, the field relationships and the data reveal that with the small amount of Roxana deposited upon the Sangamon surface, the soil formation did not appear to stop. The added material simply developed upper solum characteristics because of its position at the paleoground surface. Even in the areas of thick Roxana, the early increments were subjected to soil-forming processes before complete burial by what appears to be a rapid increase in the deposition rate.

In landscape positions such as Cisne, profile 3, the Roxana can be distinguished from the Sangamon A2 horizon, but it has upper solum characteristics instead of the characteristics of an unaltered deposit. The soil horizons are confounded here. The zone II is too thick to be a Sangamon A1 horizon, but the lower part may have been. The modern solum is influencing this zone by superposing lower B horizon characteristics on it, such as argillans. Therefore, in the present circumstances, it is treated as one horizon that developed much of its characteristics during the formation of the Farmdale Soil.

In the accretionary environments, such as Shiloh, profile 1, the Roxana is masked by the soil-forming processes that continued during the accumulation of the Roxana. In many places, no physical feature can be observed other than the silty nature of the upper solum that is a part of a pedologic continuum downward into the till-derived horizons. Rarely can A horizon characteristics, such as organic matter enrichment and granularity, be observed in the upper part of the till-derived accretionary material because the superposed B horizon characteristics, principally reformed blocky peds, mask the former features. Also, organic matter is degraded in a B horizon environment.

Therefore, for purposes of identifying geomorphic surfaces in this study, the Farmdale Soil surface is recognized to be the contact between the Peoria Loess and the early Wisconsinan-aged, silt-enriched deposits that have A and/or B horizon characteristics. Even though this contact has been interpreted to be the top of the Sangamon Soil in a general sense, the Sangamon Soil surface is recognized here in the strict sense to be the contact between the Wisconsinan-age deposits and the Sangamonian, Illinoian or older deposits that also have A or B horizon characteristics. The estimates of the ages of the Farmdale and Sangamon surfaces are 24 000 to 26 000 yr BP for the Farmdale (McKay 1979) and about 75 000 yr BP for the Sangamon (Frye *et al.* 1974).

Sangamon and Farmdale surfaces

The Cisne–Shiloh catena represents a typical soil sequence in the flat Illinoian till plain of south-central Illinois. A common variation in the pattern is Hoyleton on the summits and Ebbert in the depressions and drainageways. In the preparation of the soil map, it was found that the variations in the soil patterns, except for Huey, directly relate to the geomorphic features of the area. Also in the study of the 59 soil sites, it was found that the relief on the Farmdale and Sangamon surfaces is greater than the modern surface, with the relief on the Sangamon being the greatest (Fig. 6.6). The Sangamon surface was constructed by an Illinoian glacier and then locally modified by erosion. All of the sediments produced in the Cisne–Ebbert segment (Fig. 6.6) must have been deposited in the Shiloh area because of it being in a closed depression. However, this sediment had to be identified by its composition. It has no morphological distinction from the overlying sediments because of blending by soil-forming processes.

Erosion continued through Sangamonian time but at a slow enough rate that normal, well-developed soil horizons formed. Then, during and following the deposition of the Roxana, about one-quarter to one-half of the Roxana on the Cisne–Ebbert slope (Fig. 6.6) was eroded and redeposited in the depression. Soil formation during the Farmdalian time developed a well-expressed, dark-colored IIAb horizon at the Ebbert and Shiloh sites, and a weakly expressed A horizon at the Cisne site. Because the sand has the same composition through all of the profiles, pedogenic mixing

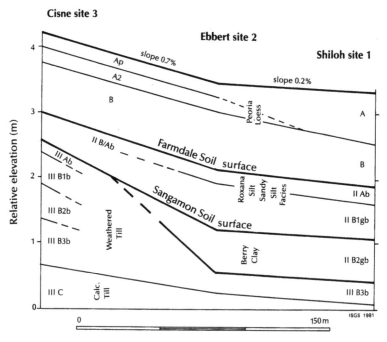

FIGURE 6.6
Cross-section through the Cisne–Shiloh catena.

must have been active at this time to explain the 14% to 25% sand content in the Roxana zone. The base of the Peoria Loess is distinct at these sites and was observed to be somewhat blurred to quite distinct in all of the other 56 profiles studied. This indicates that the Peoria was deposited rapidly in comparison to the rate of pedogenic mixing at this time. Erosion during and following the deposition of the Peoria left 1.17 m of Peoria at the Cisne site and produced 1.42 m of loess and accreted deposits at the Shiloh site. This suggests that the Cisne site had a net loss of about 0.12 m and the Shiloh site had a net gain of about 0.12 m. In summary, the net result of the Wisconsinan erosion and deposition in the region has been to make a significant contribution to the flatness of the Illinoian till plain.

Sangamon Soil map

The Sangamon and Farmdale Soils are buried and cannot be examined in the same detail as the soils on the present surface. Another limiting factor for mapping is they do not appear to vary as much as the modern soils because the chemical and color characteristics appear to have converged, making fewer separations possible. However, drainage and related characteristics were found that distinguished three mappable Sangamon Soil units. Therefore, a map of the Sangamon surface was prepared for Section 6 where the largest amount of information was available (Fig. 6.7). The Sangamon Soil units on summits of swells in some places are moderately well to well drained, but on the smaller and lower swells that are underlain by till the Sangamon is most often somewhat poorly drained.

The somewhat poorly drained units are *in-situ* soils developed in till and are podzolic in nature (Hapludalfs or Hapludults), having A2 and argillic horizons. The zones I, II and III are distinct at these locations. The poorly drained unit forms a continuum of gleyed *in-situ* soils developed in till, which commonly grade upwards into the Roxana. The Sangamon here appears to range from a podzolic soil (Ochraqualf) to a Humic Gley (Haplaquoll) that lost its humus. The very poorly drained units are strongly gleyed and are primarily accretion-gley soils that have till-derived horizons at the base of the weathering profile. The accretion–till boundary is commonly gradational but is sometimes sharp, distinct and overlain by a lense of sand and gravel. The very poorly drained unit is commonly found in small, isolated spots outcropping along the eroded area. Therefore, the poorly and very poorly drained units are contained in the modern Blair soils that occupy the slopes in the eroded area and the Hoyleton soils occupy the flat areas next to the incised streams. The accretionary Sangamon Soil appears to be a Humic Gley (Cumulic Haplaquoll) that lost its humus.

Sangamon geomorphic surface

The modern surface on the thin loess-covered Illinoian till plain in south-central Illinois reflects to a high degree the spatial attributes of the Sangamon surface. An allowance must be made for the levelling effect caused by the burial and localized redistribution of the Roxana Silt and Peoria Loess. This causes the Sangamon surface to be nearer to the surface under the swells than in the closed depressions. In the Casey area, the relief on the Sangamon surface is estimated to be about 1 to 3 m greater than on the modern surface between corresponding high and low points.

The modern soil types and their microtopographical position serve as dependable guides as to the type and depth to the Sangamon Soil in most cases. The general relationship between the modern and Sangamon Soil types follows a simple pattern. In the study area, Cisne and the adjacent

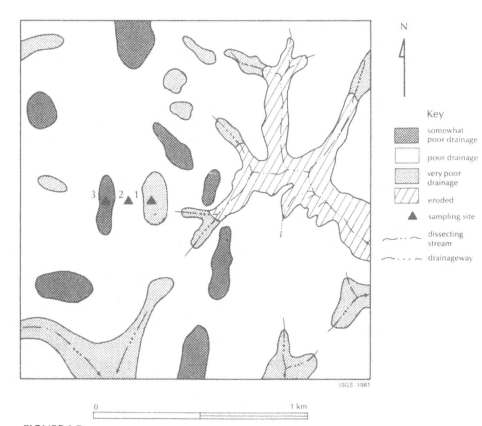

FIGURE 6.7
Map of Sangamon Soil of Section 6, T. 9 N., R. 14 W., Clark County.

Newberry and Huey soils most often overlie the poorly drained *in-situ* Sangamon Soil that is about 1.5 to 1.6 m below ground surface. Ebbert and Shiloh occur over the very poorly drained, accretionary Sangamon Soil that ranges from about 2 to 3 m below ground surface. Hoyleton and Tamalco always occur over a somewhat poorly or better-drained Sangamon, except for the Hoyleton near the dissecting streams where it is gleyed. The depth to the Sangamon under these soils is about 1.4 m, but may be locally as shallow as 1 m. The drainage class of the Sangamon and the modern soil tends to correspond most of the time. However, the modern, poorly drained soils tend to be slightly more extensive than the Sangamon poorly drained unit, which is a consequence of the levelling effect. For example, a change in the drainage class at a location can be observed at site 3, where the poorly drained Cisne overlies a somewhat poorly drained Sangamon. A guide for determining the relation of the modern soil to a relative depth to the Sangamon Soil is given in Table 6.3. The relative depth is largely a function of the landscape position and the average loess thickness covering the area. In other parts of the Illinoian till plain where the average loess thickness is greater, the depth values must be increased, but the general relationships remain about the same. Therefore, a reasonably accurate appraisal of the Sangamon surface can be made where modern Soil Survey maps are available within the region of the Illinoian till plain.

TABLE 6.3
Guide for the determination of the Sangamon surface.

Modern soil	Location on landscape	Sangamon drainage class	Relative depth to Sangamon (m)
Blair	valley slopes	poor–very poor	<1.0
	convex divides	moderate–well	1.1
	slightly convex divides	somewhat poor	1.4
Hoyleton	valley borders	poor–very poor	1.0
	convex divides	moderate–well	1.1
Tamalco	slightly convex divides	somewhat poor	1.4
	slightly convex divides	somewhat poor	1.5
	nearly level flats	poor	1.6
Cisne	depression borders	poor	2.0
	slightly convex divides	somewhat poor	1.5
Huey	nearly level flats	poor	1.6
	nearly level flats	poor	1.6
Newberry	slight depressions	poor	2.25
	slight depressions	poor–very poor	2.25
Ebbert	depressions	very poor	2.5
Shiloh	depressions	very poor	3.0

CONCLUSIONS

The spatial and temporal attributes of the Sangamon surface are difficult to assess in a precise manner. The recognition and mapping of the Sangamon surface require careful field studies and laboratory analysis. For a general characterization, the maximum error in the determination and accurate mapping of the elevation of the Sangamon surface is represented by the thickness of the sandy silt facies of the Roxana Silt. This is an early Wisconsinan deposit that in general appraisals has been included with the Sangamon Soil where thin, because it is pedologically altered and forms a pedological continuum with the Sangamon horizons developed in older deposits below.

On the basis of pedological criteria, in combination with the correlation studies of the parent materials, the top of the Sangamon is recognized to be the uppermost Sangamon Soil horizon that is in contact with an A or B horizon that has developed in early Wisconsinan sediments. By this definition, the Sangamon Soil is recognized as the product of the soil-forming event that preceded the Wisconsinan glaciation. This is in keeping with the original intent of the definition. The soil-forming event that followed is recognized in the region to have produced the Farmdale Soil, which can be physically correlated from the Casey area to the soil developed on thick Roxana to the west. However, even in the thick deposits of Roxana, the lower 1–2 m of the sequence has soil characteristics that are comparable to those in the sandy silt facies. This indicates that soil formation continued into the Wisconsinan and was not significantly reduced until local depositional rates of the Roxana exceeded a certain threshold. In the thin, loess-covered Illinoian till plain, this threshold was not reached and each increment of Roxana was incorporated into the developing soil.

The general effect of the loess deposition in the Casey study area was to inhibit the development of surface drainage and to contribute to the levelling of the till plain. Comparing the Sangamon surface to the modern surface in the area, the Sangamon has more relief by about 1 to 3 m; the Sangamon has more better-drained soils and more very poorly drained soils. The Sangamon is dominated by soils developed in Illinoian till and ice-contact deposits. Only localized erosion and deposition effected the Sangamon surface on the flat upland portions of the area.

The principal geomorphic process that operated during Sangamonian time and has continued up to the present in the area is a headward erosion of entrenching streams that are progressing into the low areas underlain by the very poorly drained, accretionary soils on the Sangamon surface. This process is linking the buried Sangamonian accretion–gleys which then outcrop on the eroding slopes in a discontinuous pattern that reflects the distribution of depressions on the original surface.

ACKNOWLEDGEMENTS

The main study was supported by the Department of Agronomy, University of Illinois at Urbana-Champaign. The extension of the original study to its present form was supported by the Illinois State Geological Survey.

REFERENCES

Butler, B. E. 1959. *Periodic phenomena in landscapes as a basis for soil studies.* CSIRO Australia, Soil Publ. 14.

Daniels, R. B., E. E. Gamble and J. G. Cady 1971. The relation between geomorphology and soil morphology and genesis. *Adv. Agron.* **23**, 51–88.

Fehrenbacher, J. B., J. L. White, H. P. Ulrich and R. T. Odell 1965. Loess distribution in southeastern Illinois and southwestern Indiana. *Soil Sci. Soc. Am. Proc.* **29**, 566–72.

Follmer, R. 1970. *Soil distribution and stratigraphy in the Mollic Albaqualf region of Illinois.* Unpubl. PhD dissertation. Urbana: University of Illinois.

Follmer, R. 1978. The Sangamon Soil in its type area – a review. In *Quaternary soils,* W. C. Mahaney (ed.), 125–65. Norwich: Geo Abstracts.

Follmer, R., E. D. McKay, J. .A. Lineback and D. L. Gross 1979. *Wisconsinan, Sangamonian, and Illinoian Stratigraphy of Central Illinois.* Friends of the Pleistocene, Midwest Section, 26th Ann. Field Conf., Illinois Geol. Surv. Guidebook 13.

Frazee, C. J., R. T. Odell and J. B. Fehrenbacher 1967. Hydraulic conductivity and moisture regimes in solonetzic and associated soils in south-central Illinois. *Soil Sci.* **105,** 362–8.

Frye, J. C., H. B. Willman and H. D. Glass 1960b. *Gumbotil, accretion-gley, and the weathering profile.* Illinois State Geol. Surv. Circ. 295.

Frye, J. C., P. R. Shaffer, H. B. Willman and G. E. Ekblaw, 1960a. Accretion gley and the gumbotil dilemma. *Am. J. Sci.* **258,** 185–90.

Frye, J. C., R. Follmer, H. D. Glass, J. M. Masters and H. B. Willman 1974. *Earliest Wisconsinan sediments and soils.* Illinois State Geol. Surv. Circ. 485.

Horberg, C. L. 1953. *Pleistocene deposits below the Wisconsin drift in northeastern Illinois.* Illinois Geol. Surv. Rept. Inv. 165.

Johnson, W. H., L. R. Follmer, D. L. Gross and A. M. Jacobs 1972. *Pleistocene stratigraphy of east-central Illinois.* Illinois State Geol. Surv. Guidebook Series 9.

Leighton, M. M., G. E. Ekblaw and L. Horberg 1948. *Physiographic divisions of Illinois.* Illinois Geol. Surv. Rept. Inv. 129, reprinted from *J. Geol.* **56,** 16–33 (1948).

Leighton, M. M. and P. MacClintock 1962. The weathered mantle of glacial tills beneath original surfaces in north-central United States. *J. Geol.* **70,** 267–93.

Leverett, F. 1899. *The Illinois glacial lobe.* U.S. Geol. Survey Monograph 38.

MacClintock, P. 1929. *Physiographic divisions of the area covered by the Illinoian drift sheet in southern Illinois.* Illinois State Geol. Surv. Rept. Inv. 19.

McKay, E. D. 1979. Wisconsinan loess stratigraphy of Illinois. In *The Wisconsinan, Sangamonian, and Illinoian Stratigraphy in Central Illinois,* 95–108. Friends of the Pleistocene, Midwest Section, 26th Ann. Field Conf., Illinois State Geol. Surv. Guidebook Series 13.

Ruhe, R. V. 1956. Geomorphic surfaces and the nature of soils. *Soil Sci.* **82,** 441–55.

Ruhe, R. V. 1959. Stone lines in soils. *Soil Sci.* **87,** 223–31.

Ruhe, R. V. 1969. *Quaternary landscapes in Iowa*. Ames: Iowa State University Press.

Ruhe, R. V. 1974. Sangamon paleosols and Quaternary environments in midwestern United States. In *Quaternary environments*. W. C. Mahaney (ed.), 153–67. Series in Geography, Geographical Monographs 5. Toronto: York University.

Ruhe, R. V. and C. G. Olson 1980. Soil welding. *Soil Sci.* **130,** 132–9.

Simonson, R. W. 1954. Identification and interpretation of buried soils. *Am. J. Sci.* **252,** 703–32.

Thorp, J., W. M. Johnson and E. C. Reed 1951. Some post-Pliocene buried soils of the central United States. *J. Soil Sci.* **2,** 1–19.

Willman, H. B. and J. C. Frye 1970. *Pleistocene stratigraphy of Illinois*. Illinois State Geol. Surv. Bull. 94.

Willman, H. B., H. D. Glass and J. C. Frye 1966. *Mineralogy of glacial tills and their weathering profiles in Illinois. Part II – Weathering profiles*. Illinois State Geol. Surv. Circ. 400.

7

Spatial and temporal variations in karst solution rates: the structure of variability

D. C. Ford and J. J. Drake

INTRODUCTION

This essay upon the karst geomorphic system focuses on the spatial and temporal validity of process data alone. It is not concerned with, for example, morphometric data, although this is collected by students of karst landforms as it is by others (e.g. Williams 1972). The process of interest is the aqueous solution of minerals. This predominates within the karst system, so distinguishing it from others. Halite, gypsum and anhydrite, calcite and dolomite are the significant karst minerals. Our discussion is limited to the two latter, the constitutents of limestone and dolostone, because these are the most widespread and host the greatest wealth and variety of karst landforms. Dissolution of calcite and dolomite in natural circumstances is dominated quantitatively by aqueous reactions with CO_2, producing the class of waters that is designated 'bicarbonate type' by hydrochemists. At the global scale, this is the most abundant type of fresh water found in the surficial and shallow subterranean domains.

In the presence of water there is no dynamic threshold of calcite or dolomite solution (Ford 1980, p. 348); reactions may proceed in a static pool or at any stage of flow in a channel. In terms of the theme of this present Symposium, one consequence is that the karst erosion system is more dampened or stable than most others, i.e. it is less prone to the impact of extreme events because there is no critical threshold or series of thresholds of stress to be exceeded before significant morphologic work may be done. Compared to the fluid mechanical complexities of clast erosion, the carbonate solution system is comparatively simple within itself, as well. The findings of theoretical and experimental work are more readily transferable to the natural environment as a result. Langmuir (1971) and Plummer and Wigley (1976) provide recent instances of the applicability of laboratory-determined equilibrium constants to the descriptions of chemical equilibria in karst waters; Blumberg and Curl (1974) and Glew and Ford (1980) show that theoretical models, direct hardware simulations, or the two together may offer satisfactory quantitative explanations of the kinetic circumstances that create particular karst landforms.

Note that 'the presence of water' is a necessary threshold condition for karst solution. If the faucet is turned off, for example by introduction of a very arid or cold phase to an area, different erosion processes will become dominant and introduce the complexities of polygenetic development to any previous karst landform assemblages. In a majority of karst areas that are actively developing today, it is generally possible to assert that the entire postglacial period at least (10 000 years or more) has constituted a single phase of metastable erosion, in the sense of Schumm (1977, p. 5). This can often be demonstrated by a class of evidences that is quite independent of the kinds of modern solution and precipitation studies described below. For example, at the close of the last glaciation, erratic blocks of insoluble rock were deposited directly onto ice-scoured limestone surfaces in Yorkshire, England (Sweeting 1973, p. 99) and elsewhere. They have functioned as umbrellas, shielding the limestone beneath so that it now stands 20 cm or more proud of a solution surface. In many cases there is no evidence of such limestone plinths being damaged by competing processes, so that net solutional erosion must have prevailed since the withdrawal of the ice. At a timescale of decades, Goodchild (1875) showed that such solutional stability had prevailed by investigating the weathering of dated inscriptions in marble headstones in the north of England. Similar studies of the limestone facing blocks on the pyramids at Giza indicate such stability for a timespan of 4000 yr in the very different climatic circumstances of lower Egypt.

There are now many sources of solutional data of local or regional (basin) scale available to the karst geomorphologist. From an early data set, Corbel (1959) published a famous study, *La vitesse de l'érosion en terraine calcaire*, which contended that limestone solution rates were greatest in cold regions. As this appeared to contradict the morphological evidence of maximum solution occurring in humid tropical settings, it stimulated many karst geomorphologists to commence regional measurement of solution rates, e.g. Gerstenhauer and Pfeiffer (1966). As a consequence, detailed process studies commenced in karst some years before they became the fashion in many other branches of geomorphology, so that there are a greater variety and longer records to inspect. In addition, there are long, unbroken data sequences from governmental water quality monitoring agencies. It can be said that data coverage of cold and temperate regions is now quite good (see reviews in Smith & Atkinson 1976, Drake 1981). Unfortunately, it remains sparse for tropical humid environments.

In this paper we shall discuss features of limestone solution data sets at the local, basin and global scales. But before doing so, we introduce a new feature into analytical reviews of this kind by considering the converse of limestone solution, i.e. calcite precipitation that occurs at sites within the karst cascading system or basin. The work described is recent and ongoing, and it has not been focused upon the determination of precipitation rates. However, it is hoped to show the reader that it offers the possibility of rather accurately determining the duration of metastable phases and the positions of thresholds (extrinsic and intrinsic) between them during the last few hundred thousand years in karst systems in a wide range of spatial settings.

THE PRECIPITATION OF CALCITE IN LIMESTONE CAVES

The limestone groundwater basins that we study are, in almost all cases, in a net erosional condition because there is net discharge of solutes at the boundary springs. Where the basins are soil mantled and vegetated there is consensus that most of the solution (70 + %) takes place at the soil/bedrock interface or above it on carbonate clasts in the soil. From the interface, a myriad of threadlike channels drain infiltration water into the lower aquifer, where they amalgamate to discharge at the springs. A small proportion of these channels may be intercepted by air-filled caves, where a part of the dissolved load that they transmit may be precipitated as calcite containing trace amounts of other elements that are in solution. The principal morphologic forms adopted by the precipitates are the gravity forms – stalactites, stalagmites and flowstones. At a site, accumulations may total many meters. Rates of accumulation of 'hard' calcite (i.e. non-evaporitic kinds) with which we are concerned appear to range between 0.1 and 10.0 cm per 1000 yr. As they grow, these accumulations are recording something of erosional conditions at the soil/bedrock interface above.

Over the lifespan of a given calcite deposit (e.g. a stalagmite), one of three different conditions may prevail at different times.

(a) Water is supplied from a microfissure and there is net deposition of calcite under metastable conditions. These conditions may be of strictly constant deposition, or of continuous deposition but with variation in rate that may be of seasonal origin, or deposition with seasonal cessation and (perhaps) minor erosion, and/or deposition with a periodic cessation caused by inundation of the cave.

(b) Water is supplied from the same fissure but now dissolves the deposit in net terms – this is re-solution. A threshold has been passed and a new metastable condition exists. The threshold may be of extrinsic origin, in which case a radical change of conditions at the soil/rock interface is indicated; or it may be an effect intrinsic to the aquifer, occurring where solvent water from a different ultimate source has been routed into the feeder channel. At a site where many stalagmites with different point sources of water all display net re-solution, the operation of extrinsic factors may reasonably be supposed. Re-solution of cave calcite is a common phenomenon.

(c) No water is supplied from the source. Calcite growth or re-solution ceases. This may be due to extrinsic factors of regional scale which shut off the faucets; or it may be a consequence of sealing or diversion of the feeder channel within the aquifer, which is analogous to river capture in a surface basin. This is now a third metastable condition of very slow weathering by atmospheric water vapor plus (in many instances) the accretion of a veneer of aerosols.

Figure 7.1 provides a real example of a sequence of depositional and re-solutional periods in stalagmite growth at Bone Cave, West Virginia. Figure 7.2 generalizes real data to give model growth histories for stalagmites in four differing environments which are compared to the dynamic metastable equilibrium curve for model river channels given by Schumm (1977). Over the total timespan recorded in the figure, there is net channel erosion and net calcite deposition. Curve T is for a stalagmite from a perhumid tropical environment where deposition is continuous with little change of rate over spans of tens or hundreds of thousands of years. Curve M represents a possible history for a deposit in a temperate, extraglacial environment, such as the Mendip Hills of southwest England (Atkinson *et al.* 1978). Metastable periods of accumulation alternate with shorter periods of re-solution or cessation which may be correlated with cold phases when soil CO_2 was drastically depleted or the water shut off entirely by growth of discontinuous permafrost. Curve A is for an alpine stalagmite; shorter spells of growth are succeeded by major re-solution phases and times of cessation. Curve B displays periods of rapid growth and complete cessation that are rather precisely out of phase with those of curves M and A. It represents a stalagmite in a cave that is now shallowly submerged by the sea, as at Bermuda (Harmon *et al.* 1978) or the Bahamas (Gascoyne *et al.* 1979); subaerial calcite growth is thus limited to periods of emergence during glacials. The deepest example that we have studied is from − 44 m at Andros Island in the Bahamas.

Cave calcite deposits may be dated radiometrically by the ^{14}C method (effective limit ~38,000 yr BP), or by U-series methods where the effective limit at time of writing is 350 000–400 000 yr BP). Application of U-series

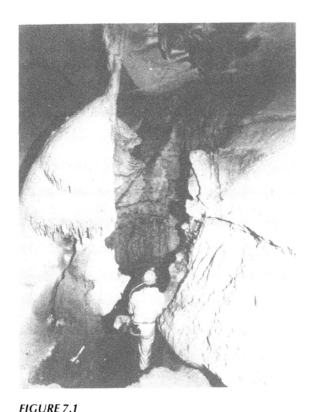

FIGURE 7.1
Eroded stalagmite in Bone Cave, Pocahontas Co., West Virginia. *A minimum of three successive phases of metastability can be seen in this picture. (i) Supersaturated waters from an overhead fissure deposited a calcite boss 4 m in diameter and 1–3 m in height. The left half of it is preserved and, in detail (not apparent in this view), reveals that there were many cessation and minor re-solution events during its accumulation, which probably extended for some hundreds of thousands of years. (ii) The fissure waters became aggressive w.r.t. calcite (a threshold was passed) and drilled a cylindrical solution shaft of 3 m diameter through the stalagmite and 2.5 m into the limestone floor beneath it. (iii) The waters reverted to a supersaturated condition and deposited stalactites and bosses of calcite, seen at the rear of the shaft. Accumulation continues today. At the least, this latest depositional phase probably extends through all of the Holocene.*

dating to terrestrial carbonates was pioneered by Cherdyntsev (1955, cited in Cherdyntsev 1971), but most of the investigation of cave deposits has been carried on at the McMaster laboratories, where more than 1000 specimens have now been successfully dated. Schwarcz (1978) provides a recent review. In addition, we have demonstrated that some stalagmites and flowstones carry a natural remanent magnetism of measurable intensity (Latham *et al.* 1979). Four specimens known to be older than 350 000 yr have recorded the Brunhes/Matuyama magnetic reversal at 730 000 yr BP. In samples known to exceed 730 000 yr in age, assumptions concerning the ^{234}U : ^{238}U ratios at the time of deposition allow us to infer whether they are younger or older than ~10^6 yr (Ford *et al.* 1981). We are able, therefore, to determine the duration of metastable periods of cave calcite deposition

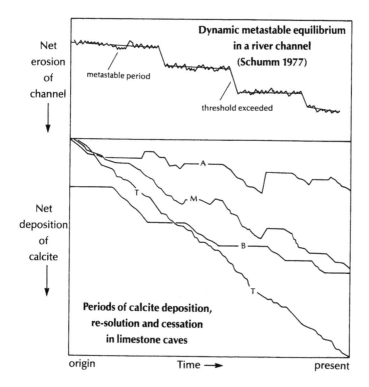

FIGURE 7.2
Model growth histories for four hypothetical stalagmites growing in different environments, compared to the model curve for dynamic metastable equilibrium in a river channel (Schumm 1977). See text for discussion.

(= enhanced subsoil limestone solution) and the ages of deposition, re-solution, and cessation thresholds in the karst system with good accuracy to >350 000 yr BP and, in broader terms, to about one million years in the past. Two examples of ancient North American deposits that have been dated are shown in Figure 7.3.

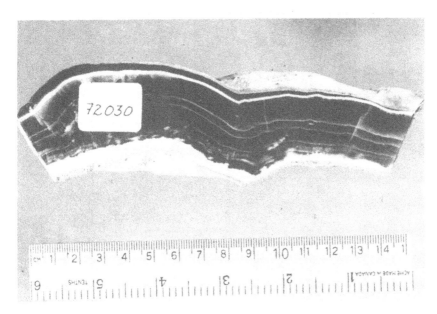

FIGURE 7.3
(a) A flowstone specimen displaying an alternation of phases of calcite growth and cessation. *It was recovered, in situ but fractured by frost action, in Grotte Valerie, latitude 61 1/2°N in the Mackenzie Mountains of Canada. The site is permafrozen today. The specimen displays a main phase of growth ~ 3 cm in thickness, plus a later phase that is separated from it by the bright weathering horizon that is seen immediately above the label. There is also prominent weathering at the base, where moisture evidently penetrated the calcite–bedrock contact. The base of the main phase of calcite growth is dated at 320 ± 26 x 10³ yr by U-series methods, and its top at 280 ± 16 x 10³ yr. It is interpreted as growing for several tens of thousands of years in an environment warmer than the present; growth was halted for brief spells at four times that are marked by thin weathering zones within the phase. The later phase is dated at 191 ± 21 x 10³ yr. The intervening weathering horizon therefore represents 60 000 yr or more of very slow decomposition. The weathering zone on the very top of the specimen, 1–3 mm in thickness, represents the last 170 000 yr at the site. (From the work of R. S. Harmon 1975.)*

FIGURE 7.3
(b) Severely eroded stalagmite massif
that grew on a boulder pile in the First
Fissure passage, Castleguard Cave,
Rocky Mountains of Alberta (Ford et al.
1976). The stalagmite is being destroyed
by waters falling from the source that
originally created it – an overhead shaft.
Today, these waters sink directly from
the sole of a glacier fringing the
Columbia Icefield. The youngest
surviving calcite is dated at 144 000
± 19 000 yr BP by U-series methods. The
oldest (basal) calcite is older than the
limit of these methods; recent analysis
has shown that it is magnetically
reversed and so older than 730 000 yr BP.
Between the oldest and youngest
material, the stratigraphy reveals many
sequences of calcite growth, cessations
or re-solution.

Within a given period of calcite deposition in a cave, something of the nature of short-period oscillations and longer-term trends within the solution–deposition system may be learned from the stable isotope ratios of the oxygen in the mineral. In suitable conditions the $^{18}O : {}^{16}O$ ratio may yield direct regional paleotemperature information (see review by Schwarcz & Ford 1981). Hendy and Wilson (1968) first described this application, and it has been followed up intensively at McMaster (Thompson 1973, Harmon 1975, Gascoyne 1980). Independent estimates of regional palaeotemperatures may sometimes be obtained from the partition of certain trace metals in the calcite lattice (Gascoyne 1981), and potentially from contained colloidal organic matter.

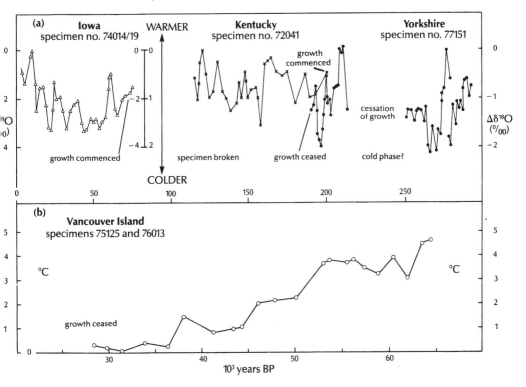

FIGURE 7.4
(a) Examples of $\delta^{18}O$ isotope records from dated stalagmites collected in Iowa, Kentucky and Yorkshire (England). Change of ^{18}O permil is normalized to zero points at the peaks (inferred warmest positions) in the records, to facilitate comparison. (From the work of R. S. Harmon 1975 and M. Gascoyne 1980.) (b) A $\delta^{18}O$ stalagmite record converted to the calculated change of mean annual air temperature during the period of growth, 64 000–28 000 yr BP at a cave site in Vancouver Island, BC. (From the work of M. Gascoyne 1980.)

The nature of ^{18}O : ^{16}O data produced from stalagmites, etc. is shown in Figure 7.4a, a representative sequence from 290 000 yr BP to the present day that is compiled from samples from three temperate environments that have strongly contrasted upper Quaternary histories (see subscript of Fig. 7.4a for details). There is a close analogy to the oscillations within a metastable erosion period in Schumm's modelling (Figure 7.2). Figure 7.4b shows the trend of mean annual temperature calculated from such data for a further contrasting temperate karst site in Vancouver Island (Gascoyne *et al.* 1980).

The U-series datings and oxygen isotope curves from calcites measured at McMaster University have so far been applied (i) to classical archaeologic, geomorphic and stratigraphic dating exercises; (ii) to date threshold events; and (iii) to set up palaeoclimatic chronologies for karst areas. Our posture has been to look outwards from the karst solution–deposition system that provides the sample material to general chronologic and environmental questions. But, in principle, this can be reversed to look from the measured deposit to the source solutional activity. The view will be clouded by effects intrinsic to the intervening aquifer, but general subsoil solution estimates of some precision in space and time are feasible, using the temperature dependencies of the solution process that are outlined below. Given a datable specimen that is also fast growing, these estimates are resolvable to 50-yr running means or better. Finally, fission-track mapping of uranium in the calcite lattice reveals an apparent seasonal habit in some instances (Truscott & Schwarcz, unpublished). This opens up the prospect of a tree-ring scale of resolution of limestone weathering activity to − 350 000 yr in the ideal case.

Metastable solution of limestone

The karst solution process is not subject to dynamic thresholds. At the basin scale, extrinsic thresholds are generally large-scale climatic events, and the whole of Holocene time is therefore potentially a single metastable karst erosion phase. The extent to which this potential is realized depends on thresholds intrinsic in a particular karst system. Examples of these intrinsic thresholds are the eventual removal of all carbonate material in the regolith by solution which causes a change in the solution system from coincident to sequential (Drake 1981); the agricultural liming of a residual soil which has the reverse effect; and the complete removal of the soil cover which causes solution to proceed at rates governed by the lower atmospheric PCO_2 rather than by a higher soil air PCO_2. Soil removal may be due to natural processes, but in Holocene times it is more likely to be anthropogenic, following upon deforestation; Drew (1981) describes an excellent example of this phenomenon in Eire. Provided that areas in which such transitions have occurred can be identified from historical or geomorphological evidence, the assertion that Holocene time forms a single karst metastable phase is true.

Spatial and temporal variations in erosion rates within a metastable phase are greatly dependent upon the space- and timescales considered. At the microscopic scale, the rates of solution of the various shell fragments in a magnesian calcite are different, depending upon their particular mineralogy (Plummer & Mackenzie 1974). At the intermediate scale of the local aquifer, such microscale variations of mineralogy are insignificant, but variations in lithology resulting from gradations in the original depositional environment may be of significance. At the macroscale, where various karst areas of the world are compared, variations of karst lithology may be insignificant in comparison to the gross regional variations of climate. The variability of erosion rates may depend similarly on timescale. An exposed limestone face may erode at very different rates from instant to instant, depending on the instantaneous rainfall intensity, while the variation from year-to-year may be quite small. There is evidently a space and time averaging effect that is a reflection of the **central limit theorem** – the erosion of a particular karst feature or area over some period is the accumulation of an infinity of instantaneous, infinitesimal erosion events. Church (1980) provides a more general discussion of this consideration.

Karst erosion rates can only be estimated at particular scales of area and time, and they may be estimated directly or indirectly. In general, the estimates are of either point or instantaneous rates: for example, the erosion at a point over a span of time can be estimated by comparing the elevation of the point at different times, or the erosion rate of an area at an instant in time can be estimated by measuring the instanteous transport rate of dissolved limestone out of a basin. Point rates tend to be measured directly by erosion meters (High & Hanna 1970) or by the weight loss of small tablets of rock (Trudgill 1977), and instanteous rates tend to be estimated indirectly through measurements of transport rates.

The problem of the validity of karst erosion data is, therefore, the problem of transforming unique values into values representative of some defined space and time, and the problem of the variability of the data is the problem of determining the relative power (in the spectrum analysis sense) of the variance of the data at the specified scales of space and time and at other scales.

The concentration of ions from the solution of limestone in a particular water depends on two sets of factors. Equilibrium factors determine the mineral solubility – primarily PCO_2, temperature and the nature of the dissolution system. Kinetic–dynamic factors determine the extent to which a water has attained chemical equilibrium – these include time and flow hydrodynamics and geometry (Drake & Wigley 1975). Within any aquifer, there may be great differences in both the mean and the variance of concentration data of waters of different hydrologic types (Fig. 7.5). In general, the least variance is exhibited by types which represent the greatest degree of spatial and time averaging. Well waters, which represent the general aquifer characteristics of slow-flowing waters, are generally saturated and show relatively constant concentrations, while streams flowing on to the limestone are unsaturated and have highly variable concentrations. The different types of waters may, therefore, be used to study variations in erosion rates at different areal and temporal scales.

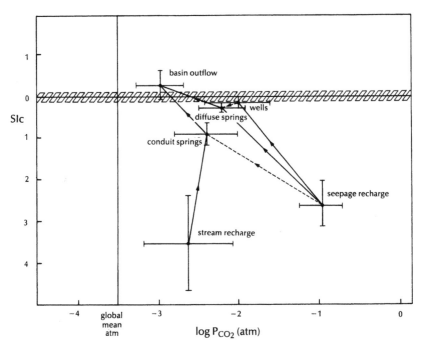

FIGURE 7.5
Mean and standard deviation of the saturation index with respect to calcite (SIc) and equilibrium partial pressure of carbon dioxide (PCO₂) for different types of waters in central Pennsylvania karst basins. The types are distinct, with the exception of the almost saturated wells and diffuse springs. These both sample the long-resident groundwater that is recharged mainly by seepage through the soil. Conduit springs, which are the outlets of discrete underground systems fed by discrete stream recharge, show considerably more variability. All springs combine in the basic outflow stream, which is oversaturated in the free atmosphere PCO₂. (After Drake & Harmon 1973, Fig. 4.)

We will discuss the nature of the variability of estimates of karst erosion rates (as represented by dissolved solids concentrations) in three typical situations: (i) differences between various types of water in a given area, (ii) annual variations at the basin scale, and (iii) differences between regions.

Intraregional variability The karst literature abounds with detailed studies of solution in local karst areas, in which the investigator has sought to differentiate waters playing distinct morphogenetic roles (e.g. soil waters, doline waters, springs, etc.). Although there is no doubt of the validity of the data for the particular time and site studied, the extent to which they are valid at a higher level of aggregation is dependent on the situation. Waters flowing in distinct underground conduits (as opposed to true groundwaters) may show extreme seasonal or short-term (storm-response) variations, as described by Shuster and White (1972), and also great between-site variation (Fig. 7.6, Bakalowicz 1979). Equally, they may not. Ford (1971) shows that a considerable diversity of concentration val-

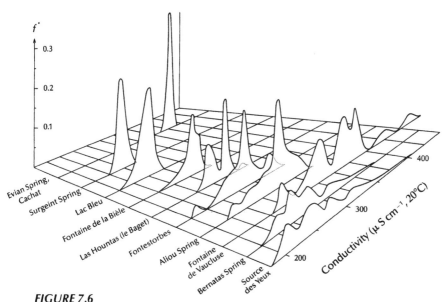

FIGURE 7.6
Frequency distributions (vertical scale) of specific conductivity (SpC)
of repeated samples from a number of springs in southern France.
Evian Spring is a seepage spring, and Surgeint Spring and Lac Bleu are
fissure-flow springs. Each type has a relatively constant and distinct
SpC. The other springs are fed by various proportions of pore and
fissure flow. (After Bakalowicz 1979, Fig. 3.60.)

ues exists between sites in the same locality and between localities in the Canadian Rockies (Fig. 7.7a). In contrast, there is little range of concentration values between apparently very diverse sites sampled during exceptionally wet and exceptionally dry summers in the karst of Newfoundland (Karolyi 1978, Fig. 7.7b).

Order can be brought to such apparent chaos by analysis of the chemical equilibrium state of the individual water samples. Although the karst solution process does not have a dynamic threshold, there is a limit, which is the mineral solubility. The geomorphic significance of this limit is discussed by Ford (1980). A particular water sample may exhibit undersaturation, saturation or, in the collecting environment, supersaturation. The capability to determine the state of a karst water with respect to its theo-

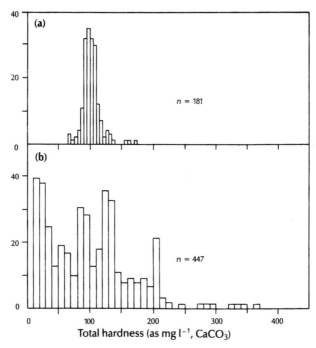

FIGURE 7.7
Histograms of the total hardness (Ca^{2+} + Mg^{2+} expressed as mg/l $CaCO_3$) of water samples from carbonate regions of (a) Newfoundland and (b) the Rocky Mountains and Selkirk Mountains, Canada. Both data sets sample waters at differing stages in many different environments. (From Ford 1971 (Rocky Mountains) and Karolyi 1978 (Newfoundland).)

retical thermodynamic equilibrium state is a powerful explanatory aid that is not available in comparable degree in other geomorphic systems. Much of the variability between sites or over time at a site is a consequence of non-saturation of the water and is governed by the kinetic–dynamic factors that are particular to a situation. Data derived from a given situation are therefore only valid for it. Where the equilibrium limit is attained, data will be generally valid for all similar sites in the same environment, the prevailing factors being the equilibrium ones. Detailed hydrologic and chemical studies are necessary to be able to place a particular situation, and individual or random sets of values are of little general use because the variance due to kinetic–dynamic effects is overpowering.

Annual variations at the basin scale   In the past 25 yr numerous studies of karstified basins have provided data on the year-to-year and within-year variability of dissolved load transport rates (e.g. Douglas 1964, Drake & Ford 1976a). Erosion rates have usually been estimated by the classic erosion formula, either as originally developed by Corbel (1959):

$$X = \frac{4ET}{100} \tag{7.1}$$

where X = erosion rate (mm 1000 yr^{-1}), E = annual runoff (dm) and T = average Ca + Mg concentration (mg 1^{-1} CaCO$_3$) or, as subsequently modified, to include the fraction of basin area underlain by karst rocks and a rock bulk density of other than 2500 kg m^{-3}. However, as shown by Drake and Ford (1976b), this form of equation gives correct estimates only if the covariance (or correlation) of discharge and concentration is zero. In most streams there is an inverse relationship between these variables, because the dissolved solids are contributed largely by a high concentration groundwater flow component which is progressively diluted at higher discharges by a low concentration surface water component (Fig. 7.8).

The relationship between discharge and concentration for streams in all karst regions of Canada can be well modelled by:

$$C = a\,Q^{-1} + b \tag{7.2}$$

from which the annual total dissolved load can be calculated as:

$$\Sigma L = a + b\,\Sigma Q \tag{7.3}$$

where a and b are constants, C = concentration, Q = discharge and ΣL and ΣQ are annual total dissolved load and discharge. Typically, over 50% of the concentration variance is explained by discharge variations. This type of analysis provides a means of extending the few years of concentration data available from karst studies to the several decades or, in some instances, centuries covered by discharge measurements, and can therefore partially fill the gap between the short-term evidence and the long-term evidence from speleothem analyses summarized above.

The lack of a threshold for karst erosion and the inverse concentration–discharge relationship imply that the dissolved load is relatively evenly distributed over the entire range of discharge experienced on a stream, and that importance of short periods of very high discharge is small. In consequence, as noted in the Introduction, the relationship can be well defined by samples that are distributed uniformly over the range of discharge. This is in marked contrast to the solids transport regime, where a few short periods of discharge that exceed some intrinsic threshold may be responsible for most of the annual net transport and which must, therefore, be defined by sampling during those periods. The long-term validity of a few years' dissolved load data is significantly higher than that of a solid load data record of comparable length because of both its better definition and the lesser impact of larger-than-observed events.

Inter-regional variability   Comparisons of erosion rates in different karst regions of the world (e.g. Corbel 1959, Sweeting 1973, Smith & Atkinson 1976) have generally shown that the greatest part of the observed variability is due to regional variations of runoff. At this scale, differences of climate are generally paramount because annual precipitation between regions may vary by several orders of magnitude while the solubility

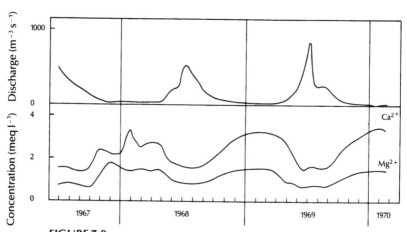

FIGURE 7.8
Seasonal variations in discharge and Ca and Mg concentrations in the North Saskatchewan River, Rocky Mountain House, Alberta. Overall erosion rate is approximately 11 mm 1000 yr^{-1}.

limited concentration varies by not more than one. There are, however, important differences in groundwater concentrations that result from different soil types and recharge regimes which are reflected in differences in the locus of most solution, and hence in landform, development (e.g. Trudgill 1976). In many early instances, these studies have not compared similar hydrochemical environments, but more recently the various components of inter-regional differences have been separated by comparing only sets of data from well samples that show saturation, thus eliminating the kinetic–dynamic factors, and which are gathered over a wide area and more than one year, thus eliminating local lithologic and seasonal variability.

The most important factors in inter-regional differences of limestone groundwater concentrations are the average temperature and the evolution system. The first determines the average rates of biogenic CO_2 in the soil, and a simple, biologically reasonable model is given by:

$$\log PCO_2^* = -2 + 0.04T \tag{7.4a}$$

$$PCO_2 = \{(0.21 - PCO_2)/0.21\} PCO_2^* \tag{7.4b}$$

where PCO_2^* is the potential partial pressure of CO_2(atm), PCO_2 is the actual value and T is the mean annual air or groundwater temperature (°C). The first equation expresses the temperature dependence of biological activity in the soil (Drake & Wigley 1975), and the second expresses the inhibition of soil respiration caused by depressed values of PO_2 at high PCO_2 (Drake 1980). The second factor, the evolution system, determines the equilibrium concentration of dissolved limestone for a given PCO_2. In general, the system may be coincident or sequential, with the former having higher values. Coincident systems are those where an atmosphere of elevated PCO_2, limestone and water are spatially and temporally coincident. Typical situations are a carbonate-rich till, or a very porous bedrock (e.g. chalk) beneath a thin soil cover. Sequential systems are those where water first equilibrates with an atmosphere of elevated PCO_2 in the absence of limestone and then with limestone in the absence of an atmosphere. Typical situations are where a non-carbonate unit intervenes between the soil and the limestone aquifer, or where there is a very thick non-carbonate regolith. Data from many published sources, including Smith and Atkinson (1976) and Trainer and Heath (1976), appear to fit this scheme well (Fig. 7.9).

The 'coincident' and 'sequential' evolution systems correspond to the geochemical 'open' and 'closed' systems in the limit where the atmosphere is present in infinite volume. The effect of a restricted air supply due to a waterlogged soil is to reduce the dissolved mineral concentrations: this effect is pronounced only for areas which follow the coincident system (Drake 1981), which have low temperatures and which have highly seasonal recharge. In the limit, this situation leads to a mimicking of sequential system evolution and is a possible explanation of the fact that many Arctic and alpine areas appear to show sequential system equilibrium.

Regional mean Ca concentrations in groundwaters generally fall in the range 30–140 mg l^{-1}. Inter-regional differences of the order of 100 mg l^{-1} are attributable to the presence of a soil with an elevated PCO_2, of 70 mg l^{-1} to the nature of the evolution system, and of up to 80 mg l^{-1} to highly seasonal recharge in cold areas. The global range of regional groundwater Ca concentration values, a factor of 5, is dwarfed by the global range of regional annual runoff values. Even within Canada, annual runoff in karst regions varies from over 3000 mm in parts of Vancouver Island to less than 25 mm in southern Alberta – a factor of 120. It is therefore not surprising that studies of the regional variations in erosion

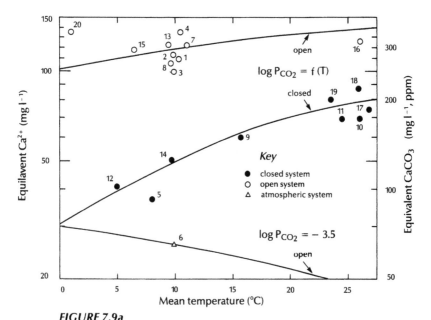

FIGURE 7.9a
The lines show the equilibrium Ca^{2+} concentration for the coincident and sequential systems with a PCO_2 given as a function of temperature by Equation 7.4, and for the open atmosphere system where $PCO_2 = 10^{-3.5}$ atm. Data are mean values for sets of samples selected from the literature such that they represent true mean groundwater conditions, and show a general concordance with the lines. (After Drake and Ford 1981.)

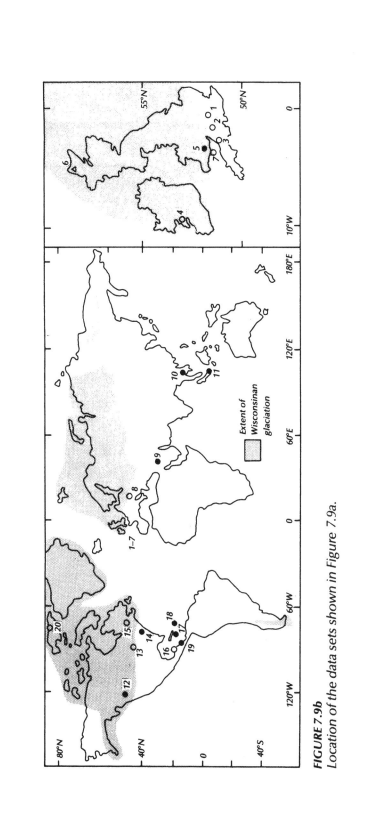

FIGURE 7.9b
Location of the data sets shown in Figure 7.9a.

rate, estimated from the dissolved solids transport rates, show high correlations between erosion rate and precipitation (or runoff). The data of Lang (1976), for example, show that 97% of the variance in regional erosion rates is explained by variation in regional precipitation. Even though a part

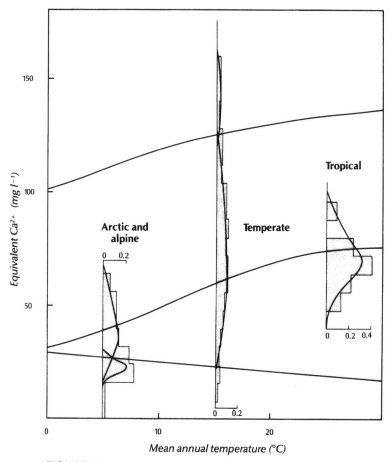

FIGURE 7.9c
Equilibrium lines as in Figure 7.9a with the data of Smith and Atkinson (1976) shown as histograms with the normal approximations superimposed. There is general concordance with the lines, but a considerable spread of values due to there being included data from many water types. The apparent tendency for Arctic and alpine data to show sequential evolution may result from mimicking by the coincident system in conditions where most recharge occurs through waterlogged soils.

of this explanation is due to the correlation of precipitation with temperature, and of temperature with concentration, it is evident that inter-regional concentration differences play only a minor role in determining the inter-regional differences in erosion rates. They are, however, important in that the factors that govern them also govern, in large part, the location of solution within a given regional aquifer (see, for example, Drake & Ford 1981); that is, whether most of the total erosion is of soil zone carbonates (and therefore does not produce or develop karst forms), of the uppermost zone of the aquifer (producing subsoil karren), of the main body of the aquifer (producing subsurface karst forms) or of subaerial rock (producing surface karren). Studies of overall regional erosion rates, however accurate the data might be, do not therefore immediately have validity in the understanding of the processes of and the progression of landform development – the true aim of geomorphic process studies.

CONCLUSIONS

Limestone solution and the landforms that it is creating today have been studied at sites ranging from the Equator to within 12° latitude of the North Pole. A comparable range of global runoff regimes has been investigated, from the perhumid to both hot arid and cold arid environments. Valid radiometric age and stable isotope data for cave calcite precipitates are available, at the time of writing, from sites ranging from the equatorial to the sub-polar zone of discontinuous permafrost. In many of these study areas it is reasonably established that the entire Holocene period may be treated as a single phase of karst metastable equilibrium.

The limestone solution system lacks intrinsic thresholds and is damped by other effects, so that a few years of solute load record in a given basin may be validly extended to calculate karst erosion rates over timespans of decades, or even centuries where the basic discharge record is sufficiently long. Radical disturbance by deforestation or urbanization, however, can invalidate this approach. The system also possesses equilibrium limits and, where waters at those limits can be segregated from data sets, they may be validly compared between basins or larger regions. This permits study of the determinants of global variability of solution rates, which appear to be the familiar variables of runoff regime, temperature and lithology. A constraint placed upon such investigation at the present time is the fact that in many data sets saturated (equilibrated) waters cannot be separated from others because of incomplete chemical analysis. Adequate data from tropical humid and perhumid environments are especially needed.

In principle, the varying oxygen isotope ratios of stalagmites and flow-stones in suitable caves may be resolved to 50-yr running means of change, although we have not attempted to do better than c. 250 yr in our work thus far. The calcite that we measure derives from waters that are saturated at or close to the subsoil environment of maximum solvent potency and, being at the saturation limit, may be compared between regions. At shorter timescales, an overlap is obtained with the decade–century long extended solute transport records. At greater timescales, parameters of the karst solution system may be inspected through metastable periods and across thresholds back to 350 000 yr BP and, with much poorer temporal resolution, to one million years ago.

ACKNOWLEDGEMENTS

Funding for much of our work has been provided by individual grants-in-aid of research from the Natural Sciences and Engineering Research Council, Canada. We are grateful for the stimulus and aid provided by our faculty and graduate colleagues at McMaster University and elsewhere over many years.

REFERENCES

Atkinson, T. C., R. S. Harmon, P. L. Smart and A. C. Waltham 1978. Paleoclimatic and geomorphic implications of ^{230}Th./^{234}U dates on speleothems from Britain. *Nature* **272**, 24–8.

Bakalowicz, M. 1979. *Contribution de la géochimie des eaux a la connaissance de l'aquifère karstique et de la karstification*. Unpublished D.ès Sc. thesis. Paris: Université Pierre et Marie Curie.

Blumberg, P. N. and R. L. Curl 1974. Experimental and theoretical studies of dissolution roughness. *J. Fluid Mech.* **75**, 735–42.

Cherdyntsev, V. V. 1971. *Uranium 234*. Translated by J. Schmorak. Jerusalem: Israel Program for Scientific Translations, 234 pp. Originally published in Russian as *URAN-234* (Moscow: Atomizdat, 1969).

Church, M. 1980. Records of recent geomorphological events. In *Timescales in geomorphology*, R. A. Cullingford, D. A. Davidson & J. Lewin (eds), 13–30. Chichester: Wiley–Interscience.

Corbel, J. 1959. Erosion en terrain calcaine. *Annal. Géog.* **68**, 97–116.

Douglas, I. 1964. Intensity and periodicity in denudation processes with special reference to the removal of material in solution by rivers. *Z. Geomorph.* **8**, 451–73.

Drake, J. J. 1980. The effect of soil activity on the chemistry of carbonate groundwaters. *Water Resources Res.* **16**, 381–6.

Drake, J. J. 1981. Geomorphic and seasonal controls of the chemistry of carbonate groundwater. *J. Hyd.* (in press).

Drake, J. J. and D. C. Ford 1976a. Solutional erosion in the southern Canadian Rockies. *Can. Geog.* **20**, 158–70.

Drake, J. J. and D. C. Ford 1976b. The dissolved solids regime and hydrology of two mountain rivers. *Int. Speleology,* Proc. 4th Congr. Speleol. Union Proc. **4**, 103–14.

Drake, J. J. and D. C. Ford 1981. Karst solution: a global model for groundwater solute concentrations. *Jap. Geomorph. Union* (in press).

Drake, J. J. and R. S. Harmon 1973. Hydrochemical environments of carbonate terrains. *Water Resources Res.* **9**, 949–57.

Drake, J. J. and T. M. L. Wigley 1975. The effect of climate on the chemistry of carbonate groundwater. *Water Resources Res.* **11**, 958–62.

Drew, D. P. 1981. Human intervention in an upland karst: The Burren, Co. Clare, Western Ireland. *J. Hydro.* (in press).

Ford, D. C. 1971. Characteristics of limestone solution in the southern Rocky Mountains and Selkierk Mountains, Alberta and British Columbia. *Can. J. Earth Sci.* **8**, 585–609.

Ford, D. C. 1980. Threshold and limit effects in karst geomorphology. In *Thresholds in geomorphology,* D. L. Coates & J. D. Vitek (eds), 345–62. London: George Allen & Unwin.

Ford, D. C., H. P. Schwarcz, J. J. Drake, M. Gascoyne, R. S. Harmon and A. G. Latham 1981. On the age of the existing relief in the southern Rocky Mountains of Canada. *Arct. Alp. Res.* **13**, 1–10.

Ford, D. C., R. S. Harmon, H. P. Schwarcz, T. M. L. Wigley and P. Thompson 1976. Geohydrologic and thermometric observations in the vicinity of the Columbia Icefield, Alberta and British Columbia, Canada. *J. Glac.* **16**, 219–30.

Gascoyne, M. 1980. *Pleistocene climates determined from stable isotope and geochronologic studies of speleothem.* Unpublished PhD dissertation. Hamilton, Ontario: McMaster University.

Gascoyne, M. 1981. Trace element partition coefficients in the calcite–water system and their palaeoclimatic significance in cave studies. *J. Hydrol.* (in press).

Gascoyne, M., H. P. Schwarcz and D. C. Ford 1980. A paleotemperature record for the mid-Wisconsin in Vancouver Island. *Nature* **285**, 474–6.

Gascoyne, M., G. J. Benjamin, H. P. Schwarcz and D. C. Ford 1979. Sea level lowering during the Illinoian Glaciation: evidence from a Bahama 'blue hole'. *Science* **205**, 806–8.

Gerstenhauer, A. and K. H. Pfeiffer 1966. Beiträge zur Frage der Lösungsfreudigkeit von Kalkesteinen. *Abhandlungen Karst zur und Hohlenkunde.* Reihe A. Heft 2, 1–46. München: Verband der Deutschen Höhlen und Karstfurscher.

Glew, J. R. and D. C. Ford 1980. Simulation of rillenkarren. *Earth Surf. Proc.* **5**, 25–36.

Goodchild, J. G. 1875. The glacial phenomena of the Eden Valley and the west part of the Yorkshire Dales district. *Q. J. Geol. Soc.* **31**, 55–99.

Harmon, R. S. 1975. *Late Pleistocene paleoclimates in North America as inferred from isotopic variations in speleothems.* Unpublished PhD dissertation. Hamilton, Ontario: McMaster University.

Harmon, R. S., H. P. Schwarcz and D. C. Ford 1978. Late Pleistocene sea level history of Bermuda. *Quat. Res.* **9**, 205–18.

Hendy, C. H. and A. T. Wilson 1968. Palaeoclimatic data from speleothems. *Nature* **216**, 48–51.

High, C. and F. K. Hanna 1970. *A method for the direct measurement of erosion on rock surfaces.* Tech. Bull. Br. Geomorph. Res. Group 5.

Karolyi, M. S. 1978. *Karst development in Ordovician carbonates: Western Platform of Newfoundland.* Unpublished MSc thesis. Hamilton, Ontario: McMaster University.

Lang, S. 1976. Setting of karstic denudation in the global denudation of the Earth's surface. *7th Inter. Speleol. Cong. Proc.* 282–3.

Langmuir, D. 1971. The geochemistry of some carbonate groundwaters in central Pennsylvania. *Geochim. Cosmochim. Acta* **35**, 1023–45.

Latham, A. G., H. P. Schwarcz, D. C. Ford and W. G. Pearce 1979. Paleomagnetism of stalagmite deposits. *Nature* **280**, 383–5.

Plummer, L. N. and T. M. L. Wigley 1976. The dissolution of calcite in CO_2-saturated solutions at 25°C and 1 atmosphere total pressure. *Geochim. Cosmochim. Acta* **40**, 191–202.

Plummer, L. N. and F. T. Mackenzie 1974. Predicting mineral solubility from rate data: application to the dissolution of magnesian calcites. *Am. J. Sci.* **274**, 61–83.

Schumm, S. A. 1977. *The fluvial system.* New York: J. Wiley.

Schwarcz, H. P. 1978. Uranium-series disequilibrium dating. *Geoscience Can.* **5**, 184–7.

Schwarcz, H. P. and D. C. Ford 1981. Applications of stable isotope fractionation affects in waters, sedimentary deposits, flora and fauna. In *Geomorphological techniques*, A. S. Goudie (ed.). London: George Allen & Unwin.

Shuster, E. T. and W. B. White 1972. Source areas and climatic effects in carbonate groundwaters as determined by saturation indices and carbon dioxide pressures. *Water Resources Res.* **8**, 1067–73.

Smith, D. I. and T. C. Atkinson 1976. Process, landforms and climate in limestone regions. In *Geomorphology and climate*, E. Derbyshire (ed.), 367–409. New York: J. Wiley.

Sweeting, M. M. 1973 *Karst landforms.* New York: Columbia University Press.

Thompson, P. 1973. *Speleochronology and Late Pleistocene climates inferred from O, C, H, U and Th isotopic abundances in speleothems.* Unpublished PhD dissertation. Hamilton, Ontario: McMaster University.

Trainer, F. W. and R. C. Heath 1976. Bicarbonate content of groundwater in carbonate rock in eastern North America. *J. Hydrol.* **31**, 37–55.

Trudgill, S. T. 1976. The erosion of limestones under soil and the long-term stability of soil vegetation systems on limestone. *Earth Surf. Proc.* **1**, 31–41.

Trudgill, S. T. 1977. Problems in the estimation of short-term variations in limestone erosion processes. *Earth Surf. Proc.* **2**, 251–6.

Truscott, M. L. and H. P. Schwarcz (n.d.). *U fission track mapping in cave calcites.* Unpublished Mss. Department of Geology, McMaster University.

Williams, P. W. 1972. Morphometric analysis of polygonal karst in New Guinea. *Geol Soc. Am. Bull.* **83**, 761–96.

8

Alpine mass-wasting in contemporary time: some examples from the Canadian Rocky Mountains

James S. Gardner

INTRODUCTION

Geomorphology has been bipolar in its emphasis. One emphasis has been on the operation and functioning of present-day processes. The other has been on long-term landform evolution and landscape chronology. Recent attention to topics and concepts such as 'geomorphology and time' (Thornes & Brunsden 1977, Cullingford *et al.* 1980), 'geomorphic thresholds' (Coates & Vitek 1980), and 'dynamic metastable equilibrium' (Schumm 1977), has indicated a need for reconciling the long-term and short-term emphases in geomorphology. Process studies exist in a void apart from the historical context of reality. Pronouncements about landform evolution over great vistas of time and space make little sense, apart from a careful understanding of the processes and agents involved in change. In this paper, relatively long-term observations of present-day mass wasting events or material fluxes on mountain slopes are presented and discussed in the context of the long-term geomorphic record, so far as it is known.

The data have accumulated over 15 yr of research in the Canadian Rocky Mountains, notably the Lake Louise area (116° 15'W, 51° 25'N) and the Mt Rae area (116° 6'W, 50° 35'N). Four material fluxes are described: debris accretion, debris shift, rockfall and flood-debris flow. In the observation periods, the events observed have had little measurable impact on the gross morphology of the region or, indeed, on the morphology of individual slopes in most cases. Nonetheless, the events have been highly variable in time and space. And, at the scale of an individual rock particle, a rockfall is a 'threshold event'. At the scale of the regional morphology and its geomorphic chronology, the tempo of the observed events represents 'between threshold variation' or 'contemporary time' variation. Church has defined 'contemporary time' as follows:

'...the (present) period within which a geomorphological process has maintained a substantially constant mean – that is, has undergone no important changes due either to non-stationarity of forcing function or impingement upon some threshold' (Church 1980, p. 19).

'Contemporary time' and the variability described by the observed events must be understood in the context of the known environmental geomorphic chronology of the region.

RECENT ENVIRONMENTAL CHRONOLOGY

Schumm (1977) implies that a timespan of 10^6 yr is an appropriate temporal context for a consideration of dynamic metastable equilibrium. Within such a spectrum, one might expect significant geomorphic change to occur. The pattern of change could take one of several forms but, consistent with the threshold concept and episodic erosion, the pattern should include short periods of rapid change (threshold events), interspersed with longer periods of relative stability (between threshold variation or contemporary time variation).

A period of 10^6 yr in the Canadian Rocky Mountains represents a time of considerable environmental variability. It includes the Holocene and the latter part of the Pleistocene, a period characterized by several major and minor glacial episodes. The chronology of this period is not well developed in the literature on the region, for two reasons. First, there has been a relative lack of research due to remoteness. Second, the existing landforms and deposits are not an articulate palimpsest of Pleistocene events. Datable material is apparently not abundant and most deposits of pre-Wisconsin age were greatly modified, if not destroyed, by late-Pleistocene events.

The earliest identifiable evidence of a glaciation has been associated with the 'Great Cordilleran Advance' (Stalker & Harrison 1977) which has been assigned an Illinoian age equivalence. All subsequent glacial episodes in the Canadian Rockies were of lesser magnitude, but several are better represented in the stratigraphy. A confusing array of Wisconsin age chronologies has been developed for the region (Fulton 1977, Harris & Waters 1977, Stalker & Harrison 1977). Common to all the chronologies are two, if not three, episodes roughly coincident with Wisconsin time.

For the area of the Rockies covered by this paper, the late-Wisconsin (past 25 000 yr BP), is characterized by two glacial episodes, termed the 'Canmore' and 'Eisenhower Junction' advances (Rutter 1972). Most surfaces were directly or indirectly (i.e. through periglacial effects) affected by the former, while the effects of the latter and more recent advance are limited. In many respects, the region is a Pleistocene landscape, implying that most aspects of the present geomorphic scenery were in place at the close of the Wisconsin (10 000–12 000 yr BP). The data for the Pleistocene permit only a very coarse resolution in describing environmental variation and geomorphic change.

Resolution improves for the Holocene (past 10 000 yr) but the geomorphic impact of events in this period is spatially limited. Several authors have suggested an early-Holocene glacial episode (8 000–10 000 yr BP) referred to as the 'Crowfoot Advance' by Luckman and Osborn (1979). In many places, the geomorphic effects of this episode were obliterated by the Neoglacial advances (past 2500 yr). The most extensive of those advances was the most recent – the 'Little Ice Age' or 'Cavell Advance' which reached a zenith about 250–150 yr BP. Most glaciers have been in a more or less constant state of recession since that time. The period between the 'Crowfoot Advance' and the Neoglacial was probably warmer and drier than the preceding and subsequent periods and may represent 6 000 yr of relative geomorphic stability.

While glacial episodes indicate the major environmental–geomorphic shifts, some non-glacial landforms suggest tempo changes in fluvial and mass-wasting processes in the late-Wisconsin to present-time spectrum. Paraglacial alluvial fans (Ryder 1971) have been identified in the Bow River Valley (Roed & Wasylyk 1973), the Kananaskis Valley (Stalker 1973), and the Elbow Valley (Desloges 1980). Such deposits suggest a very active period of fluvial and/or fluvioglacial activity subsequent to the late-Wisconsin deglaciation of the major mountain valleys. From this fan-building period to the present, these deposits have been little modified, with the exception of minor flood-debris flow activity in the latter part of the Neoglacial (Gardner 1980a).

In addition, high magnitude rockfall–rockslide deposits have been identified in the Cordillera (Dishaw 1967, Cruden 1976, Eisbacher 1971, Gardner 1977). Such deposits post-date late-Wisconsin glacial deposits but degree of weathering, soil development and vegetation cover indicate that they are of considerable antiquity. One such deposit in the Mt Rae area contains pockets of Mazama tephra (6700 yr BP) on its surface. Research in the Rockies and elsewhere (e.g. Whalley 1974) suggests that high magnitude rockfall–rockslide deposits of this type could be indicative of an intense period of slope failure immediately following or coincident with valley deglaciation. Removal of lateral ice support from glacially oversteepened bedrock slopes and groundwater shifts have been suggested as causative factors. A lack of more recent deposits of equivalent magnitude suggests a decline in high magnitude rockfall–rockslide frequency in the past 6000 yr at least. Of course, the one exception is the Frank slide of 1903 which serves as a reminder that the process is not extinct in the region.

In summary, the past 10^6 yr in the Canadian Rocky Mountains has been characterized by a variable environment. Several major glaciations and numerous minor glaciations have had obvious geomorphic effects. Deglaciation would appear to be a phase of some considerable geomorphic significance, given copious meltwater and the necessity for slopes to adjust to a new subaerial regime. Some of these adjustments may not be manifest for considerable periods of time. At a regional scale, a reasonable working definition of 'contemporary time' would be the Holocene, or that period post-dating the last deglaciation of the major mountain valleys and those events deriving from that deglaciation process. Thus 'contemporary time' would encompass the past 10 000 to 12 000 yr. Within this period, there have been minor glacial fluctuations and undoubtedly fluctuations in the tempo of most geomorphic processes. But major shifts in forcing functions and identifiable threshold events leading to abrupt morphological changes have not occurred at the regional scale. As scale is reduced to individual cirques, individual slopes or streams, and individual segments of slopes and reaches of streams, this generalization becomes less valid. Clearly, the morphology of some cirques has been completely remodelled in the Little Ice Age from which the area is emerging. Locally, slope failures such as that at Frank, Alberta, are catastrophic geomorphologically and otherwise. Thus, as scale is reduced, the time spectrum of contemporary time or 'between threshold' time is reduced as well. The following data must be viewed in this context. While the data series are reasonably long from the viewpoint of geomorphic studies, they are descriptive of but an instant in the Holocene, let alone a period of 10^6 yr.

DEBRIS ACCRETION

Debris accretion describes the buildup or deposition of material on the surface of debris slopes. The assumption is that mountain debris slopes are, in general, accreting deposits under present conditions, recognizing that specific points may undergo erosion or scour from time to time. The longest record of debris accretion in the Canadian Rocky Mountains is that published by Luckman (1978) to describe the sediment transport by snow avalanches. The following data have accumulated over five years of monitoring five debris slopes in the Mt Rae area. Accretion was

measured using meter-square fabric sampling mats placed at regular intervals on the long profiles of the slopes.

The data (Table 8.1) permit examination of absolute amounts and variability of accretion rates on a year-to-year, slope-to-slope and point-to-
point basis. Considerable variation is shown for accretion rates for different points on the same slope in any given sampling period. Slope number
162B in the period 1975–76 provides an extreme example, where 5 of 6
sample points showed no accretion and the remaining sample point accumulated an estimated 40 000 $cm^3 m^2$ of debris. Point-to-point variation
is described by the coefficient of variability which, in most instances (i.e.
each slope in each sampling period), is greater than 1.00, indicating a
standard deviation in excess of the mean accretion for the slope in question. Some order is suggested in the point-to-point variation in that accretion appears to increase towards the top of each slope. This pattern is
clear on some slopes in some years (e.g. number 231 in 1979–80) whereas
a reverse pattern is the case in other years (e.g. number 231 in 1976–77).
The latter reverse pattern reflects one instance of massive downslope
deposition by a snow avalanche in the winter of 1977, a winter noted
throughout the Canadian Rocky Mountains for numerous, destructive
full-depth avalanches. When the data for all slopes are aggregated, the
relationship between accretion rate and position on the slope is not statistically significant (coefficient of determination = 0.02).

Year-to-year variation at individual sample points is also considerable.
Coefficients of variability (Table 8.1) are mostly greater than 1.00 and often
in excess of 2.00. However, the data do indicate that, through time, individual sample points do actually accumulate debris even though they may
have zero accretion in a given year or years. Obviously, the mean accretion
rates are not good predictors of accretion at any given point in any given
year. Likewise, data from one year are not a good estimate of contemporary accretion rates. Furthermore, over short periods of time one slope is
not a good estimator for all slopes in an area.

Data with the variability indicated here suggest that the agents or
processes responsible for the movement of material are discrete or highly
intermittent. Field observations suggest that much of the accretion measured in the study occurs in a single or very few deposition events. A similar
pattern is suggested by data describing debris shift.

DEBRIS SHIFT

Movements such as slow creep, rolling, sliding and subsidence result
in the downslope shift of material on the surface of talus and other debris
slopes. A variety of forces initiate the shift, including those imported by

TABLE 8.1
Debris accretion 1975 – 80 inclusive (data in cm³ m⁻² yr⁻¹).

Slope		Sample point	1975–6	1976–7	1977–8	1978–9	1979–80	Total	Mean	Standard deviation	Variability coef.
205A	down	1	0	0	1375	0	50	1425	285	609.7	2.14
		2	0	0	0	–	–	0	0	0	0
		3	0	0	160	0	250	410	82	116.7	1.42
		4	452	250	–	1125	–	1827	609	458.1	.75
		5	862	950	313	1000	7875	11 000	2200	3184.3	1.45
		6	635	500	1110	1000	2125	5370	1074	639.1	.60
		7	365	875	7400	1650	5250	15 550	3110	3067.8	.99
	up	mean	330.6	367.9	1726.3	795.8	3110				
		st. dev.	346.1	415.0	2832.8	661.5	3383.7				
		var. coef.	1.05	1.13	1.64	.83	1.09				
119B	down	1	0	0	0	0	2955	2955	591	1321.5	2.24
		2	0	0	113	25	176	314	63	78.5	1.25
		3	0	0	525	15	0	540	108	233.2	2.16
	up	mean	0	0	212.7	13.3	1043.7				
		st. dev.	0	0	276.3	12.6	1657.6				
		var. coef.			1.30	.95	1.59				
231	down	1	0	30	0	0	0	30	6	13.4	2.23
		2	0	33 600	0	0	0	33600	6720	15 026.4	2.24
		3	0	3500	100	50	0	3680	730	1549.0	2.12
		4	2	1750	0	0	0	1752	380	782.4	2.24
		5	5468	–	35	0	167	5670	1418	2697.3	1.90
		6	0	150	33	15	0	198	40	63.2	1.58
		7	3750	1945	0	325	740	6760	1352	1529.6	1.13

TABLE 8.1 (continued)
Debris accretion 1975 – 80 inclusive (data in cm³ m⁻² yr⁻¹).

Slope	Sample point	1975-6	1976-7	1977-8	1978-9	1979-80	Total	Mean	Standard deviation	Variability coef.
up	8	4650	290	235	–	210	5385	1346	2202.8	1.64
	9	0	–	20	1750	2250	4020	1005	1166.9	1.16
mean		1541.2	5895.0	47	267.5	374.1				
st. dev.		2380.1	12 280.7	77.5	609.3	743.5				
var. coef.		1.52	2.08	1.65	2.28	1.99				
162B down	1	0	0	1088	0	0	1088	218	486.6	2.23
	2	0	0	0	0	0	0	0	0	0
	3	0	0	0	425	0	425	85	190.1	2.24
	4	0	200	5	750	0	955	191	324.1	1.70
up	5	40 000	1625	588	115	325	42 665	8530	17 601.5	2.06
	6	0	780	5500	–	550	6800	1700	2553.1	1.50
mean		6667.8	429.2	1197.3	258.2	146.8				
st. dev.		16329.4	653.9	2153.0	325.2	236.1				
var. coef.		2.48	1.52	1.80	1.26	1.61				
162A down	1	0	0	0	0	0	0	0	0	0
	2	0	500	3	0	0	503	100	223.3	2.23
up	3	6295	1125	138	215	2150	9923	1985	2544.0	1.28
	4	2925	2375	240	700	1200	7439	1488	1130.3	.76
mean		2305.0	1000	95.3	228.8	837.8				
st. dev.		2995.8	1025.7	116.0	330.1	1041.7				
var. coef.		1.30	1.03	1.22	1.44	1.24				

such external factors as falling rocks, snow avalances, flowing water and, occasionally, animals and people. Debris shift has been described and monitored by Rapp (1960a, 1960b), Barnett (1966), Gardner (1969a, 1973, 1979), and Carniel and Scheidegger (1976). The data presented in this paper come from 11 yr observation in the Lake Louise area and have been discussed previously (Gardner 1979) in the literature. Individual rocks of varying size and shape were taken from the slope, painted and replaced in their original positions along transverse lines (transects) across the debris slopes. Movement of the marked stones was measured from these original positions, which were subsequently located using fixed points in nearby bedrock.

Debris shift on talus and other high-gradient debris slopes may be seen as a progressive flux of material from the top to the bottom of the slope (Thornes 1971). Gross slope morphology need not be changed by this re-ordering. Threshold events, such as massive slope failures or slumps brought on by intrinsic changes (e.g. gradual oversteepening through deposition), or by extrinsic changes (e.g. accelerated erosion of material at the slope toe due to glacial or fluvial activity), can change slope morphology radically. No such changes were observed in 11 years of study. The data, therefore, probably describe 'between threshold' activity, recognizing that such activity may contribute to gradual steepening and massive failure. Like the accretion data, the shift data suggest events that are highly intermittent and discrete in time and space.

Even in the aggregated form shown in Table 8.2, the data suggest intermittency and variability. Variation in mean rates of shift is apparent between measurement periods, from one slope to the next, and between transects (sample points) on the same slope. Mean annual movement rates are given by transect and they are considerable (13–88 cm yr^{-1}) in comparison to slow mass movements in general. Mean movement rates for individual transects show variability when compared through time and when compared with one another in the same measurement period. For example, Slope A, Transect 1 had mean movement value of 381 cm in the 1966–7 period, but in 1975–6 the value was 0. The 1966–7 value was inflated by the fact that one of the 25 marked rocks moved 45.4 m downslope. In the same time period, many of the remaining markers did not move. This intratransect variation is described by a high coefficient of variability of 2.6. The 1975–6 value of 0 movement on Transect 1 of Slope A must be viewed in the context that only 24% (6 of 25) of the original markers were recoverable, the others having been buried in the shifting of debris over 11 yr. Intratransect variation is shown by high variability coefficients throughout Table 8.2 and this is largely a result of movement involving single marked rocks rather than groups of markers within the transect or all the markers on the transect. One marker may display movement of

Table 8.2
Debris shift data by measurement, period 1965–76.

Slope transect	Original markers	Markers recovered 1976	1965-6 X̄ᵃ	1965-6 CVᵇ	1966-7 X̄	1966-7 CV	1967-70 X̄	1967-70 CV	1970-2 X̄	1970-2 CV	1972-4 X̄	1972-4 CV	1974-5 X̄	1974-5 CV	1975-6 X̄	1975-6 CV	Annual X̄	Annual CV
A 1	25	6	302	2.5	381	2.6	70	0.6	24	1.2	53	2.2	16	2.4	0	0	76.8	1.72
2	51	15	569	2.3	49	3.7	64	1.0	60	1.2	125	0.8	64	4.0	37	3.0	87.9	1.83
3	85	8	318	1.9	163	2.1	193	1.5	40	1.6	103	4.2	40	3.6	54	3.0	82.7	1.05
C 1	30	10	50	3.3	5	1.8	23	1.9	98	2.7	64	5.0	19	6.3	3	2.9	23.9	0.80
2	41	10	4	4.5	4	1.4	10	1.4	185	2.3	13	3.9	11	4.7	8	3.9	21.4	1.66
3	50	9	11	4.0	45	4.3	22	0.9	30	1.2	11	4.4	10	3.4	19	2.9	13.4	0.85
D 1	35	15	10	1.8	27	1.0	59	1.6	45	1.6	63	4.5	22	3.9	34	3.8	23.7	0.29
2	46	35	7	1.8	20	2.1	22	0.9	20	0.6	33	5.9	29	6.1	8	5.0	12.5	0.58
3	41	18	10	2.2	16	1.9	49	0.7	16	2.5	122	1.0	342	5.8	16	4.3	51.8	1.90
F 1	21	7	–		43	1.6	119	2.1	41	2.0	8	2.8	0	0	11	2.7	22.2	0.78
2	14	7	–		15	1.2	30	0.6	12	1.2	74	2.6	47	2.6	19	2.6	18.6	0.87
Between transects			142	1.44	70	1.67	60	0.90	52	0.96	61	0.70	55	1.76	19	0.84		

ᵃ X̄ = mean values in cm.
ᵇ CV = coefficient of variation.

several meters, whereas rocks, marked and otherwise, immediately adjacent often showed no movement whatsoever within one measurement period. It was rare (two transects in two measurement periods) for no movement to occur within a transect over one measurement period.

Despite year-to-year variation and variation between transects and slopes over the long term (11 yr), downslope movement was general. All markers shifted. Many became buried by material deposited from above. Thus, what appears as a very staccato pattern of flux in the short term, appears as a more general flux as the observation period increases.

ROCKFALLS

Low-magnitude, high-frequency rockfall is an important material flux on the free faces in high mountain environments (Rapp 1960a, Gardner 1970, Luckman 1976, Church et al. 1979). Such rockfalls are contributors to the debris accretion and debris shift discussed earlier. Rockfalls of varying magnitude in the Mt Rae area have been described by Gardner (1980b), using data collected over a three-yr period. The data presented here represent a continuation of that study over the period 1975–80.

The rockfall data were collected using an inventory approach and continuous observation while in the field. The Mt Rae study area covers approximately 100 km². At a scale of 1 : 50 000, a grid with cells of the equivalent of 500 m to a side was superimposed on this study area. Each cell was given an identification number so that any location could be identified according to grid cell. When in the field, it was sometimes possible to observe slopes located in several grid cells at once. The time during which each cell was under observation was recorded, providing a continuous record of observation time. If a rockfall occurred, and it was usually sensed first by sound, its cell location, time and date of occurrence, and rough size characteristics (number of rocks and their size) were recorded. Specific attempts were made to observe in all weather conditions and during the night. To date, an insufficient sample has been accumulated for the night and, therefore, the data discussed here include only the hours between 07.00 and 20.00 (Table 8.4). Moreover, data collection has been confined to the late spring and summer seasons. Therefore, the presentation covers June to September (Fig. 8.1).

Observed rockfalls were all of the low magnitude variety or debris fall type (Whalley 1974), involving less than 1 m³ of material. From 1975 through 1980, 1042 rockfalls were recorded in 1943 hours of observation, giving an average frequency of 0.62 rockfalls per hour of observation (Table 8.3). For comparative purposes, a frequency value of 0.70 was found during a three-year study in the Lake Louise area (Gardner 1970) and Luckman

TABLE 8.3
Rockfall occurrences in the Mt. Rae area, 1975–80.

Year	1975	1976	1977	1978	1979	1980	Total	$\bar{X}$[a]	CV[b]
observation hours	434	324	235	480	204	266	1943	325	NA
observed rockfalls	98	257	472	111	39	65	1042	174	0.95
hours with rockfalls	–	135	81	80	35	34	365	73	NA
rockfalls/ hour obs.	0.23	0.79	2.01	0.23	0.19	0.24	3.69	0.62	1.16
rockfall probability	–	0.42	0.34	0.17	0.17	0.13	1.23	0.25	0.48

[a] $\bar{X}$ = mean value 1975–80.
[b] CV = coefficient of variability (standard deviation/mean).

TABLE 8.4
Annual 'average day' rockfall frequencies, 1975–80.

Year	Hours of the day (07.00–20.00 hours)														$\bar{X}$	CV
	7	8	9	10	11	12	13	14	15	16	17	18	19	20		
1975	0.16	0	0	0.03	0.38	0.58	0.40	0.33	0.78	0.46	0.08	0	0.07	0.06	0.24	1.04
1976	2.80	0.44	0	0.16	1.28	1.37	1.07	1.15	0.73	0.01	0.51	0.30	0.07	0.50	0.74	1.03
1977	0	1.00	4.04	3.85	3.43	2.17	3.04	1.81	1.93	0.79	0.29	0	0	0	1.59	0.96
1978	0	0	0.03	0.14	0.38	0.38	0.39	0.35	0.34	0.39	0	0	0.05	0.05	0.18	1.00
1979	0.50	0	0	0.16	0.16	0.20	0.16	0.44	0.25	0	0	0	0.20	0.25	0.17	0.94
1980	0	0.17	0.14	0.03	0.50	0.49	0.32	0.25	0.17	0.26	0.20	0	0.25	0	0.20	0.80
$\bar{X}$	0.58	0.26	0.70	0.72	1.03	0.87	0.90	0.72	0.70	0.32	0.18	0.05	0.11	0.14		
CV	1.91	1.54	2.34	2.13	1.20	0.87	1.22	0.86	0.93	0.94	1.11	2.40	0.91	1.43		

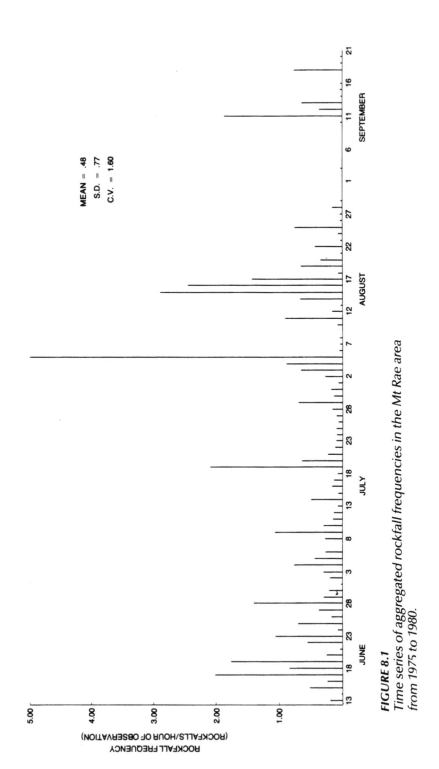

FIGURE 8.1
Time series of aggregated rockfall frequencies in the Mt Rae area
from 1975 to 1980.

(1976) published a value of 0.66 for the Surprise Valley area of Jasper Park. The data give a probability of 0.25 of observing one or more events in any given hour. These summarizing statistics do not give a good indication of the variability found in rockfall occurrences. For example, Table 8.3 shows considerable variation from one year to the next in numbers of rockfalls observed (coefficient of variability 0.95).

A probability value of 0.25 indicates that all the observed rockfalls occurred in 486 of the total 1943 hours of observation. Thus, in 1457 hours no rockfalls occurred, suggesting that events were concentrated in time. Over the study period, 269 cells were monitored for varying periods of time. Of these, 23 probably have no potential for low magnitude rockfall, since they do not contain, and are located some distance from, potential rockfall source areas (i.e. freefaces). A further 173 cells recorded no rock-fall occurrences, meaning that observed rockfalls were recorded in 73 cells (or about 30%) of the cells with rockfall potential. Within this 30% of cells, the majority of rockfalls was restricted to a few cells. For example, only 13 cells (5%) had rockfall frequencies of greater than 1.00 per hour of observation and 19 cells (8%) recorded more than 10 rockfalls in the study period.

Using aggregated data for the 1975–80 period, the day-to-day variation of rockfall frequency over the June to September period is considerable (Fig. 8.1). Because data are restricted to the late-spring through summer period, it is not possible to assess seasonal patterns in rockfall occurrence. Within the observation period shown in Figure 8.1, there is little if any pattern or trend apparent. The coefficient of variability describing this data array is 1.60 and there is no clear temporal autocorrelation. That is, because one day has a high rockfall frequency, it does not follow that the next day will too. The mean frequency is not a good predictor of the frequency value for any given day. Given that rockfall occurrences appear to be concentrated in a few sites within the study area, the location of observations on any given day will influence the possibility of observing high rockfall frequencies.

A pattern emerges when the rockfall data are aggregated according to hours of the day and 'average day' or diurnal data sets are produced (Table 8.4). The mid-day period is clearly most prone to rockfall activity. This pattern is evident in most previously published data as well. However, frequency values for given hours of the day vary considerably from one year to the next (Table 8.4). This variability is least during the mid-day period (CV = 0.86) and greatest in the early morning and evening (CV = 1.9–2.4).

The variability and patterns of observed rockfalls point to some envi-ronmental associations under contemporary conditions. About 75% of the rockfalls, besides being restricted to a small part of the total study area, occurred on freefaces with a north to northeast exposure. These are the exposures which carry snow and ice patches throughout the summer period in the Mt Rae area. Invariably, these are scarp slopes as well, with the numerous bedding planes characteristic of Paleozoic sediments of the region being exposed to the atmosphere. Coupled with joints, the bedding planes provide geological conditions suitable for the production of rockfall material via physical weathering. At altitudes above 2500 m diurnal freeze–thaw activity can be expected even during the summer, especially in the presence of snow and ice (Gardner 1969b). Thus, friable bedrock, fluctuating temperatures causing freeze and thaw, the resulting meltwater and the steep slopes provide the necessary conditions for frequent, low-magnitude rockfall occurrence.

The contemporary intensity of rockfall activity and the volumes of material produced by this process do not account for the large volume of debris found at the base of most freefaces in the study area. Talus and other types of debris slopes are the principal non-glacial deposits and depositional landforms in the Mt Rae area. If the presence of snow and ice in the source areas is an important factor in rockfall, as many studies have suggested, it is apparent that, other things being equal, rockfall intensity could have been less during the long interlude between the early Holocene glacial episode (Crowfoot Advance) and the Neoglacial (Cavell Advance).

The volume of material in existing debris slopes demands alternative explanations. Clearly, one alternative would lie in more intense low magnitude rockfall activity in previous periods, possibly during the earlier part of the Neoglacial, during the early Holocene glacial episode and, most likely, during the late-Wisconsin deglaciation. Another alternative lies in the possibility that debris slopes in the area rely on more than rockfall for their nourishment. Snow avalanches and surface runoff of snowmelt water and rainfall are known to contribute debris in several locations in the Mt Rae area. A third alternative is that talus and other debris slopes contain remodelled glacial drift, probably that derived from former lateral moraines. Large quantities of fines are present in most debris slopes beneath a surface veneer of coarse debris (Bruckle *et al.* 1974) and this is certainly the case in the Mt Rae area. Some of the fine-sand and silt fraction could be derived from glacial drift. In areas of recent glacial recession in the Canadian Rocky Mountains, it is not uncommon to see lateral moraines being inundated with debris from upslope and being remodelled by a variety of slope processes. All alternatives could play a role in explaining the debris slopes of the area and radical departures from contemporary rockfall intensity need not be invoked.

FLOOD-DEBRIS FLOW

In the steep, ephemeral channels on mountain slopes, floods or high magnitude discharges and debris flows are often part of the same event. Floods are transformed into debris flows through the incorporation of debris from the channel bed and sides. A fluvial process becomes a mass-wasting process involving a grain-dispersion type of motion (Mears 1979). This has been an effective mode of material flux on the slopes of the Canadian Rocky Mountains. Indeed, the paraglacial alluvial fans identified by Roed and Wasylyk (1973) are, in part, a result of flood-debris flow activity.

My studies of flood-debris flow events in the Canadian Rocky Mountains have spanned only two years. Thus, the discussion which follows relies less on direct observation, as in the case of the previously discussed fluxes, and more on secondary data sources. These include historical records, old photographs, sequences of airphotos, and botanical (dendrochronology) information. Through these data sources and fieldwork, sites through the central Canadian Rockies have been identified and chronologies to illustrate the tempo of the process are being developed.

In this section, flood-debris flow frequency at three sites is examined. The sites are: Cathedral Crags, near Field, British Columbia, for which a good historical record exists; Mt Hector, which is nearby and for which few historical data exist; and Mt Elpoca, for which few historical data exist. The former two sites are in the Lake Louise region and the latter is in the Mt Rae area. At these sites, flood-debris flow events start in response to snow-cover melt, glacier melt and rainfall, or combinations of these.

The Cathedral Crags site has attracted considerable media and scientific attention (Jackson 1980). In September 1978, approximately 0.5×10^6 m^3 of water and debris rushed down an ephemeral stream course, damaging the Trans Canada Highway and Canadian Pacific Railway and delaying traffic on both. The water source for this event was a jokulhlaup from a small, high altitude glacier atop Cathedral Mountain. Glacier melt and rain contributed to the accumulation of water in a basin on the glacier which suddenly drained through a subglacial channel leading to the ephemeral stream course. This was the fourth destructive flood-debris flow at this site in the past 55 yr. Historical records and photos indicate that the others occurred in 1962, 1946 and 1925. However, old photographs and direct observations in the past 2 yr indicate that numerous smaller scale, non-destructive and non-jokulhlaup flood-debris flow events have occurred at this site in the past 80 yr. One such event was observed on July 10, 1980. Direct observations also suggest that small volumes of snow-melt runoff occupy the channel, or parts of it, every year, though the geomorphic effects of this may be minor.

A photo taken in 1910 shows deposits of numerous small debris flows high on the slope at the Cathedral Crags site. There is no evidence in the photo of large levées and an incised channel extending to near the valley bottom, as was the case following the 1978 event. A photo taken in 1923 reveals significant levées downslope, but these do not extend to the CPR tracks and thus the event did not register in the historical record. A photo taken in 1925 clearly shows the extent of the high magnitude 1925 flood-debris flow. This signalled the beginning of the series of large-scale, destructive debris flows over the past 55 yr.

This period coincides with the major part of the Little Ice Age recession of the glacier which today serves as the jokulhlaup source area. The 1910 photo shows the terminus of a small outlet glacier extending from the summit area of Cathedral Crags into the upper part of the ephemeral channel in question. By 1925, this tongue of ice had receded approximately 300 m and had thinned. Today, it has virtually disappeared. The lack of any morainic features more extensive than the Little Ice Age maximum suggests that the ice shown in the 1910 photo may represent the most extensive glaciation of this slope since the late-Wisconsin deglaciation. If this is the case, the Little Ice Age advance and the 1910-to-present recession is a very significant geomorphic event at the site. The recession may be a significant factor in providing both large quantities of water and debris to generate high magnitude flood-debris flows and thereby increase the tempo of this activity over the average for the Holocene.

The well-documented history of the Cathedral Crags site provided an opportunity to test the utility of dendrochronology in dating flood-debris flow events generally (Alestalo 1971, Shroder 1978). Eighteen trees were sampled in 1979 to test for correlations between growth perturbations and known flood-debris flow events. Comparisons of growth curves from upslope and downslope radii of trees growing in affected areas (see Fig. 8.2 for examples) show the 1962 and 1946 events. Samples taken higher on slopes indicate the 1925 occurrence, but analysis of these is not yet complete. The 1978 event is too recent to have registered in samples obtained in 1979 or 1980. In conifers, the usual impact on tree growth of inundation by flood-debris flow material is downslope deformation followed by accelerated growth, usually involving reaction wood, on the downslope radius and decelerated growth on the upslope radius. These patterns are apparent in Figure 8.2. The initial results being favorable, the technique was applied at the other two sites.

The Mt Hector site is located 20 km north of Lake Louise. Morphological evidence on the southwest slopes of Mt Hector suggested that flood-debris flow activity occurs in the ephemeral channels on the slope. Indeed, in July 1977, traffic on the Banff–Jasper Highway, which crosses the slope, was delayed by a flood-debris flow resulting from intense rainfall.

Other water sources include snowmelt runoff and combinations of the same with rain. At this site, water is collected in a steep bedrock basin above treeline. It passes through a gorge on to a colluvium and till-mantled slope below treeline where debris is incorporated into the flow. An incised and levéed channel leads to widespread flood and debris flow deposits on the lower valley slopes. A well-defined fan is not present.

The 1977 event is the only observed occurrence at the Mt Hector site. However, comparison of 1951 and 1977 airphotos shows evidence of several occurrences at this and nearby sites in the interim. Cores from 23 trees were sampled in 1979. Most of these trees date from the mid-nineteenth century, presumably following extensive forest fires in the region. Sir James Hector, who travelled in the area in 1858, described the forest as having been devastated by fire. Preliminary results from the core samples suggest occurrences in 1966, 1941 ± 2 yr, 1912, 1887 ± 3 yr and 1876 in addition to the 1977 event. This would give a frequency of about 1 in 20

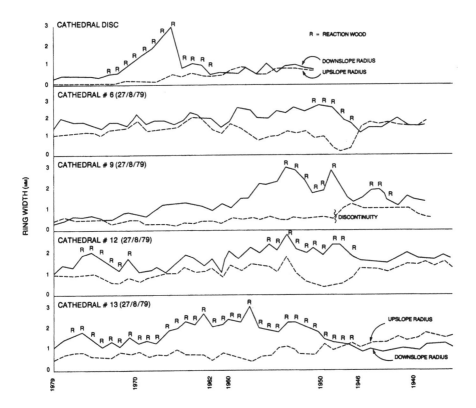

FIGURE 8.2
Examples of radial growth curves from trees sampled at Cathedral Crags.

yr for major out-of-channel flood–debris flow events at the Mt Hector site. Again, some snowmelt discharge occurs in the channel for a short period on an annual basis.

The Mt Elpoca site is located in the front ranges near Mt Rae. As in the other examples, a highway (Alberta 40) crosses the ephemeral channel. Snowmelt and rainfall are the primary water sources and some baseflow discharge is evident in a short reach of the channel through most summers. Morphologically, the site is similar to the site at Mt Hector. The upper catchment is composed of steep bedrock surfaces above treeline. The mid-section cuts through a suite of soft, erodable Mesozoic shales which provide copious debris to the system. The lower third of the site is composed of a paraglacial fan deposited over till and colluvium.

Flood-debris flow occurrences which obstructed the road were recorded in the summers of 1975 and 1979. These resulted from intense rainfall. The 1975 event followed a storm with an intensity of 80 mm/24 hours which has a return period of slightly more than 1 in 25 yr in this part of the Rockies.

Being in an isolated location, very little historical data exist for the Elpoca site. Therefore, dendrochronology was used to assess flood-debris flow occurrences in the recent past. The observed 1975 event is evident in growth series, indicating that the trees are sensitive to disturbance at this location. Three samples indicate growth anomalies and reaction wood in the mid-1950s. Subsequent examination of airphotos taken in the fall of 1954 and summer of 1955 indicate the occurrence of a very extensive flood-debris flow event in the intervening months. Field studies indicate both tree scarring without debris inundation as well as tree deformation with debris inundation, which suggests that the event involved a mixture of debris, water and snow indicating a spring thaw–rain combination. Tree growth patterns prior to 1955 indicate little disturbance, and deposits on the fan suggest little out-of-channel deposition of the 1955 magnitude. Fire destroyed part of the forest cover at this site in 1920, precluding dendrochronology beyond that date. Airphotos taken in the 1940s indicate little disturbance between then and 1920. Further sampling is required in the unburned part of the fan.

Data from the three locations indicate that combinations of morphological, botanical and historical information can be used to develop chronologies for the flood-debris flow mode of mass transfer. Over the past 2 yr, nine other sites have been examined using the same methods. As yet, all the data have not been systematically analyzed. However, preliminary impressions in combination with the results from the three sites discussed above suggest a pattern.

Flood-debris flow activity is a common occurrence in the Canadian Rocky Mountains under contemporary environmental conditions. Major out-of-channel occurrences have a frequency of 1 in 15 yr to about 1 in 25 yr. The Little Ice Age recession at the Cathedral site has generated a more frequent occurrence than that experienced prior to the recession. Extensive out-of-channel levées and lobe deposits at Hector and Elpoca indicate events of greater magnitude in the past 35 yr than in the previous century. The pattern is upheld at other sites. At three other sites, a detailed stratigraphy is exposed in fan deposits. The sections reveal surprisingly little (i.e. 0.75–1.50 m) deposition over Mazama ash which dates from 6700 yr BP. The fans themselves are typical postglacial or paraglacial deposits (Roed & Wasylyk 1973). Such a sequence suggests late-Wisconsin or early Holocene fan-building activity with flood-debris flow processes of far greater magnitude than at present. This was followed by a period of relative quiescence to the Neoglacial. The Little Ice Age appears to coincide with increased flood-debris flow frequency.

CONCLUSIONS

The material fluxes described in this paper all represent geomorphic work. However, they do not engender marked changes in morphology at the regional or individual slope scale. In terms of landform change at this scale, the individual rates of debris accretion and debris shift, and individual rockfalls and flood-debris flows do not represent 'threshold events'. The debris flows at the Cathedral Crags come closest to this category. Yet, even they do not radically alter the slope morphology, even though they can be highly destructive and disruptive to man-made structures in the area. It is a reasonable conclusion that the variability outlined here is within the variation and rate of operation typical of 'contemporary time'. There is some suggestion that the intensity of rockfall and flood-debris flow activity may be greater than it was over a large part of the Holocene. Yet, the morphological context in which these fluxes have operated is essentially that which has existed throughout the Holocene. Other morphological and stratigraphic evidence in the form of paraglacial fans, debris slope volumes, high magnitude rockfall deposits and the near-surface position of tephra of Mazama origin in a variety of deposits, indicate that present and Holocene rates of activity are less than that experienced during the late-Wisconsin deglaciation, for example. The paraglacial flood-debris flow fans and, possibly, the high magnitude rockfall deposits are seen as being of late-Wisconsin origin. Thus, the late-Wisconsin deglaciation may be seen as a relatively short period of radical geomorphic change, and subsequent 10 000–12 000 yr as a period of relative stability or variation in the context of dynamic metastable equilibrium.

The observational data describing this 'between threshold' variability clearly point out that variation in material fluxes from year to year and from place to place is considerable. Mean values are not good predictors for a given place and time. Yet, even the extreme fluxes noted in the data do not produce radical geomorphic changes, except at the scale of individual rock particles and small segments of ephemeral stream courses. At this scale, virtually any flux becomes a 'threshold event'. The material fluxes described here are all highly intermittent in their behavior. This seems to be a characteristic of alpine mass-wasting processes generally. The forcing functions and agents – freeze-thaw, presence of water, snow avalanches – produce intermittent stresses. Thus, it follows that responses should be intermittent occurrences and produce the illusion of geomorphic thresholds.

ACKNOWLEDGEMENTS

This research is funded by Natural Sciences and Engineering Research Council of Canada Grant A9152. I would like to thank D. Sauchyn, M. Sauchyn, J. Desloges, D. Smith, P. Kershaw, L. Kershaw and S. Hunter for their field assistance.

REFERENCES

Alestalo, J. 1971. Dendrochronological interpretation of geomorphic processes. *Fennia* No. 105.

Barnett, D. M. 1966. Preliminary field investigations of movement on certain arctic slope forms. *Geogr. Bull.* **8,** 377–82.

Bruckle, E. F. K. Brunner, E. Gerber, and A. E. Scheidegger 1974. Morphometrie einer schutthalde. *Mitt. Osterr. Geogr. Ges.* **116,** 79–96.

Carniel, P. and A. E. Scheidegger 1976. Mass transport on an alpine scree cone. *Rev. Ital. Geofisica* **3,** 82–4.

Church, M. 1980. Records of recent geomorphological events. In *Timescales in geomorphology,* R. A. Cullingford, D. A. Davidson & J. Lewin (eds), 13–29, New York: Wiley.

Church, M., R. F. Stock and J. M. Byder 1979. Contemporary sedimentary environments on Baffin Island, N.W.T., Canada: Debris slope accumulations. *Arct. Alp. Res.* **11,** 371–402.

Coates, D. R. and J. D. Vitek, (eds) 1980. *Thresholds in geomorphology.* London: George Allen & Unwin.

Cruden, D. M. 1976. Major rockslides in the Rockies. *Can. Geotech. J.* **13,** 8–20.

Cullingford, R. A., D. A. Davidson and J. Lewin (eds) 1980. *Timescales in geomorphology.* New York: Wiley.

Desloges, J. R. 1980. *Fluvial morphology and sedimentation in a high mountain drainage basin, southwestern Alberta.* Unpublished BFS thesis. University of Waterloo.

Dishaw, H. E. 1967. Massive landslides. *Photogramm. Engng.* **33**, 603–08.

Eisbacher, G. H. 1971. Natural slope failure, northeastern Skeena Mountains. *Can. Geotech. J.* **8**, 384–90.

Fulton, R. J. 1977. Late Pleistocene stratigraphic correlations, western Canada. In *Quaternary glaciations in the northern hemisphere,* D. J. Easterbrook & V. Sibrava (eds), 204–17. Prague: IUGS–UNESCO Int. Correlation Program.

Gardner, J. 1969a. Observations of surficial talus movement *Z. Geomorph.* **13**, 317–23.

Gardner, J. 1969b. Snowpatches: Their influence on mountain wall temperatures and the geomorphic implications. *Geogr. Annaler* **51** (A), 114–20.

Gardner, J. 1970. Rockfall: A geomorphic process in high mountain terrain *Albertan Geographer* **6**, 15–20.

Gardner, J. 1973. The nature of talus shift on alpine talus slopes: an example from the Canadian Rocky Mountains. In *Research in polar and alpine geomorphology,* B. D. Fahey & R. D. Thompson (eds), 95–106. Norwich: Geo Abstracts.

Gardner, J. 1977. High magnitude rockfall–rockslide frequency and geomorphic significance in the Highwood Pass area, Alberta. *Great Plains–Rocky Mountain Geog. J.* **6**, 228–38.

Gardner, J. 1979. The movement of material on debris slopes in the Canadian Rocky Mountains. *Z. Geomorph.* **23**, 45–57.

Gardner, J. 1980a. Dendrochronological evidence for flood/debris flow frequency in the Canadian Rockies (Abstract). *Abstr. Paper Can. Assoc. Geogrs.* Annual Meeting 1980, Montreal, Canada.

Gardner, J. 1980b. Frequency, magnitude, and spatial distribution of mountain rockfalls and rockslides in the Highwood Pass area, Alberta, Canada. In *Thresholds in geomorphology,* D. R. Coates & J. D. Vitek (eds), 267–95. London: George Allen & Unwin.

Harris, S. A. and R. W. Waters. 1977. Late-Quaternary history of southwestern Alberta: a progress report. *Can. Petrol. Geol. Bull.* **25**, 35–62.

Jackson, L. E. 1980. A catastrophic glacial outburst flood (jokulhlaup) mechanism for debris flow generation at the Spiral Tunnels, Kicking Horse River Basin British Columbia. *Can. Geotech. J.* **16**, 806–13.

Luckman, B 1976. Rockfalls and rockfall inventory data: some observations from Surprise Valley, Jasper National Park, Canada. *Earth Surf. Proc.* **1**, 287–98.

Luckman, B. 1978. Geomorphic work of snow avalanches in the Canadian Rocky Mountains. *Arct. Alp. Res.* **10**, 261–76.

Luckman, B. and G. D. Osborn 1979. Holocene glacier fluctuations in the middle Canadian Rocky Mountains. *Quat. Res.* **11**, 52–77.

Mears, A. I. 1979. Flooding and sediment transport in a small alpine drainage basin in Colorado. *Geology* **7**, 53–57.

Rapp, A. 1960a. Recent development of mountain slopes in Karkevagge and surroundings, northern Scandinavia. *Geogr. Annaler* **42**, 73–200.

Rapp, A. 1960b. Talus slopes and mountain walls at Templefjorden, Spitzbergen *Norsk Polarinst. Skr.* **119**, 1–96.

Roed, M. A. and D. G. Wasylyk 1973. Age of inactive alluvial fans – Bow River Valley, Alberta. *Can. Earth Sci.* **10**, 1834–40.

Rutter, N. W. 1972. Geomorphology and multiple glaciation in the area of Banff, Alberta. *Geol. Survey Can. Bull.* **206.**

Ryder, J. M. 1971. The stratigraphy and morphology of paraglacial alluvial fans in south-central British-Columbia. *Can. J. Earth Sci.* **8**, 279–98.

Schumm, S. A. 1977. *The fluvial system.* New York: Wiley–Interscience.

Shroeder, J. F. 1978. Dendrogeomorphological analysis of mass movement. *Quat. Res.* **9**, 168–85.

Stalker, A. Macs. 1973. *Surficial geology of the Kananaskis Research Forest and Marmot Creek basin region of Alberta.* Geol. Survey Can. Paper 72-51, 1–25.

Stalker, A. Macs. and J. E. Harrison. 1977. Quaternary glaciation of the Waterton-Castle River region of Alberta. *Can. Petrol. Geol. Bull.* **25**, 882–906.

Thornes, J. B. 1971. State, environment and attribute in scree slope studies. In *Slopes: Form and process,* D. Brunsden (compiler) 49–63. Inst. Br. Geogs. Spec. Pub. 3.

Thornes, J. B. and D. Brunsden. 1977. *Time and Geomorphology.* London: Methuen.

Walley, W. B. 1974. The mechanics of high-magnitude, low frequency rock failure and its importance in a mountainous area. *Geographical papers* **27**. University of Reading.

9

Spatial variation of fluvial processes in semi-arid lands

William L. Graf

INTRODUCTION

Within the past decade, a great deal of attention has been given to the concept of geomorphic thresholds and abrupt changes in the operation of fluvial systems (summarized by Schumm 1977). This work has been instrumental in focusing attention on the alteration of process rates through time, and several models have been proposed to describe the temporal characteristics of abrupt fluvial adjustment (e.g. Graf 1979c). These concepts and their models have differed significantly from earlier works which emphasized equilibrium states with little, or at least relatively smooth, change (summarized by Leopold *et al.* 1964). There is, however, an additional perspective that may be valuable in interpreting, explaining and predicting the operation of fluvial processes. In placing emphasis on temporal variation, geomorphologists may recently have unnecessarily slighted the analysis of spatial variation of process, assuming, perhaps, that the precepts of hydraulic geometry and drainage network laws are adequate. The theses of this paper are that the spatial variation of energy is a fundamental component of the explanation of fluvial processes, that the distribution of energy is not simply related to the classic measures of network position or discharge, and that major adjustments in system operation are manifestations of shifts in the spatial distribution of energy.

From the spatial perspective, periods of system change can be defined most easily by the spatial patterns of energy that are observed before or after the change. Thus, in the periods when the system crosses critical thresholds of operation the patterns are difficult to define, but during relatively stable periods the patterns are more readily discernible. There-fore, this paper will concentrate on the analysis of systems that might be characterized as being in some kind of steady state equilibrium, with a minimal rate of changes – that is, between thresholds. Chorley and Kennedy (1971) provide a review of types of equilibrium, and Schumm and Lichty (1965) provide a temporal framework for their analysis.

The specific questions to be addressed are:

(i) what is the basis of structural and external controls on spatial variation of fluvial processes?

(ii) how does energy vary spatially in channel networks?

(iii) what are the geomorphic consequences of the spatial variation of fluvial processes?

The investigation of these questions is most easily accomplished in semi-arid lands where system adjustments are known to occur abruptly and are separated by periods of relatively little change (Cooke & Reeves 1976). The shallow soils of such regions also readily permit identification of geologic variations. Finally, the products of fluvial system adjustment, arroyo–gully development and massive sedimentation–erosion are among the most remarkable aspects of semi-arid valleys and have long excited scientific curiosity (Bryan 1925, Johns 1976).

For the purpose of the research reported below, geomorphic processes are interactions between forces and resistances at the Earth's surface (modified from Embleton & Thornes 1979, p. 1). These interactions are usually difficult, if not impossible, to measure in the field, but in many cases geomorphic processes can be measured as changes in observable system variables (Thornes & Brunsden 1977, p. 96). Lack of change does not necessarily imply lack of process, since the net result of interactions over a restricted period of observation might be zero. Measurements of processes may be approximated by using known or estimated impacts of energy compared to known or estimated resistances.

Much fluvial geomorphic research concerning spatial variation has been restricted to channel forms, and this is a source of some problems when attempts are made to analyze fluvial processes in semi-arid lands. Arroyos and gullies, for example, cannot be analyzed using classic hydraulic geometry because the forms are not channels, they are trenches with channels at the bottom. Estimates of the 10-, 50-, and 100-yr flows in nearly 200 cross-sections of arroyos and gullies in Colorado and Utah indicate that in only a few instances do these flows exceed 'bank full' capacity, and in no case do the stages correspond to morphological characteristics of the overall forms. Since arroyos and gullies are not channels, discharge is not an effective predictor of width or depth, except in very general terms.

Presumably there are two types of spatial variation in fluvial systems – random and systematic. Random variation is expectable because of the complexity of the fluvial system operating within an environmental matrix. This variation insures that even sophisticated models accounting for large numbers of variables will be unlikely to reproduce synthetic duplicates of natural streams, and simplified models will be likely to be gross approximations to reality. The frequently observed scatter about regression lines relating fluvial geomorphic variables is a reflection of this random variation (Leopold *et al.* 1964, p. 274, Patton & Schumm 1975), as are the departures from expected conditions outlined in the models below.

Systematic variation in fluvial processes comes from two sources – that which is the product of the spatial structure or network within which processes operate, and that which is imposed by external constraints such as surficial materials, vegetation or man-made structures. These two types of spatial controls compose the spatial system or stage where the interactions of forces and resistances take place, resulting in fluvial processes.

The sections that follow are an exploration of some aspects of spatial variation of energy and process in the streams of an area that is generally representative of many semi-arid lands. After an overview of the study area, individual sections address spatial variation imposed by structure, variation imposed by external controls, the distribution of energy in the form of force differentials, and the geomorphic consequences of these distributions.

THE HENRY MOUNTAINS AS A STUDY AREA

The Henry Mountains region of south central Utah provides an ideal setting for the analysis of spatial variation of fluvial processes in semi-arid lands. Geomorphic studies have been conducted there for more than 100 yr (though not continuously), so that the operation of the fluvial systems in this remote and relatively undisturbed environment is known for an extended period. The first scientific observations were by A. H. Thompson, an associate of the Powell Survey in 1872 (Gregory 1939), and Gilbert's work there in 1875-6 insured the lasting fame of the region in the field of geomorphology (Gilbert 1877). In the 1880s, intensive and exceptionally detailed Government Land Office surveys provided valuable quantitative information (Ferron 1883). About 50 yr after Gilbert, Hunt conducted an intensive analysis of the geology and geography of the area in the period 1933-9 (Hunt *et al.* 1953). Further geomorphic research was accomplished in the area by several of Hunt's students, with some published results (Everitt 1979, 1980, Godfrey 1980a, 1980b). Almost half a century after Hunt's work, Graf (1980) has conducted additional geomorphic research in the area, sponsored by the National Science Foundation.

The Henry Mountains are a series of five lacolithic mountains arranged in a roughly north–south chain stretching some 85 km (50 miles) through south central Utah (Fig. 9.1). Porphry cores are exposed in the centers of the lacolithic domes, with sharply upturned sedimentary rocks around each core (Williams & Hackman 1971, Williams 1972). Pediments sweep down from the steep mountain slopes across gently inclined to flat-lying sedimentary rocks to terminate in the badlands, table lands and canyon-lands associated with the surrounding major streams. The stream systems used in the present work lie on the north side of the Henry Mountains, generally between 1280 m (4200 ft) and 1700 m (5600 ft).

The climate of the Henry Mountains is highly variable and strongly dependent on elevation. In the river valleys surrounding the region, rainfall averages only 15.4 cm (6 in) per year, while on the highest mountain slopes it exceeds 76.8 cm (30 in) (Covington & Williams 1972). Most precipitation in the region falls as rain from summer thunderstorms, but there is some

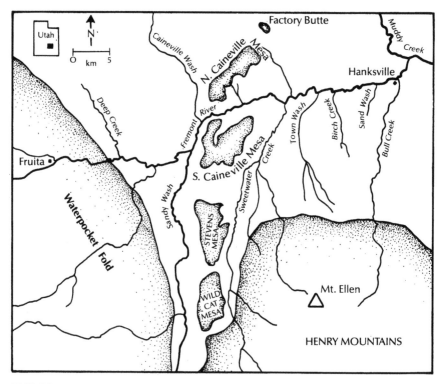

FIGURE 9.1
Location map for the Henry Mountains regions, Utah. Stippled boundaries enclose highland areas.

snowfall as well. Winters are cold with average January temperatures in the valleys of − 5.6°C (22°F), and summers are hot with average July temperatures of 25°C (77°F) (Hunt *et al.* 1953). In most of the elevational envelope containing the streams analyzed in the present research, the climate can be characterized as semi-arid in Köppen system (Trewartha 1954), with shrubs and grasslands (Hackman 1973). It is also semi-arid in Peltier's morphogenetic system (Peltier 1950).

Valley floors are partially buried by alluvial materials, and stream channels lie at the bottoms of continuous arroyo systems. In a few cases discontinuous systems also exist, especially high on pediments near the mountain fronts. In their upper reaches, the trenches are small − 1 m (3 ft) in depth − but part way between the mountains and the major streams they reach depths of 13.7 m (45 ft). The Fremont River is entrenched into its alluvial fill about 6.4 m (21 ft). Some gullies existed in the area as early as 1872 (Gregory 1939), but most cutting did not begin until a large flood on the Fremont River in 1897 (Hunt *et al.* 1953).

According to residents interviewed by Hunt and as recorded in his field notes (stored by the US Geological Survey in Denver), the 1897 flood caused drastic entrenchment of the Fremont River, and initiated headward erosion on the tributaries. An extensive system of continuous arroyos developed, reaching many kilometers from the Fremont. By 1917 most arroyos had partially refilled, only to be subjected to erosion again by the 1920s. Some refilling has occurred in some places since the 1920s.

SPATIAL VARIATION IMPOSED BY STRUCTURE

Stream processes operate within a hierarchical network of channels, an arrangement that pre-ordains many spatial characteristics of the resulting processes. A considerable amount of work has already been accomplished on the network properties of streams (e.g. Playfair 1802, Horton 1945, Shreve 1966, 1967, Woldenberg, 1969, Abrahams 1975). An example result of the influence of network arrangements is the movement of water and sediment. Generally, larger streams in a network carry larger quantities of sediment because of greater amounts of water available for work: smaller streams in the network carry larger sizes of sediment because of steeper gradients, and vice-versa. In a hypothetical, perfectly tuned fluvial system, the transport capability of each segment is slightly greater than the segment upstream and slightly less than the segment downstream.

That this is not the case with natural streams of semi-arid environments in general, however, is indicated by the development of discontinuous arroyo systems in many regions. That it is not the case in the Henry

Mountains is indicated by the temporary storage of sediments on the floodplain of the Fremont River during the period of Gilbert's work (1875–6), and their partial removal by the time of Hunt's work (1933–9). At present (1980), the sediments appear to be accumulating again. This series of events is strongly reminiscent of the model of episodic erosion wherein systems adjust quickly as thresholds are crossed, triggering complex responses (Schumm 1975). Assuming that the Henry Mountains' systems are not at a threshold (except for restricted local cases) and are operating in a period between thresholds, it might be possible to define the spatial distribution of processes that are concentrating sediments in one place in the network (along the Fremont River) and evacuating them from other places (intermediate-sized tributaries).

One measure of process operation is depth of erosion at sample cross-sections through the stream networks. The depth of erosion is the product of the amount of energy available as represented by the discharge of a representative event, and the variation imposed by the spatial structure within which the flow occurs. The location of the cross-section is critical because, if it is close to the main stream of the system where erosion was initiated, it will experience longer and perhaps deeper incision than a cross-section that has the same discharge but that is located far from the point of initial erosion.

Consider Figure 9.2 as an example of the role of distance or location in fluvial processes. The figure represents a simple hypothetical drainage system with three monitored cross-sections: numbers 1 and 2 with some known and equal 10-yr discharge, and number 3 with a larger 10-yr discharge. At time 1, as in the upper diagram, an arroyo is eroding headward, but has not yet reached the monitored sections. At time 2, the lower diagram, the erosion has reached two of the sections. Although sections 1 and 2 have similar discharges, they exhibit dissimilar conditions, one with an arroyo and the other with a pre-existing relatively shallow channel. Discharge is not a useful explanatory variable for predicting channel geometry in this case.

Even after the headward erosion has affected both of the Q-magnitude cross-sections, their characteristics are not likely to be the same, since one will have been subjected to cutting for a longer period. Relative location or distance from the source of disruption is, therefore, a significant variable related to the spatial structure of the system, which must be taken into account in explaining channel and near-channel conditions such as arroyo geometry or sediment movement and storage.

The most general spatial law available for the description and analysis of processes operating within a structural framework as measured by distance is the spatial interaction law (Taylor 1975). This law specifies that the interaction (x_{12}) between two entities is directly proportional to the magnitude of the two entities (m_1 and m_2) and inversely proportional to the distance between them (d_{12}), or:

$$X_{12} = k_1 \frac{(m_1)^{b_1}(m_2)^{b_2}}{(d_{12})^{b_3}} \qquad (9.1)$$

where k_1 and $b_{1,2}$...are constants. The potential in a system (y_{ij}) is the sum of all the interactions, and is defined as

$$y_{ij} = k_2 \frac{(m_i)^{b_4}(m_j)^{b_5}}{(d_{ij})^{b_6}} \qquad (9.2)$$

where k_2 and $b_{4,5}$...are constants, $b_6 = b_3 - 1$, and $i,j = 1,2...$

Spatial interaction laws have been successfully applied in the spatial analysis of economic, social and political problems where, for example, the entities are cities, the distances between the entities are road distances, and the interactions are numbers of trade transactions, communications exchanges, or migrations (Morrill 1963, Olsson 1970). The most basic application of spatial interaction laws, which can serve as an illustrative ex-

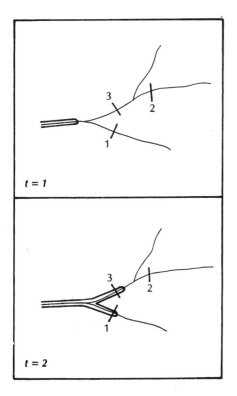

FIGURE 9.2
The development of continuous arroyos by headcut migration. Labels indicate relative amounts of discharge at sample cross-sections.

ample, is from physics and celestial mechanics – the gravity law (Feynman *et al.* 1965, Resnick & Halliday 1977). In this particular application, the interaction term, x_{12} is the gravitational attraction between two bodies in space, the constant k_1 is the universal gravitational constant, the magnitudes of the two entities (m_1 and m_2) are their respective masses, and the distance term (d_{12}) is the straight-line distance between the entities. The potential of the system (y_{ij}) is the sum of all the gravitational interactions between pairs of entities in the system. In this application, the constants $b_{1,2,4,5,6} = 1.0$, while $b_3 = 2.0$.

The definitions of the variables are different in the application to the fluvial system, but the function forms and principles remain the same. Considering the development of an arroyo system with the erosion and evacuation of valley-fill sediments, the interaction between two entities or cross-sections (x_{12}) is the erosion of material, and the potential for erosion at a given site (y_{ij}) is the observed final depth of erosion at site j, after a system has adjusted to some initial disruption at i. The magnitudes (m_1 and m_2) of the entities or cross-sections are for some event of reference, either an observed magnitude or a calculated one such as an arbitrarily selected return interval of interest. The distance term (d_{12}) is the network distance from one site to the other along the channel system. As with the gravitational system, the values of the constants must be determined empirically.

Since almost all the streams in the Henry Mountains region appear to be in a period between thresholds, the depth of each arroyo cross-section probably represents the influence of process and spatial structure. The depth of each cross-section is the product of erosional processes controlled by the amount of discharge available for geomorphic work plus the influence of the spatial structure as controlled by the distance between the cross-section and the point of disruption. If systems are tributaries to the Fremont River, the sums of all the interactions or the erosion potential (y_{ij}) can be calculated for each system if numerous observations are available. A total of 88 units of observation (one cross-section with its associated measurements is a unit of observation) in seven tributary systems are available from field data collected in 1979 and 1980. Measurements include the depth of the arroyo, which was measured in the field as the maximum vertical distance between the top of the valley fill to the floor of the channel. The magnitude of the site was measured as the probable 10-yr discharge, as estimated from previously established regional models (Berwick 1962). The magnitude of the other entity was measured as the probable 10-year discharge of the Fremont River where the tributary system joins it. The distance variable was measured as the channel distance between the site in question and the junction of its tributary system and the Fremont River. Most distance measures were made on aerial photography.

The quantitative observations from a single tributary system can be used as input for the linear form of Equation (9.2):

$$\log y_{ij} = \log k_2 + b_4(\log m_i) + b_5(\log m_j) - b_6(\log d_{ij}) \qquad (9.3)$$

A least-squares solution of this function produced values of the constants, with the entire process being repeated for each tributary system. In the present discussion, only the value of b_3 (which is defined by $b_6 + 1$) is of interest because it represents the distance factor in the interaction. The solutions provide an indication of the significance of the spatial structure by assessing the increased statistical power of the function when the distance variable is added to the calculations, as opposed to using only the discharge variable, and the values of the distance interaction parameter (b_3) provide a useful comparison among tributaries.

When Equation (9.3) is solved for all seven tributary systems without using the distance variable, the mean coefficient of determination is 0.54,

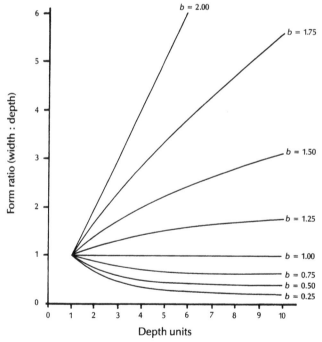

FIGURE 9.3
The effect of various b values of the arroyo cross-section model that relates width to depth. Low values for the form ratio indicate relatively deep narrow channels, high values indicate broad, shallow ones. A form ratio of 1.0 indicates a perfectly rectangular cross section.

but when the distance variable is added, the mean coefficient of determination increases to 0.76. One tributary system showed no increase in the coefficient, but it was already at 0.96 and it is a small system where distance effects have little chance to operate. The greatest increase in the coefficient realized by the additional consideration of distance was Town Wash, where the value increased from 0.13 to 0.77. Although the mere addition of a random variable might also increase the coefficient of determination, the substantial increase observed here indicates the important role that spatial structure plays in the process of erosion and sediment transport. Clearly, some spatial variation of process is the result of the spatial structure within which the process operates.

The distance interaction parameter (b_3) shows how important distance is in accounting for change in the depth of erosion. If the value is less than 1.0, distance is of relatively minor importance, but as its values approach 2.0, distance plays an increasingly important role in the system operation. The mean value of b_3 for all seven systems considered is 1.71 – an indication that distance is an important consideration in fluvial processes, but that it is slightly less influential than in gravitational systems. Distance exerted a stronger influence in some systems than in others, since the values of b_3 ranged from 1.01 to 2.92. The system-to-system variation appears to be a function of surficial materials, an aspect of spatial variation that is discussed in the following section.

SPATIAL VARIATION IMPOSED BY EXTERNAL FACTORS

Spatial variation in fluvial processes, imposed by structural considerations such as the hierarchical channel networks or by distance from points of disruption, is superimposed on an environmental backdrop of the external factors of climate and surficial materials. Climate determines large regions with similar geomorphic processes on a subcontinental scale (Tricart & Cailleux 1972). Within each morphogenetic region, however, geology, through the medium of surficial materials, contributes systematically to spatial variation, to erosion and deposition. In semi-arid lands, geology is especially prominent because of shallow soils and the dominance of mechanical weathering which insures close spatial association between weathering debris and the parent rock. Boundaries between rock types are usually sharp and distinct, and are likely to be significant process boundaries as well.

In order to define the impacts of variation in surficial materials it is necessary to compare the behavior of a number of streams representative of each type of material. No single limited study area could provide such wide-ranging data, but the northern flank of the Henry Mountains provides a useful sample. Some watersheds are developed wholly within shale, others in sandstone, and still others in a mixture of the two.

Any comparative tool that is used to explore the impact of external spatial controls must be essentially scale-free in order to avoid the problem of comparing streams of disparate sizes. The tool also must be able to take into account entire stream systems so that the comparison can be accomplished on a network-to-network basis rather than on a site-to-site basis. Finally, the tool must be able to accommodate a parameter that can be linked to process. Allometric models serve as the comparative tool in this study and meet these criteria.

Allometric models describe the rate of change of a part of a system as compared to the rate of change of another part of the same system or to the rate of change for the system as a whole (Woldenberg 1968, Faulkner 1974, Bull 1975). In biologic applications, for example, an allometric model might describe the rate of growth of arteries as a ratio to the rate of growth of the heart muscle. In the arroyo application, the model relates rate of depth change (dD/D) to rate of width change (dW/W).

$$\frac{dW}{W} = b\frac{dD}{D} \tag{9.4}$$

where b is a constant of proportionality (Graf 1979a). Integration of both sides of the equation and taking antilogs produces the simple mathematical expression of the allometric law that the rate of depth change and the rate of width change may change from time to time, but they maintain a constant ratio. One is a non-linear function of the other:

$$W = a\,D^b \tag{9.5}$$

where a is an additional constant.

The application of allometric principles and models is especially appropriate for arroyos, because their cross-sectional form is the product of two processes, deepening by fluvial entrainment and widening by mass movement. It is unlikely that these processes would operate at exactly the same rate, but they might be related by some constant ratio, as expressed in the value of b in Equations (9.4) and (9.5). If $b = 1.0$, the two rates of change are equal, and the cross-sectional form does not change from one time or place to another. As shown in Figure 9.3, if $b < 1.0$, when the cross-section changes in size through time or across space, it changes shape. An increase in size produces a form that is relatively deeper and narrower. If $b > 1.0$, when the cross-section changes in size, it also changes in shape: an increase in size produces a form that is relatively wider and shallower. In all cases, the value of a is an indicator of scale if models with similar b values are compared.

Allometric models are ideal tools to compare arroyo networks developed on different lithologies or with different vegetation characteristics. The *b* values in particular describe the nature of the relative rates of change within each network, so that the network as a whole is considered and compared rather than a single site with another single site, neither one of which might be typical of its network. If one type of change in a network can be associated with particular geologic or biotic conditions, valuable additional information can be provided on spatial variation, since the spatial distribution of the control factor can be mapped. Drainage networks superimposed on these controls can then be viewed as having processes that are controlled by spatial properties of the networks as well as the environment where the network happens to be located.

The 88 sites used earlier in the spatial interaction tests also provided data for allometric models. Arroyo width was measured in the field horizontally from the top of one edge of the arroyo to the other. Depth, also measured in the field, was the maximum vertical distance from the top of the valley fill to the floor of the channel. These data, collected from seven watersheds in a variety of geologic settings, were used in ordered pairs to solve by standard least-squares methods the linear form of Equation (9.5):

$$\log W = \log a + b(\log D) \tag{9.6}$$

The results are given in Table 9.1, and the geologic materials of the region are reviewed in Figure 9.4.

Those stream systems with arroyos excavated in fine-grained sediments (derived mostly from shales) have power function models with $b < 1.0$. As larger cross-sections are considered, they become relatively deeper and narrower, probably because, although the fine materials are readily excavated by fluvial processes, the walls of the arroyos are stable and do not retreat rapidly – width increases more slowly than depth. The middle reaches of Sweetwater Creek, the tributaries of Sandy Creek, and the main stem of Sandy Creek (despite its name) flow through silty sediments and exemplify this behavior.

Stream systems with arroyos excavated in sandy materials derived from sandstones have power function models with $b > 1.0$. As larger cross-sections are considered, they become relatively wider and more shallow, probably because the walls are unstable and collapse readily in sandy materials. Width increases more rapidly than depth. The north bank tributaries of the Fremont River and the lower reaches of Sweetwater Creek flow through terrains dominated by sandstones and their weathering debris, and have power function models with $b > 1.0$.

Town Wash and the south bank tributaries of the Fremont River flow through arroyos excavated in valley fills that have a mixture of sedimentary sizes, and interbedded rock units are especially common in these basins.

The *b* values of their power function models are less than 1.0, but are intermediate in relationship to the extreme cases cited above. Sweetwater Creek, taken in total, has a compound function, with a low *b* value for the fine-grained middle reaches and a high *b* value for the coarse-grained lower reaches.

All the *b* values are near 1.0 or below, indicating that in the restricted tributary systems being considered there is little tendency for the rapid development of wide and shallow channels. This is probably because the arroyos are continuous rather than discontinuous and so lack areas of deposition with attending widening, and because none of the data sets includes the arroyo of the Fremont River, which has a width:depth ratio exceeding 20. If the scale of analysis were to be broadened to a regional perspective to include the Fremont and Dirty Devil Rivers, compound functions might result. When subregional basins are considered, arroyos tend to be slot shaped or rectangular, irrespective of size, and only the largest streams have wide, shallow forms that represent large amounts of sediment removed from the valley cross-sections.

The results of the allometric analysis show a clear partitioning of the backdrop over which drainage networks are superimposed. This partitioning is directly related to surficial materials, which control in part the nature

TABLE 9.1
Solutions for power functions as allometric models.

Data group	Geologic material	a	b	r	s	p
Fremont tributaries, north bank	sandstone and shale	1.66	1.14	0.93	0.39	0.01
Fremont tributaries, south bank	sandstone and shale	6.23	0.94	0.81	0.33	0.10
Sandy Creek tributaries, west	sandstone	0.83	0.40	0.31	0.69	0.62
Sandy Creek tributaries, east	fine alluvium	9.25	0.40	0.58	0.31	0.17
Sandy Creek	fine sands	20.27	0.55	0.68	0.42	0.09
Sweetwater Creek	sandstone and shale	2.05	1.17	0.57	0.30	0.15
Town Wash	fine alluvium	3.80	0.90	0.68	0.59	0.03

Notes: *a* = power function coefficient; *b* = power function exponent; *r* = coefficient of determination; *s* = standard error; *p* = level of significance.

of the change in arroyo morphology, so that systematic variation in fluvial processes is introduced. Narrow confining arroyo walls result in deep flows, and deep flows produce high levels of tractive force and unit stream power, further enhancing erosion potential. This positive-feedback system of erosion, deep forms and further erosion is most likely to occur in terrains dominated by fine-grained surficial materials, so that the distribution of surface materials becomes a major concern in explaining the distribution of process intensity.

FORCE DIFFERENTIALS

Spatial variation in fluvial processes imposed by external considerations such as surficial materials and by structural considerations such as the hierarchical drainage network and distance to points of disruption sees its most significant expression in the distribution of fluvial energy in the system. The fluvial energy, measured as tractive force or as stream power, accomplishes work by entraining and transporting sediment (see Bull (1979) for discussion). In the discussion that follows, only the simpler case of tractive force will be considered. Tractive force is important because it represents the sheer force available on the channel bed, and it determines the competence of the stream. Higher amounts of tractive force move larger particles. If insufficient tractive force is available to move particles on the bed, no erosion takes place.

Tractive force is defined by the DuBois equation as

$$\tau_o = \gamma R S \tag{9.7}$$

Where τ_0 is tractive force in dynes, γ is unit weight of the fluid in gm cm^{-3}, R is hydraulic radius in cm, and S is dimensionless energy slope (Leliavski 1966, Graf 1971, Bogardi 1974). Values of tractive force in natural rivers range from below 1.0 dyne to above 10 dynes for deep, restricted flows. In the Henry Mountains region, the values range from less than 1.0 dyne to about 5.0 dynes. Small boulders up to 25 cm in diameter occur in many of the channels, and they require about 3–4 dynes for movement by fluvial processes. Smaller particles, of course, are moved by lesser amounts of tractive force.

Theoretical considerations lead to the conclusion that tractive force has a non-linear distribution because it is controlled primarily by two variables, hydraulic radius and slope, which change in opposition to each other when larger streams are considered. Very small streams with limited drainage areas are likely to have low values of tractive force, because their hydraulic radii (closely allied with depth) are very small. The steep gradients of these small streams do not make up for the lack of depth, and low values of tractive force result.

At the opposite end of the scale, large streams have deep flows, which would lead to increased tractive force, but high values of force do not occur because of the gentle gradients involved. A trade-off occurs between the variables of hydraulic radius and gradient, so that the large streams also have low values of tractive force. Somewhere between the two extremes, at moderate sizes of hydraulic radii and moderate values of gradient, maximum values of tractive force occur.

A plot of tractive force against drainage basin area, then, would show a function with its low values at both ends of the drainage area scale, and a 'hump' in the middle where the hydraulic radius and gradient combine in the most favorable way to produce maximum force value, as in Figure

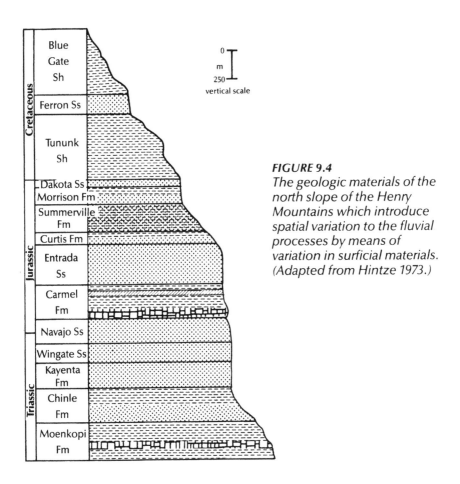

FIGURE 9.4
The geologic materials of the north slope of the Henry Mountains which introduce spatial variation to the fluvial processes by means of variation in surficial materials. (Adapted from Hintze 1973.)

9.5. The spatial implications of this circumstance are obvious: when proceeding through a drainage network, the most forceful streams are those in the middle. The change in values of tractive force along a single channel or within a hierarchical group of watersheds is here defined as a force differential. A positive force differential occurs when tractive force increases in the downstream direction or when larger watersheds are considered. A negative force differential occurs when tractive force decreases in the downstream direction or when larger watersheds are considered.

Erosion and sedimentation can modify force differentials as well because they alter the shape of the channel in a feedback process. If the channel cross-section is filled and becomes shallow and wide, tractive force values decline because of declining hydraulic radius and gradient. The reverse occurs in an erosional situation. In semi-arid lands, short-term fluctuations on a scale of a few years might also occur as a result of spatially restricted flood events. Thunderstorm cells may produce intensive precipitation over only a limited portion of channel network, so that channel adjustments during the resulting flood may affect only a portion of the network in question. Presumably, over a period of a century or more the events would have occurred throughout the area, and all channels would be adjusted.

If force differentials can be used to characterize streams between periods of significant alteration, they might also be used as a perspective

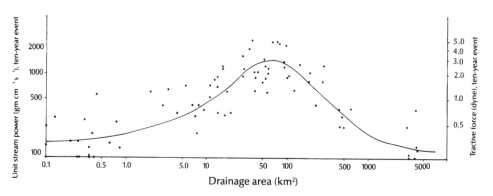

FIGURE 9.5

Tractive force and unit stream power at steam cross-sections in the northern Henry Mountains, plotted against drainage areas above the cross-sections. A positive force differential exists for streams with drainage areas ranging from near 0 to about 70 km^2; a negative force differential prevails for streams draining areas greater than 70 km^2.

on thresholds or catastrophic change. The locations of positive and negative force differentials are probably different for different stable periods, and the 'hump' in the plot of force vs. drainage basin area may move during system-wide adjustment. Maximum values of tractive force might now occur at channel cross-sections with 100 km^2 drainage areas upstream, whereas in the past the maximum values may have occurred at channel cross-sections with only 50 km^2 drainage areas.

As a representative of typical semi-arid lands, the Henry Mountains region provided data to test the concept of force differentials. At 88 cross-sections (used previously), channel and near-channel forms were measured in the field to determine hydraulic radius and gradient. The unit weight of flood waters was assumed to be 1.15 gm cm^{-3} to account for the debris and high suspended-sediment concentrations in flood waters. Hydraulic radius and gradient were used to estimate the tractive force for the 10-yr flood event (arbitrarily selected for comparative purposes) at each cross-section, by the method outlined by Graf (1979a, 1979b), and then plotted against the drainage area above the cross-section. Magnitudes of 10-yr events were calculated using regional relationships previously developed by Berwick (1962). The resulting plot (Fig. 9.5) demonstrates the existence of positive and negative force differentials which produce maximum tractive force values in streams draining about 50–90 km^2.

The particular location of the maximum values of tractive force on the landscape and on the plot is a function of the influence of the hierarchical aspects of the channels (with increasing discharge, increasing radius, and decreasing gradient in the downstream direction), but other factors also play a role. The development of a continuous arroyo system after the 1897 flood on the Fremont River has led to a headcut migration into tributary basins, and especially deep trenches extend into those basins as small as 30 km^2. Many smaller basins are too far away from the source of disruption and, following spatial interaction laws, they were not trenched at all or only slightly. Where the arroyos exist, they constrict channel flow, causing relatively great depths of flow and high values of tractive force.

Geologic or surficial materials are also important since basins of intermediate size in this region have floors dominated by fine-grained alluvium which supports high arroyo walls. Once the vertical erosion had taken place in the medium-sized basins, the geologic materials insured the maintenance of relatively deep and narrow forms, which resulted in constrictions and deep flow with high values of tractive force. An unequal distribution of force and the attending existence of positive and negative force differentials in their present arrangement, therefore, is an expectable outcome of spatial controls on the processes.

CONSEQUENCES OF SPATIAL VARIATION

The most important consequences of the spatial variation in fluvial energy is the uneven movement and temporary storage of sediments. Under present conditions in the Henry Mountains region, for example, the distribution of force suggests that sediment is most likely to accumulate in small streams and in the largest ones, while those of intermediate size (with drainage areas upstream of 50–90 km^2) are most likely to be erosion sites with little sediment accumulation. A field examination of the channel networks reveals this to be the case.

After the development of the extensive arroyo networks by headward erosion after the 1897 flood, some refilling and recutting occurred. Visits to sites noted or photographed by Hunt during the 1930s show that the Fremont River has stablized and is aggrading slightly since the period. Sediment also partially fills small channels that were without filling in the 1930s. But in the intermediate areas along Sandy Creek, Sweetwater Creek, and Town Wash (see Fig. 9.1 for locations), intensive cutting continues with only minor amounts of desposition.

Further support for the effects of force differentials is offered by investigation of stream junction areas within the region. The cross-sections of the trenches cut during the late 1800s provide a reference, and the percentage of those cross-sections that have been refilled provides a guide to the accumulation of sediment under present system conditions. The behavior variable at stream junctions is the amount of sediment accumulated in the main channel, and it is influenced by two control factors: the force available in the tributary to deliver sediment, and the force available in the main channel to carry the sediment further downstream. Following the precepts of catastrophe theory outlined in an earlier paper (Graf 1979c), a graphical representation of the interaction of the control and behavior variables is a cusp-shaped surface. Figure 9.6 is a smoothed version of the data collected from 28 junction areas in the Henry Mountains region. The figure demonstrates a basic aspect of the spatial variation of fluvial processes in the example area – differential movement of sediments.

At any given junction, when tributary force is low, little sediment is added to the main channel, so there is little sedimentation, regardless of force levels in the main channel. When tributary force is high, more sediment is poured in to the main channel. Much of it remains there if the main-channel force levels are low; less remains if the main-channel force levels are high. Between the extremes there are two possible conditions, where intermediate force levels for both main and tributary channels might produce either large or small amounts of sediment in the main channel. This dual behavior is in force levels for both main and tributary channels from 1 to 3 dynes. It is to be expected, given the inherent nature of systems controlled by two major control factors (Graf 1979c), and probably results in part, in this case, from uneven distribution of floods. Some tributaries have experienced especially large floods in the past 80 yr and have flushed sediment into the main channel. Junction areas of this type would plot on the upper-center surface of Figure 9.6. Other junctions have not experienced large floods from tributaries and, therefore, do not yet have major deposits in the main valleys. They would plot on the lower-center surface of Figure 9.6.

Therefore, under present conditions, sediment is being temporarily stored in the largest and the smallest channels. When a threshold is crossed and system adjustment occurs on a massive scale, this pattern will change and temporary storage locations will shift. An adjustment of this type might occur if the largest channels fill with sediment, causing gradient changes and eventual filling of the entire arroyo system. Then an entirely new distribution of fluvial energy would develop, force differentials would probably be different from those at present, and sediment transport and deposition would, likewise, have different patterns. Widespread evidence that this has happened in many semi-arid lands is found in numerous well-documented sequences of cutting and filling in channel networks.

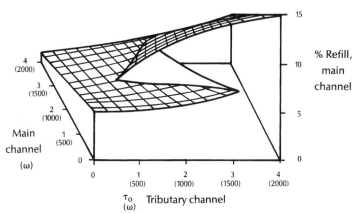

FIGURE 9.6
Conditions at stream junctions in the northern Henry Mountains. The horizontal axis shows tractive force in the main and tributary channels; the vertical axis shows the percent refilling of the arroyo in the main channel at each junction. Surface is smoothed from 28 points.

CONCLUSIONS

The nature of this paper is clearly exploratory, and there are many aspects of the issues raised here that require further evaluation before all the aspects of spatial interaction laws, allometry and force differentials can be fully known. For example, the characteristics of the sedimentary materials as the resistance part of the process has been totally ignored. The complete picture of the process requires a blending of the energy concepts outlined here and more conventional views of sediment characteristics. The role of vegetation has not been included. Channel vegetation influences hydraulics, and upland vegetation influences runoff patterns, both with important implications for the distribution of energy and process. The influence of human activities is a factor to be further evaluated, since near-channel structures play a critical role in the channel processes of semi-arid regions. Climatic change and its ability to adjust altitudinal zones of precipitation in semi-arid regions, and thus to adjust the distribution of energy, has yet to be adequately addressed. Although the role of single intense rainfall events was alluded to above, the spatial properties of the events and their relationship to fluvial processes has not been intensively investigated.

These reservations aside, data from the Henry Mountains region provide confirmation of preliminary answers to the original research questions posed in this paper. First, the basis of spatial–structural controls on fluvial processes includes, in addition to general hierarchical considerations, spatial interaction laws. The laws specify that the impact at any site in the system as a result of a disruption is directly proportional to the magnitude of the disruption and inversely proportional to the network distance from the point of initiation. The basis of external controls on the fluvial processes is variation in resistance and adjustments in the rates of change among channel characteristics. Allometry is a useful tool for analyzing relative rates of change.

Second, tractive force as one representative of fluvial energy is unevenly distributed throughout the stream systems. The maximum values of tractive force occur in intermediate-sized streams in response to the influences of hydraulic radius and channel gradient, which in turn are influenced by the structural and external spatial controls. Positive and negative force differentials occur in the downstream direction or when watersheds of increasing size are considered.

Third, the geomorphic consequences of the spatial variation of fluvial energy include uneven but predictable distributions of channel instability with respect to arroyo development. Also, erosion and transportation of sediment in high-force reaches and watersheds, and stability or deposition of sediments in low-force reaches or watersheds result in an unequal but predictable distribution of sediments which changes from time to time. Sediments move from and through high-force zones to be deposited and temporarily stored in low-force zones (Figs. 9.7 & 9.8). They are remobilized when the distribution of fluvial energy changes due to complex responses from internal adjustments or external interferences in natural system operation.

The study of channels in the semi-arid portions of the Henry Mountains also has some general implications, including the observation that thresholds are defined by the system conditions on either side of the threshold. Fluvial geomorphic research of the 1950s and 1960s emphasized equilibrium conditions, while through the 1970s emphasis on change in systems was prominent. A blending of these philosophies would appear to be most successful in the study of channels in semi-arid lands. Periods of cata-

FIGURE 9.7
Photo of the channel of the Fremont River just west of South Caineville
***Mesa** (North Cainville Mesa in the background, photo looking*
downstream or northeast), showing its appearance in the late 1930s.
Channel has entrenched 6.4 m (21 ft) below the general valley surface,
and channel floor is unstable. (Photo by Dr A. L. Inglesby, courtesy of
the Utah State Historical Society.)

strophic adjustment are clearly defined in the sedimentary and morphologic record, but what has occurred between these periods of adjustment cannot be ignored. Ager (1973) has stated that the development of the stratigraphic column 'has been a very episodic affair, with short happenings interrupting long periods of nothing much in particular.' But to eliminate the periods between the interruptions from geomorphic research would be to exclude from consideration almost everything of interest in terms of fluvial processes.

Finally, the operation of fluvial processes in the Henry Mountains shows rich spatial variety, and efforts to investigate spatial variation in semi-arid fluvial processes seem especially appropriate. Much previous effort has been expended on temporal variation, and it is proper that this work continues. Knowledge of events through time is hopelessly narrow, however, unless supported by equally sound knowledge of processes across space.

FIGURE 9.8
Photo of the channel of the Fremont River from the same perspective as in Figure 9.7, showing its appearance in 1980. Channel has partially filled and stabilized because the main stream, now obscured by cottonwoods, is less powerful than tributaries which continue to deliver sediments. (Photo by author.)

ACKNOWLEDGEMENTS

Research on the fluvial processes of the Henry Mountains region was made possible by the financial assistance provided by the Earth Sciences Program of the National Science Foundation (Grant No. EAR-7727698) and by the Research and Exploration Committee of the National Geographic Society. C. B. Hunt provided numerous helpful suggestions during the research and in the field. S. A. Schumm and W. B. Bull provided useful commentary on an early draft of the present work.

REFERENCES

Abrahams, A. D. 1975. Topologically random channel networks in the presence of environmental controls. *Geol. Soc. Am. Bull.* **86,** 1459–62.

Ager, D. V. 1973. *The nature of the stratigraphical record.* London: Macmillan.

Berwick, V. K. 1962. *Floods in Utah: magnitude and frequency.* U.S. Geol. Survey Circ. 457.

Bogardi, J. L. 1974. *Sediment transport in alluvial streams.* Budapest: Akademai Kaido.

Bryan, K. 1925. Date of channel trenching (arroyo cutting) in the arid Southwest. *Science* **62,** 338–44.

Bull, W. B. 1975. Allometric change of landforms. *Geol. Soc. Am. Bull.* **89,** 1489–98.

Bull, W. B. 1979. Threshold of critical power in streams. *Geol. Soc. Am. Bull.* **90,** 453–64.

Chorley, R. J. and B. A. Kennedy 1971. *Physical geography – a systems approach.* London: Prentice-Hall.

Cooke, R. U. and R. W. Reeves 1976. *Arroyos and environmental change in the American South-west.* Oxford: Clarendon Press.

Covington, H. R. and P. L. Williams 1972. *Map showing normal annual and monthly precipitation in the Salina Quadrangle, Utah.* U. S. Geol. Survey, Salina Folio, Map I-591-D.

Embleton, C. and J. Thornes (eds) 1979. *Process in geomorphology.* New York: Wiley.

Everitt, B. L. 1979. The cutting of Bull Creek arroyo. *Utah Geol.* **6,** 39–44.

Everitt, B. L. 1980. Vegetation and sediment migration in the Henry Mountains region, Utah. In *The Henry Mountains symposium,* M. D. Picard (ed.), 209–15. Salt Lake City: Utah Geological Association.

Faulkner, H. 1974. An allometric growth model for competitive gullies. *Z. Geomorph.* **21,** 76–87.

Ferron, A. D. 1883. Unpublished survey notes. Salt Lake City, Utah: U.S. General Land Office Survey Files, Bureau of Land Management.

Feynman, R. P., R. B. Leighton and M. Sands 1965. *The Feynman lectures on physics.* Reading, Mass: Addison-Wesley.

Gilbert, G. K. 1877. *Report on the geology of the Henry Mountains.* Washington, D.C.: U.S. Geographical and Geological Survey of the Rocky Mountain Region.

Godfrey, A. E. 1980a. Debris avalanche deposits north of Mount Ellen. In *The Henry Mountains symposium*, M. D. Picard (ed.), 171–6. Salt Lake City: Utah Geological Association.

Godfrey, A. E. 1980b. Porphyry weathering in a desert climate. In *The Henry Mountains symposium*, M. D. Picard (ed.), 189–96.Salt Lake City: Utah Geological Association.

Graf, W. H. 1971. *Hydraulics of sediment transport*. New York: McGraw-Hill.

Graf, W. L. 1979a. The development of montane arroyos and gullies. *Earth Surf. Proc.* **4**, 1–14.

Graf, W. L. 1979b. Mining and channel response. *Assn. Am. Geog. Ann.* **69**, 262–75.

Graf, W. L. 1979c. Catastrophe theory as a model for change in fluvial systems. In *Adjustments of the fluvial system*, D. D. Rhodes and G. P. Williams (eds), 13–32. Dubuque: Kendall/Hunt.

Graf, W. L. 1980. Fluvial processes in the lower Fremont River Basin, Utah. In *The Henry Mountains symposium*, M. D. Picard (ed.) 177–83. Salt Lake City: Utah Geological Association.

Gregory, H. E. 1939. Diary of Almon Harris Thompson. *Utah Hist. Q.* **7**, 37–83.

Hackman, R. J. 1973. *Vegetation map of the Salina Quadrangle, Utah*. US Geol. Survey, Salina Folio, Map I-591-P.

Hintze, L. F. 1973. *Geologic history of Utah*. Brigham Young Univ. Geology Studies, vol. 20.

Horton, R. E. 1945. Erosional development of streams and their drainage basins: hydrophysical approach to quantitative morphology. *Geol. Soc. Am. Bull.* **56**, 275–370.

Hunt, C. B., P. Averitta and R. L. Miller 1953. *Geology and geography of the Henry Mountains region, Utah*. U.S. Geol. Survey Prof. Paper 228.

Johns, E. L. 1976. Sediment problems in the Mohave Valley – a case history. In *Proceedings of the Third Federal Interagency Sedimentation Conference, 4/64–4/75*. Washington: Water Resources Council.

Leliavski, S. 1966. *An introduction to fluvial hydraulics*. New York: Dover.

Leopold, L. B., M. G. Wolman and J. R. Miller 1964. *Fluvial processes in geomorphology*. San Francisco: W. H. Freeman.

Morrill, R. L. 1963. The distribution of migration distances. *Reg. Sci. Assn. Papers* **11**, 75–84.

Olsson, G. 1970. Explanation, prediction, and meaning variance: an assessment of distance interaction models. *Econ. Geog.* **46**, 223–31.

Patton, P. C. and S. A. Schumm 1975. Gully erosion, northern Colorado: a threshold phenomenon. *Geology* **3**, 88–90.

Peltier, L. 1950. The geographical cycle in periglacial regions as it is related to climatic geomorphology. *Assn. Am. Geog. Ann.* **40**, 214–36.

Playfair, J. 1802. *Illustrations of the Huttonian theory of the Earth*. Facsimile reprint, 1964. New York: Dover.

Resnick, R. and D. Halliday 1977. *Physics*. New York: Wiley.

Schumm, S. A. 1975. Episodic erosion, a modification of the geomorphic cycle. In *Theories of landform development*, W. N. Melhorn & R. C. Flemal (eds), 69 –85. London: George Allen & Unwin.

Schumm, S. A. 1977. *The fluvial system*. New York: Wiley.

Schumm, S. A. and R. W. Lichty 1965. Time, space, and causality in geomorphology. *Am. J. Sci.* **263**, 110–19.

Shreve, R. L. 1966. Statistical law of stream numbers. *J. Geol.* **74**, 17–37.

Shreve, R. L. 1967. Infinite topologically random channel networks. *J. Geol.* **75**, 178–86.

Taylor, P. J. 1975. *Distance decay models in spatial interactions.* Norwich: Geo Abstracts.

Thornes, J. B. and D. Brunsden 1977. *Geomorphology and time.* New York: Wiley.

Trewartha, G. T. 1954. *An introduction to climate.* New York: McGraw-Hill.

Tricart, J. and A. Cailleux 1972. *Introduction to climatic geomorphology.* (trans. by C. J. K. deJonge). New York: St. Martin's Press.

Williams, P. L. 1972. *Map showing types of bedrock and surficial deposits, Salina Quadrangle, Utah.* U.S. Geol. Survey, Salina Folio, Map I-591-H.

Williams, P. L. and R. J. Hackman 1971. *Geology, structure, and uranium deposits of the Salina Quadrangle, Utah.* U.S. Geol. Survey Misc. Geol. Invest. Map I-591.

Woldenberg, M. J. 1968. Open systems – allometric growth. In *The encyclopedia of geomorphology,* R. W. Fairbridge (ed.), 776–8. New York: Reinhold Book Corporation.

Woldenberg, M. J. 1969. Spatial order in fluvial systems: Horton's laws derived from mixed hierarchies of drainage basin areas. *Geol. Soc. Am. Bull.* **80**, 97–112.

10

Interrelationships among geomorphic interpretations of the stratigraphic record, process geomorphology and geomorphic models

W. Hilton Johnson

INTRODUCTION

The accompanying papers deal primarily with spatial and temporal validity of geomorphic data and, in particular, variability in geomorphic processes that occurs during times of steady state equilibrium (Fig. 10.1). Such process-oriented studies look at physical systems over relatively short time intervals in order to determine variability as well as reactions of the system in terms of rates and nature of response. Schumm and Lichty (1965) referred to such timescales as graded time and suggested time intervals in the order of 100 to 1000 yr. Most process studies, however, are limited by observations and data extending over much shorter time intervals, generally in terms of years or at the most decades. Thus, questions arise when such data are extended to longer intervals. Are they valid? Can geomorphic concepts be based on them and to what extent can they be used in interpretation of the geologic record?

An accompanying problem concerns environmental conditions during time intervals when process observations are made. We live in a time that is characterized by climatic changes of varying frequency and magnitude, large glacioeustatic sea level changes, tectonic and isostatic adjustments of continents and ocean basins and increasing influence of man on geomorphic systems. Are rates and data collected during such a dynamic timespan representative? How widely applicable are they in the interpretation of records of earlier intervals of geologic time?

Both problems concern utilization of geomorphic data in interpretation of the stratigraphic record. Questions can also be asked as to whether the stratigraphic record can be used to extend geomorphic observation over a longer timescale, and to test and possibly substantiate geomorphic concepts or models.

In the past decade, the importance of thresholds and complex response in geomorphic systems has been emphasized and was the topic of the 9th Binghamton Symposium. Although the focus of this volume is on characteristics of systems during time intervals between thresholds, relationships between intervals of equilibrium and disequilibrium in terms of response and effect on the geologic record cannot be considered independently. Interpretation of the stratigraphic record requires consideration of which aspects are the result of variability during times of steady state equilibrium as opposed to those aspects resulting from the crossing of a critical threshold and the potential complex response in terms of lag times and spatial variability within the system.

In this paper I would like to consider both the potential role stratigraphic records play in extending process data and substantiating geo-

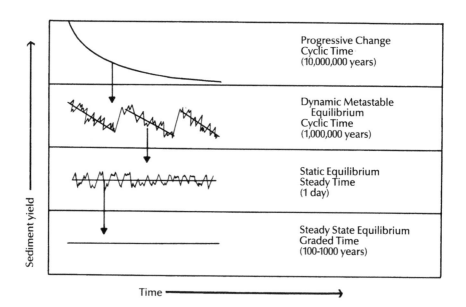

FIGURE 10.1
Variations in sediment yield with respect to Schumm's model of landscape evolution based on episodic erosion. (Modified after Schumm 1977.)

morphic concepts, particularly those encompassed in dynamic metastable equilibrium (Fig. 10.1), as well as aspects of the use of current geomorphic concepts and process data in the interpretation of the stratigraphic record.

CHARACTERISTICS OF THE QUATERNARY PERIOD

Results of modern process studies and those based on studies of Holocene sediments must be considered in a larger context with respect to the Quaternary period. This period, dating from about 1.8 million yr ago (Berggren & VanCouvering 1974), is one of the more dynamic time intervals in Earth history. Climatic change and the cyclic nature of the climatic record clearly dominate the Quaternary. The deep sea record demonstrates eight major climatic cycles in the last 730 000 yr, and similar cycles of somewhat less magnitude but greater frequency continue back beyond the Plio-Pleistocene boundary (Shackleton & Opdyke 1976). This record has revolutionized concepts of the Quaternary in terms of numbers of major climatic cycles and rapidity with which processes have operated in general (Bowen 1978). The ^{18}O record of deep sea sediments has become the standard from which to consider this interval of time (Kukla 1977). Most, or at least many, of the cycles led to widespread continental glaciation in the Northern Hemisphere which in turn resulted in major changes in sea level and isostatic adjustments of continents and ocean basins. Tectonic and other processes associated with plate movements were also active. All of these events and processes have a strong impact on most geomorphic systems and must be considered in the interpretation of Quaternary records, particularly as they might be utilized to reflect earlier and/or longer intervals of time.

Within the last complete climatic cycle (^{18}O stages 5, 4, 3 and 2) and the first part (^{18}O stage 1 or Holocene) of the current cycle, climatic cycles of greater frequency and lesser magnitude clearly are superimposed upon the larger cycle. The effect of these climatic events on geomorphic systems and resulting responses are well documented in the stratigraphic record. What must be kept in mind is that those studies based on modern processes and Holocene records are representative of conditions of a small portion of one complete cycle, i.e. the Holocene appears analogous to only about the first 10 000 yr of ^{18}O stage 5, or what is referred to as an 'interglacial' in European concepts (Kukla et al. 1972). Consequently, erosional and/or depositional rates derived from such studies cannot be extended indiscriminately over longer time intervals, even within the Quaternary. Recognition of climatic events within the Holocene and their corresponding strong effect on geomorphic processes (Wendland, this volume) only serves to magnify the possibility of misuse of data derived over a short time interval.

An accompanying problem of the Quaternary concerns the increasing influence of man's activities on geomorphic processes. Although these effects varied in importance spatially and temporally in the earlier part of the Quaternary, it is clear they have been worldwide in recent centuries. Most geomorphic systems are affected either directly or indirectly by man's activities and such influence must also be taken into account in interpretation and utilization of modern process-derived data.

These comments are not to question the value of process studies – they are fundamental to geomorphology and play an essential role in historical interpretation. They do serve to emphasize that the characteristics and controlling environmental factors of the system under study are just as important as the derived data. Likewise, independent paleoecological and paleogeographical data and interpretations are needed in historical studies if approriate use is to be made of process-derived data.

NATURE OF THE STRATIGRAPHIC RECORD

In a provocative book, Ager (1973) discusses several aspects of the stratigraphic record which are pertinent to this paper and the topic of this volume. These include the incompleteness of the stratigraphic record, the spasmodic nature of sedimentation and the importance of catastrophic events in the record. Another important aspect is the general lack of representation of some environments in the record, particularly over long time intervals.

The stratigraphic record is incomplete for two main reasons, (i) erosion, either as an inherent characteristic of the depositional process or as a result of subsequent major erosional episodes; and (ii) non-deposition. The former is widely recognized as being part of many natural systems and Earth history in general and, similarly, the latter is clearly obvious in those areas which are not located in a general depositional setting. The question concerns those areas which supposedly show more or less continuous sedimentation – do they exist? Ager argues strongly, as have others in the past, that such situations are the exception rather than the rule and that pauses in sedimentation, often reflected by bedding planes, are common. Thus, he believes that in general the record has more 'gaps' than 'record.' Schumm (1977) has noted Ager's ideas and suggested that such a record is what should be expected as a result of the inherent nature of episodic erosion in drainage basins and hence the lack of continuity in sediment yield (Fig. 10.1).

Somewhat related, but at a different scale, is Ager's view that the stratigraphic record often indicates spasmodic sedimentation. Sediments may accumulate at a more or less constant rate for some interval of time, only to be succeeded by extremely rapid sedimentation during a much shorter time interval, for whatever cause. Periods of rapid sedimentation might result from changes in the drainage basin, e.g. exceeding a critical threshold, or an exceptional or catastrophic event of brief duration in addition to a number of possible changes within the environment and in the mode of sedimentation.

The importance of catastrophic (low frequency but large magnitude) events in geomorphology and geology in general has become widely recognized and examples need not be discussed here. What is important, however, is the nature of the record of such events in terms of sediment dispersal, character and preservation, and how such events may be recorded in the stratigraphic record. This is clearly one area where detailed study of extraordinary events occurring today and those of historical and geologically recent occurrence (Quaternary) are essential if similar events are to be recognized and interpreted in the stratigraphic record.

Preservation of depositional environments in the stratigraphic record varies, and terrestrial environments, of paramount concern to geomorphologists, are the ones least likely to be preserved, particularly in the long term. This is especially true for those located at high elevations; even at low elevations, Church (1980) notes that only lacustrine and marine environments are relatively stable and likely to be preserved over the long term. Exceptions occur and we are all familiar with examples of glacial, fluvial and eolian sediments in most parts of the geologic column. Deposits of other terrestrial environments stand little chance of being preserved in the rock record (except rarely) and the stratigraphic record is of limited value in better understanding these environments and related processes. Process studies in these environments will be of most use to the interpretation of the stratigraphic record as they relate to sediment conditioning, availability, and dispersal to other environments of sedimentation.

The stratigraphic record is complex and often difficult to interpret in terms of short-term events and geomorphic controls. However, our abilities to better interpret it must come from our better understanding of geomorphic processes and responses in the short term and our use of 'ideal' stratigraphic sequences, where they exist, to extend our understanding over longer timescales.

TIME AND TIMESCALES

Time is and has always been a fundamental concept in geology and stratigraphy. It has also played an important role in geomorphic thinking, often in the context of William Morris Davis' use of 'stage' with respect to the geographic cycle. The importance of differing timescales in geo-morphic investigations and thinking was brought out in a classic paper by Schumm and Lichty (1965). They clearly showed differing roles for 'time' in geomorphic systems when viewed over long intervals, termed cyclic time, as opposed to much shorter intervals, termed graded or steady time (Fig. 10.1). The concern in this volume is with the latter, but from a stratigraphic standpoint both must be considered. Relationships and con-trasts among and between different timescales in geomorphology are the subject of an excellent new book edited by Cullingford *et al.* (1980).

Graded time

Church (1980)' has considered the records of geomorphic processes over contemporary time (the last few hundred years) in terms of charac-teristics of geophysical event sequences. Summarizing the work of others, he notes that geomorphic processes do not behave as simple stochastic processes, but illustrate an intermittency (the Hurst phenomenon) such that short-term observation of a process will not sample all scales of variability of the process. This characteristic, coupled with the probability of 'exceptional' events and the complex nature of geomorphic systems (extrinsic and intrinsic thresholds and complex response) led Church to conclude that short-term observations may provide a misleading impres-sion of long-term variability of a process, and records made within con-temporary time are too short to indicate the pattern of landscape evolution.

If this is the case, and I see no reason to question it, then the strati-graphic record must play an essential role in geomorphological research. With respect to graded timescales, such research is going on, particularly through studies of the record of the more recent part of the Quaternary period. If the stratigraphic record is to be utilized to gain further insight into geomorphic processes over relatively short timescales, it is essential that a reliable chronology be established. The most common means of doing this are through either seasonally distinctive sedimentation patterns or chronometric dating.

Rhythmically bedded sediment is the result of a regular repetitive pattern of sedimentation. If the timescale of the pattern is known or can be inferred, a chronology can be developed. The best known and most useful are those rhythmic sediments which are varved, i.e. the result of seasonal variations in the physical, chemical or biological conditions of sedimentation such that an annual deposit (usually a couplet) can be delineated. Although such deposits are best known from the glaciolacus-trine environment, they are not unique to that environment (Sturm 1979), but they generally are restricted to quiet water conditions in deeper or protected parts of lake or marine basins. In most cases, they have been used either to develop a chronology (e.g. DeGeer 1912, Schove 1979) or to characterize environmental conditions and determine the cause of varving (e.g. Anderson & Kirkland 1969, Ludlam 1979). Demonstration that a particular deposit is varved is not easy and care must be taken because some annual deposits may contain several varve-like accumulations (Lambert & Hsu 1979), but, once established, both temporal and spatial variability within the former environment of deposition can be ascertained.

Ashley (1975) demonstrated considerable spatial variability in upper varved deposits of Glacial Lake Hitchcock, Massachusetts–Connecticut. The varves were subdivided into three main groups, based on relative thickness of silt and clay layers within a couplet. Varve character was generally similar at a locality but varied spatially. Spatial variations were interpreted to be the result of differences in basin topography, types of sedimentation (deltaic, density underflow and suspension) and proximity of inflowing rivers. From these relationships Ashley deduced that, in areas of the lake away from major sediment sources, the 'classic' winter layer actually accumulated during most of the year (Fig. 10.2).

Through microtextural studies within a glacial varve, it may be possible to determine spatial and temporal variations of short-term (days) origin (Peach & Perrie 1975). These can then be used for both intrabasinal and

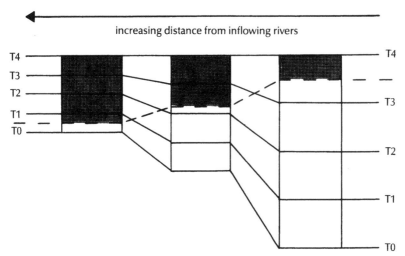

FIGURE 10.2
Interpreted spatial and temporal variations in varves of Glacial Lake Hitchcock. Time-lines T0 through T4 enclose one varve. Dashed line traces contact between classical 'summer' layer and overlying classical 'winter' layer. Multiple graded beds of subseasonal events not shown in 'summer' layer. (After Ashley 1975.)

extrabasinal varve correlation, so that, eventually, events that may be local in effect (slumping, stream shifts, etc.) can be differentiated from those that are more regional in effect (seasonal variations in climate such as increased ablation, storminess, etc.).

Studies of temporal characteristics of varve sequences have used harmonic analysis of variations in varve thickness to determine the presence of other periodicities (Anderson & Koopmans 1963). Relatively weak periodicities of varying length have been observed and have been interpreted in terms of climatic factors – either precipitation or temperature, depending on the origin of the varves – as well as possible solar control. Anderson (1964) has analyzed varved sediments (ancient and nonglacial) in order to calibrate larger-scale stratification associated with them. He noted that, except for variations in thickness, most varves do not show evidence of subseasonal weather or other events. Where such events were recorded, such as the Oligocene Florissant lake beds, the varves had characteristics of glacial varves. They were many times thicker than associated normal varves and contained one or more graded beds (Fig. 10.3). The graded beds were thought to represent turbidity current events and were interpreted to be the result of an occasional slump within the basin and severe storms and times of increased storm activity. Likewise, occasional sandy layers within varves from a Quaternary lake in Spain have been interpreted as resulting from influxes of sediment during times of flooding by the main inflowing river (DeDeckker *et al.* 1979).

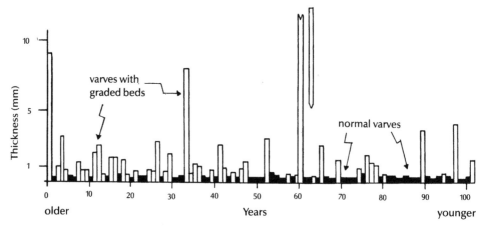

FIGURE 10.3
Graphic plot of distribution of normal varves and varves with seasonal or subseasonal graded beds in sequence of Florissant lake deposits. (After Anderson 1964.)

Varve sequences are unsurpassed in terms of time control and, where preserved in the stratigraphic record, offer an ideal situation for short-term temporal and spatial studies over graded or longer timescales. With respect to geomorphic implications, they are limited in value because they most commonly occur in environments that are spatially separated from subaerial environments, and ties between drainage basin processes, controls and sediment yield cannot be demonstrated. This does not mean that more 'ideal' situations do not exist, and some of the ancient varved sequences described by Anderson (1964), as well as others, should be reconsidered in terms of intrinsic drainage basin control of sediment influx as well as the more usual climate control. Other rhythmic sediments may also be useful in establishing chronologic control, and Bull (1980) describes unusual microlaminated clastic cave deposits in South Wales which are unique because of the close tie to the subaerial environment. Although neither the timescale of laminations nor geomorphic controls on sedimentation have been firmly established, the potential for interpretations of paleoenvironmental conditions exists. This is especially true because the clastic deposits are associated with speleothems which can be radiometrically dated and also may provide independent paleoclimatic information through oxygen isotope analyses (Hendy 1971).

The other areas where stratigraphic studies are contributing to geomorphic models on relatively short timescales are those where it is possible to obtain closely spaced chronometric dates within a sequence of deposits with or without buried soils. Most such studies have concentrated on records of the last part of the Pleistocene and the Holocene where radiocarbon dating is applicable. Other means of dating are possible and techniques and problems associated with dating Quaternary deposits have recently been reviewed by Andrews and Miller (1980). In most stratigraphic sections, dating either is not possible, except in a relative sense, or only a few dates are available and they are used to date the deposits and to make correlations. To be useful in evaluating the variability of geomorphic processes, a sequence of dates is required, and even then inherent errors in dating and the lack of continuous materials suitable for dating limit interpretations. In most cases, the end result is an average rate of sedimentation for some interval of time, and further interpretations of variability within the sequence of sediments must come from sedimentologic and/or paleopedologic investigations.

Among the most useful studies are those in Holocene terrestrial environments where sedimentation has been episodic and soil formation occurred during times of decreased or no sedimentation. Alluvial, colluvial and eolian environments are often characterized by these conditions; lacustrine and bog environments are similar except they often lack the buried soils which mark breaks in sedimentation. In the latter environments, organic carbon is often preserved, which allows radiocarbon dating, and variations in the amounts of organic carbon and types of sediments can often be interpreted in terms of changing conditions within the environment or with respect to the influx of sediments into the depositional basin (e.g. Walker 1966). If pollen or other ecological indicators are also preserved, data from these sources can be used to reconstruct vegetation and other environmental conditions from which climatic inferences can be made. Likewise, geomorphic observations and interpretations associated with the sediment and soil records may allow placement of the stratigraphic record in a broader and precise spatial context. Such background data are essential if contemporary process data are to be used in interpreting the sediment record and if the sedimentary record is going to be used to extend short-term process observations over a longer time interval. Although some circularity in interpretation is possible, both types of studies are clearly complementary and are essential if both are to achieve maximum potential in the formulation and substantiation of geomorphic models.

Many examples of late Quaternary studies could be cited which contribute to our understanding of both geomorphic processes and geomorphic models. I am only going to discuss one, because it illustrates most aspects of the potential as well as limitations and problems with this approach. It is the excellent study of the Little Sioux River Valley and the Cherokee Sewer Archeological Site in northwestern Iowa (Fig. 10.4) which has recently been summarized by Hoyer (1980a, b). The stratigraphic studies are primarily at the Cherokee Site, but geomorphic studies in the Little Sioux Valley are pertinent and provide a longer temporal and better spatial framework from which to consider the deposits at Cherokee.

The study sites are located in western Iowa, which consists of a variety of landscapes of different age and origin (Ruhe 1969). The Little Sioux Valley, along with most other major valleys in extreme western Iowa, carried meltwater discharging from the late Wisconsinan ice front to the Missouri River, and their characteristics reflect glacial, fluvial and erosional events during and subsequent to that time. The Little Sioux is unique among these valleys. It is deeper, has steeper valley walls, and contains five terrace systems (Fig. 10.5), as opposed to only two in most other valleys. The younger three terrace systems are complex (multiple levels), unpaired, and absent in the lower portion of the valley, whereas the older two systems are missing in the uppermost part of the drainage basin (Hoyer 1980a). These contrasts indicate a different history and are the result of a glacially related drainage diversion about 14 500 RCYBP (radiocarbon years before present) which more than doubled the Little Sioux's drainage basin. Briefly, Hoyer (1980a) interprets the geomorphology as follows.

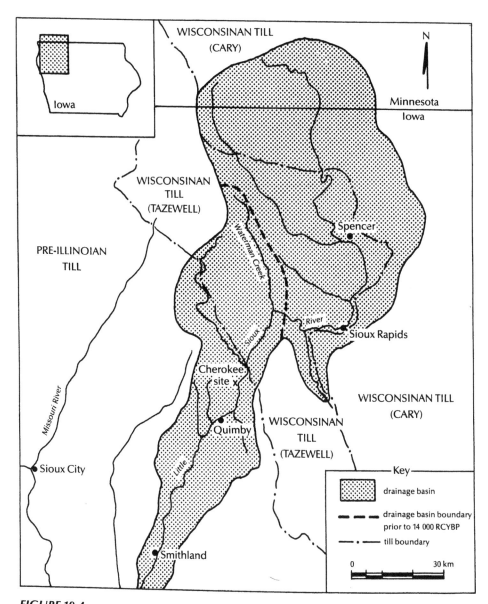

FIGURE 10.4
Location of the Little Sioux River drainage basin, and the Cherokee Sewer
Site in northwestern Iowa. (After Hoyer 1980a.)

(a) Ice advance to the Tazewell-event maximum about 20 000 RCYBP, and deposition of outwash which makes up most of the oldest terrace (T1).

(b) Glacial retreat, deposition of loess on T1 and incision to the T2 level.

(c) Ice re-advance about 14 500 RCYBP (Cary event) which dammed eastward-draining streams and formed ice-marginal lakes.

(d) Eventual spillover and diversion of eastward drainage to the westward-flowing Little Sioux drainage before or near the Cary event maximum (14 000 RCYBP).

(e) Rapid incision during and after the diversion, followed by episodic erosion and deposition in the upper and middle portions of the valley (formation of T3, 4 and 5 systems) concurrent with deposition in the lower part of the valley (Fig. 10.5).

(f) Beginning about 10 000 RCYBP, stabilization of the valley system with a regime primarily of episodic aggradation as alluvial fans and colluvial aprons encroach onto the valley bottom.

Hoyer believes the system illustrates the complex response predicted by Schumm (1973) for fluvial systems, i.e. significant spatial and temporal variability in erosional and depositional responses to the diversion event.

The Holocene part of the record is best preserved in alluvial fans and the geology of the Corrington Fan south of Cherokee was studied as part of an interdisciplinary investigation of the Cherokee Sewer Site (Hallberg et al. 1974, Hoyer 1980b). The site is located in the distal part of the fan, a short distance from where it is terminated by the floodplain of the Little Sioux River. Microstratigraphic studies identified 5 major lithostratigraphic

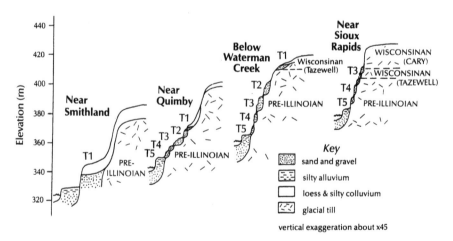

FIGURE 10.5
Sketch showing spatial variability in terrace positions at four locations in the Little Sioux Valley. Cherokee Sewer Site located in valley between Quimby and Waterman Creek. (Modified after Hoyer 1980a.)

units with many subunits (designated with Roman numerals and letter subscripts), 19 soils (designated with numbers), all of which are buried except the surface soil, and 3 main cultural horizons (Fig. 10.6). Twenty-two ^{14}C dates of charcoal, carbonized wood and organic-enriched sediment provided chronologic control from about 10 050 RCYBP through 4600 RCYBP. Sediments at the site reflect both alluvial fan and floodplain sedimentation and transition from one environment to the other changed spatially with time (Fig. 10.7). Although unconformities marked by buried

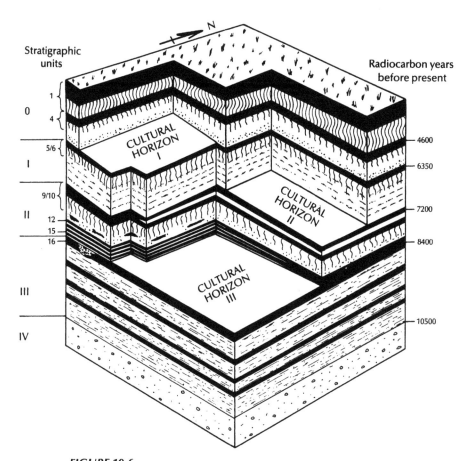

FIGURE 10.6
Schematic block diagram of Cherokee Sewer Site.
Lithostratigraphic units and key soils are labeled at the left;
radiocarbon dates are indicated at the right; cultural horizons
are indicated within the block. Dark areas are A1 horizons of
soils; wavy vertical lines indicate B soil horizon development.
(From Hoyer 1980b.)

soils are common, little evidence of erosion of sediment is apparent and, consequently, a relatively complete sediment–soil complex is preserved.

The stratigraphic sequence was divided into five major lithologic units, the lowermost (Unit IV) being relatively coarse, stratified silts, sands and gravels up to 8 m thick and interpreted to be bedload deposits of the Little Sioux River prior to 10 050 RCYBP. The upper four units (0–III) are finer-grained and consist of Holocene fan and floodplain deposits. Prominent buried soils mark the upper contacts of most of these units, but other lithologic criteria were also used to establish the stratigraphy. The main units contain several buried soils exhibiting varying degrees of development, and several sediment subunits reflecting different modes and intensities of sedimentation. Details are described by Hoyer (1980b). He notes that many of the thin alluvial subunits are graded and represent single episodes of deposition. In some cases, they have soils developed in them, indicating considerable time between depositional episodes, whereas in others no evidence of weathering and soil development is found and the depositional events are closely spaced in time. Many of the minor soils consist of A1-C profiles of varying thickness and organic carbon content, and Hoyer estimates that they represent time intervals varying from perhaps 50 to 200 or 300 yr. Some soils exhibit B horizons with structural

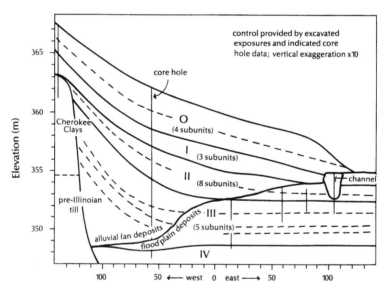

FIGURE 10.7
Cross-section of Corrington Fan showing major lithostratigraphic units and number of subunits. Selected subunit contacts dashed to show geometry of deposits. (After Hoyer 1980b.)

development and clay enrichment. Based on radiocarbon dates, degree of soil development, and comparison with studies of soil genesis, soil 1 (surface soil) and soil 4 are interpreted to represent from 2000 to 2500 yr, and soil complexes 5–6 and 9–10 about 500 yr. Thus, sedimentation at Cherokee is interpreted to represent 'intermittent episodes of rapid sedimentation followed by relatively long periods of soil formation during which insignificant deposition occurs' (Hoyer 1980b, p. 49). This interpretation, as shown by sediment yield from the 25 ha drainage basin, is illustrated in Fig. 10.8.

The nature of sedimentation at Cherokee and available radiocarbon dates allow characterization of some sedimentation events of short duration, weathering and soil formation of somewhat longer duration, and estimation of average clastic sedimentation or erosion rates on timescales ranging from about 1000 to 5000 yr. Hoyer (1980a) has considered these topics and, with respect to the latter, indicates sedimentation rates increased from an average of 51 m³/yr for the first 1600-yr interval to 83 m³/yr for the interval 6000–7200 RCYBP, and decreased to an average rate of 34 m³/yr in the last 4600 RCYBP (Table 10.1). Erosion rates calculated from the sedimentation record are minimum rates because the system is not closed, but comparisons with other studies in Iowa, some based on watershed studies of less than 10-yr duration, led Hoyer to think that they are representative and the trends in rates are in agreement with other geomorphic–stratigraphic studies.

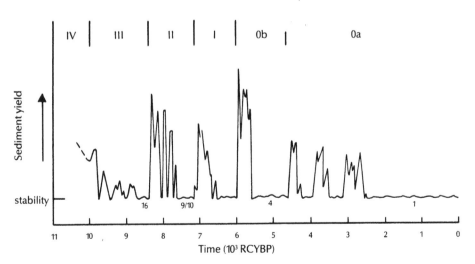

FIGURE 10.8
Holocene variations in sediment yield to Corrington Fan. *Major lithostratigraphic units shown above curve; soils formed during times of landscape stability shown below curve.*

The sedimentation rates are misleading, however, because the time intervals used to calculate the rates include intervals of soil formation when sedimentation was minimal to non-existent. Assuming little or no erosion and sedimentation on the fan during Hoyer's estimated times of soil formation, rates were recalculated on two bases. In column a of Table 10.1, only the major soil-forming intervals are deleted from the calculations; in column b, minor soil-forming intervals within times of active sedimentation are also excluded in calculating the rates. Both have the effect of giving significantly larger rates during shorter intervals of time and greater variability in rates during the entire timespan. These averages also have limitations and are probably too large because the soils are in part cumulic and time intervals of soil formation may be overestimated, but they are more representative than rates which include long intervals of stability, are more consistent with the sedimentation model, and show episodic sediment yield from the drainage basin (Fig. 10.8).

The geology can also be used to consider geomorphic models and controls on sedimentation. Hoyer (1980b) notes that the sedimentation record exhibits cycles varying by several orders of magnitude and, overall, has many of the characteristics to be expected from Schumm's (1973, 1976) ideas of complex response and episodic erosion following a threshold event. The sediment subunits which are separated by minor soils are recognized as episodes; these in turn are grouped into four main cycles and the entire sequence is considered a megacycle. The main cycles correspond approximately to the major lithologic units, with the lower two units combined to form the first cycle. Each cycle (including the megacycle and subunits) tends to be graded, has more frequent episodes of sedimentation and larger volumes of sediment in the lower part, fewer episodes and small volumes in the upper part ending with a period of minimum sedimentation and maximum soil formation. Hoyer interprets the megacycle as a delayed aggradational response to the drainage diversion which initially caused episodic cutting and filling in the Little Sioux Valley during the interval from 14 000 to 10 000 RCYBP. The cycles and subunits are thought to represent episodic behavior of the small drainage basin, as thresholds are periodically exceeded and sedimentation episodes of varying magnitude are initiated. These would correspond to the types of responses which Schumm has suggested are inherent in erosional systems. Each main cycle would correspond to intervals of graded time when the system was adjusting to achieve steady state equilibrium, and the megacycle may be considered as a small interval of cyclic time with the system exhibiting dynamic metastable equilibrium (Figs. 10.1 & 10.8).

Schumm (1976) has suggested the character of an ideal sedimentary unit derived from the denudation of a landscape over time intervals equivalent to that of a Davisian Cycle. The unit (Fig. 10.9a) consists of five cycles,

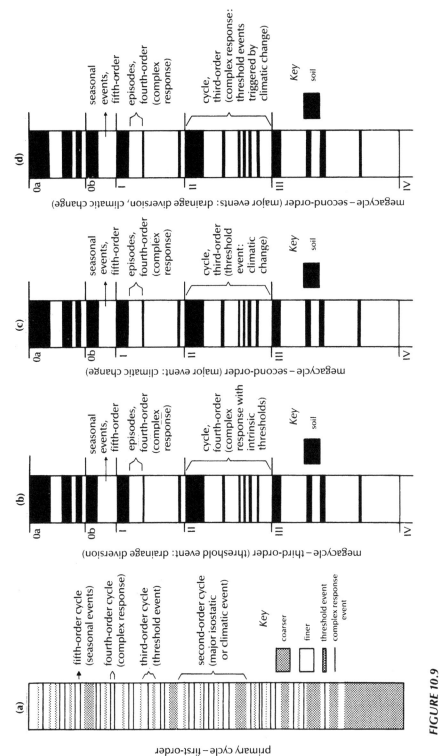

FIGURE 10.9

(a) Schumm's ideal sedimentation cycle (after Schumm 1976). (b), (c), (d) Application of ideal cycle to alternative interpretations of Corrington Fan: (b) complex response (Hoyer 1980b); (c) climatic change (Hallberg et al. 1974); (d) complex response and climatic change (Hoyer 1980b).

each of which is of increasing fineness and decreasing magnitude as one goes upward: the primary cycle related to tectonic activity and overall denudation of the landscape; second-order cycles related to isostatic adjustment and/or major climatic change; third-order cycles related to crossing of geomorphic thresholds; fourth-order cycles related to the complex response (episodic behavior) of the system to the lower orders; and fifth-order cycles related to seasonal and shorter-term events. Although not a basinal record, the Cherokee Site clearly exhibits aspects of the higher order cycles, but different interpretations of them are possible.

Individual beds within subunits correlated well to Schumm's fifth-order cycle – they represent short-term seasonal events. The cycles and subunits (episodes), as interpreted by Hoyer (1980b), correspond to Schumm's fourth-order cycles – the complex and episodic behavior of the system in response to the earlier drainage diversion. The cycles were initiated as intrinsic thresholds were exceeded in the drainage basin and sediment stored in the basin was transported to the fan. The megacycle would correspond to a Schumm third-order cycle, in this case an extrinsic threshold was exceeded as a result of the drainage diversion (Fig. 10.9b).

Other interpretations are possible and earlier interpretations at Cherokee (Hallberg *et al.* 1974) considered the main cycles to be the result of Holocene climatic change. Assuming that climate and climatic change are dominant controls on this stratigraphic record, correlations to Schumm's cycles are at a lower level (Fig. 10.9c). The main cycles would be third order resulting from extrinsic thresholds (climatic change); subunits would correspond to fourth-order cycles resulting from episodic sedimentation in response to climatic change; and individual beds within the subunits would correspond to fifth-order events. The megacycle could also have a different interpretation. Hoyer (1980b) includes both Units III and IV in his first cycle. The distinct change in sedimentation at this position suggests that the contact between them may mark a change of greater magnitude, i.e. a second-order cycle. An abrupt change from coarse to fine sediment is present in major fluvial deposits throughout Iowa, and ^{14}C dates bracket it from 10 000 – 11 000 RCYBP (G. R. Hallberg, personal communication). It corresponds broadly with climatic change and responses associated with the Pleistocene/Holocene boundary and, in terms of a Quaternary timescale, it fits a second-order cycle boundary.

Hoyer (1980b) notes that the climatic change interpretation conflicts with the complex response interpretation. However, he also notes that the timing of the main cycles corresponds closely with the regional fluvial discontinuities suggested by Knox (1976) as a result of analysis of radiocarbon dates in alluvial sediments (see Wendland, this volume). Knox interprets these discontinuities to result primarily from globally synchronous climatic changes (Wendland & Bryson 1974), and consequently Hoyer indicates that the complex response at Cherokee may be triggered by climatic change and, if so, extrinsic climatic thresholds are also part of the interpretation (Fig. 10.9d). This raises the question as to whether any of the main cycles are the result of intrinsic thresholds of the type Schumm describes for fluvial systems, or are they entirely controlled by Holocene climate and climatic change?

Further study is necessary to evaluate the various interpretations fully. The correspondence between increased sedimentation rates at the site during mid-Holocene (Table 10.1) and changes in vegetation at about the same time in northwestern Iowa (VanZant 1979) suggests that climatic change played some role. Great variability in rates within the mid-Holocene also argues for complex response. Regional and local studies of alluvial sediments in small valleys of western Iowa (Daniels & Jordan 1966, Allen 1971, Thompson & Bettis 1980) suggest broad similarities in their stratigraphy and chronology and indicate the possibility of synchronous landscape change (Thompson & Bettis 1980). Much of this record, however, postdates active fan sedimentation at the Cherokee Site and the earlier record is not dated well enough to compare and evaluate the record at Cherokee. Considerable spatial variability may exist among the records. Evaluation of the role of climatic change, as well as the complex response related to the drainage diversion, should be possible through comparable detailed stratigraphic studies of fan and floodplain sediments in the major valleys immediately northwest and southeast of the Little Sioux Valley,

TABLE 10.1
Average rates of Holocene sedimentation and erosion computed from Corrington Fan.

Time interval	Clastic sedimentation (m³/yr)			Minimal clastic erosion (t/ha/yr) (mm/yr)	
RCYBP	Hoyer 1980b[a]	Recalculated a	b[b]	Hoyer 1980b[a]	
10 000–8650	51	61	68	3.1	2.1
8650–8400		m[c]	m		
8400–7800	61	121	208	3.6	2.4
7800–7200		m	m		
7200–6500	83	143	166	4.9	3.3
6500–6000		m	m		
6000–5600	64	223	223	3.8	2.6
5600–4600		m	m		
4600–2500	34	74	141	2.0	1.4
2500–0		m	m		

[a] Combined time interval.
[b] Excludes time of estimated soil formation during the interval.
[c] m = minimum sedimentation (soil-forming intervals).

which had similar histories with the exception of the drainage diversion. Comparisons of different stratigraphic records should allow discrimination of those effects which are climatically controlled and those which are in response to the diversion. Such research would make significant contributions to understanding complexities of drainage basin responses to different combinations of extrinsic and intrinsic thresholds.

Stratigraphic sequences, such as the Cherokee Site, which contain few erosional breaks and can be dated sequentially provide excellent opportunities to examine geomorphic systems on relatively short timescales ($<10^4$ yr). Average rates can be determined and short-term events can often be interpreted through sedimentologic analysis. Where possible, paleopedologic analysis (e.g. Follmer, this volume) provides further information on temporal and spatial variability, and paleoecological data and spatial geomorphic studies allow the best interpretations of the stratigraphic sequences. Such interpretations can then be utilized to extend contemporary process data, test geomorphic models, and better interpret older stratigraphic records where less control on time and environmental factors is available.

Cyclic time

Understanding geomorphic systems over long timescales equivalent to that of an erosion cycle (10^7 yr) is not possible in terms of modern process studies or even stratigraphic studies of Quaternary sediments on timescales of 10^4 to 10^6 yr. Schumm (1976) recognized this and indicated that the stratigraphic record must be evaluated to gain long-term perspectives and to substantiate conceptual or theoretical models. His hypothesized primary sedimentary unit (Fig. 10.9a) was clearly presented as a challenge to geomorphologists and stratigraphers to be tested by careful analysis of sedimentary records. To the best of my knowledge, that challenge has not been met. Recent books in sedimentology (e.g. Reading 1978, Blatt *et al.* 1980) discuss influences on sediment yield such as relief, climate, vegetation and geology but make little or no mention of the inherent characteristics of denudation which probably produce an episodic sediment supply (Schumm 1976).

Geomorphic interpretation of ancient stratigraphic sequences is much more difficult than for late Quaternary sequences because there is usually little or no relationship to the modern landscape, and reconstruction of spatial perspectives and detailed paleogeography is difficult; chronometric dating or biostratigraphic zonation is often not possible and, even where it is, time resolution is generally in the order of 10^5 or more years; paleoecological and paleoclimatological background data are limited and/ or less accurate; and the record is usually more fragmentary and has undergone greater post-depositional modifications. In addition, because of evolutionary changes in Earth's physical, chemical and biological systems, utilization of modern process studies may not always be directly applied and in all cases must be applied cautiously. The result is that most ancient interpretations concern depositional environments and broad changes in the environment through time as a result of facies considerations. Changes in environmental conditions are then interpreted in terms of extrinsic factors such as tectonics, eustatic sea level changes, and climatic change, but little consideration is given to intrinsic factors, particularly if they are not well established. Schumm's challenge is still valid and for the long term must be evaluated through the 'ideal' stratigraphic record if concepts of complex response and geomorphic thresholds are going to be accepted widely as new paradigms in geomorphology, as suggested by Coates and Vitek (1980).

CONCLUSIONS

The stratigraphic record, in spite of inherent limitations, has a definite role to play in the formulation and substantiation of geomorphic concepts and models. Whereas process geomorphic studies are unsurpassed in understanding the mechanics and controls of geomorphic processes and short-term variability and behavior of geomorphic systems and landforms, geomorphic interpretations of the stratigraphic record document process variability over longer time intervals and responses of geomorphic systems to threshold events of varying cause and magnitude. Such interpretations are best made where the record is most complete, chronometric dating is available, ties between geomorphic systems and depositional environments can be made, and independent evidence on extrinsic controlling factors is available. These types of records are more common in the Quaternary, and geomorphic interpretations of the Quaternary record should allow for better interpretations of earlier stratigraphic records. The latter is essential to gain insight into long-term landscape evolution and to test current geomorphic models.

ACKNOWLEDGEMENTS

G. R. Hallberg, B. E. Hoyer, C. J. Mann and C. E. Thorn read an early draft and contributed to the clarity of the paper; in addition, Hoyer provided further insight into the geology of the Cherokee site. D. R. Phillips drafted the illustrations and L. L. Harbison typed the manuscript.

REFERENCES

Ager, D. V. 1973. *The nature of the stratigraphical record.* New York: Wiley.
Allen, W. H. 1971. Landscape evolution and soil formation. Unpublished PhD dissertation. Ames: Iowa State University.
Anderson, R. Y. 1964. Varve calibration of stratification. In *Symposium on cyclic sedimentation,* D. F. Merriam (ed.), 62–72. Bull. Geol. Surv. Kansas 169.
Anderson, R. Y. and D. W. Kirkland (eds) 1969. *Paleoecology of an early Pleistocene lake on the High Plains of Texas.* Mem. Geol. Soc. Am. 113.
Anderson, R. Y. and L. H. Koopmans 1963. Harmonic analysis of varve time series. *J. Geophys. Res.* **68,** 877–93.
Andrews, J. T. and G. H. Miller 1980. Dating Quaternary deposits more than 10 000 years old. In *Timescales in geomorphology,* R. A. Cullingford, D. A. Davidson & J. Lewin (eds), 262–87. New York: Wiley.
Ashley, G. M. 1975. Rhythmic sedimentation in Glacial Lake Hitchcock, Massachusetts–Connecticut. In *Glaviofluvial and glaciolacustrine sedimentation,* A. V. Jopling and B. C. McDonald (eds), 304–20. Soc. Econ. Paleontologists and Mineralogists, Spec. Publ. 23.
Berggren, W. A. and J. A. VanCouvering 1974. The Late Neogene: Biostratigraphy, geochronology and paleoclimatology of the last 15 million years in marine and continental sequences. *Palaeogeogr., Palaeoclimatol., Palaeoecol.* **16,** 1–216.
Blatt, H., G. Middleton and R. Murray 1980. *Origin of sedimentary rocks.* Englewood Cliffs, NJ: Prentice-Hall.
Bowen, D. Q. 1978. *Quaternary geology, a stratigraphic framework for multidisciplinary work.* New York: Pergamon.
Bull, P.A. 1980. Towards a reconstruction of time-scales and paleoenvironments from cave sediment studies. In *Timescales in geomorphology,* R. A. Cullingford, D. A. Davidson & J. Lewin (eds), 177–87. New York: Wiley.
Church, M. 1980. Records of recent geomorphological events. In *Timescales in geomorphology,* R. A. Cullingford, D. A. Davidson & J. Lewin (eds), 13–29. New York: Wiley.
Coates, D. R. and J. D. Vitek 1980. Perspectives on geomorphic thresholds. In *Thresholds in geomorphology,* D. R. Coates & J. D. Vitek (eds), 3–23. London: George Allen & Unwin.
Cullingford, R. A., D. A. Davidson and J. Lewin (eds) 1980. *Timescales in geomorphology.* New York: Wiley.
Daniels, R. B. and R. H. Jordan 1966. *Physiographic history and the soils, entrenched stream systems, and gullies, Harrison County, Iowa.* Tech. Bull. US Department of Agriculture, 1348.
DeDeckker, P., M. A. Geurts and R. Julia 1979. Seasonal rhythmites from a lower Pleistocene lake in northeastern Spain. *Palaeogeogr., Palaeoclimatol., Palaeoecol.* **26,** 43–71.
DeGeer, G. 1912. A geochronology of the last 12 000 years. *Proc. Int. Geol. Congr.* **1,** 241–58.
Follmer, L. R. 1982 The geomorphology of the Sangamon surface: its spatial and temporal attributes. In *Space and time in geomorphology,* C. E. Thorn (ed.), 117-46. London: George Allen & Unwin.
Hallberg, G. R., B. E. Hoyer and G. A. Miller 1974. The geology and paleopedology of Cherokee sewer site. *J. Iowa Arch. Soc.* **21,** 17–49.
Hendy, C. H. 1971. The isotopic geochemistry of speleothems – I. The calculation of the effects of different modes of formation on the isotopic composition of speleothems and their applicability as paleoclimate indicators. *Geochim. Cosmochim. Acta* **35,** 801–24.

Hoyer, B. E. 1980a. *Geomorphic history of the Little Sioux River Valley.* Iowa City: Geol. Soc. Iowa Guidebook.

Hoyer, B. E. 1980b. The geology of the Cherokee sewer site. In *The Cherokee excavations, Holocene ecology and human adaptations in north western Iowa,* D. C. Anderson and H. A. Semken Jr. (eds), 21–77. New York: Academic Press.

Knox, J. C. 1976. Concept of the graded stream. In *Theories of landform development,* W. N. Melhorn & R. C. Flemal (eds), 170–98. London: George Allen & Unwin.

Kukla, G. J. 1977. Pleistocene land–sea correlations I. Europe. *Earth-Sci. Rev.* **13,** 307–74.

Kukla, G. J., R. K. Matthews and T. M. Mitchell Jr 1972. The end of the present interglacial. *Quat. Res.* **2,** 261–9.

Lambert, A. M. and K. J. Hsü 1979. Varve-like sediments of the Walensee, Switzerland. In *Moraine and varves, origin, genesis, classification,* C. Schüchter (ed.), 287–94. Rotterdam: A. A. Balkema.

Ludlam, S. D. 1979. Rhythmite deposition in lakes of the northeastern United States. In *Moraines and varves, origin, genesis, classification,* C. Schüchter (ed.), 295–302. Rotterdam: A. A. Balkema.

Peach, P. A. and L. A. Perrie 1975. Grain-size distribution within glacial varves. *Geology* **3,** 43–6.

Reading, H. G. (ed.) 1978. *Sedimentary environments and facies.* New York: Elsevier.

Ruhe, R. V. 1969. *Quaternary landscapes in Iowa.* Ames: Iowa State University Press.

Schove, D. J. 1979. Varve-chronologies and their teleconnections, 14 000–750 BC. In *Moraines and varves, origin, genesis, classification,* C. Schüchter (ed.), 319–25. Rotterdam: A. A. Balkema.

Schumm, S. A. 1973. Geomorphic thresholds and complex response of drainage systems. In *Fluvial geomorphology,* M. Morisawa (ed.), 299–310. London: George Allen & Unwin.

Schumm, S. A. 1976. Episodic erosion: a modification of the geomorphic cycle. In *Theories of landform development,* W. N. Melhorn & R. C. Flemal (eds), 69–85. London: George Allen & Unwin.

Schumm, S. A. 1977. *The fluvial system.* New York: Wiley.

Schumm, S. A. and R. W. Lichty 1965. Time, space and causality in geomorphology, *Am. J. Sci.* **263,** 110–19.

Shackleton, J. J. and N. D. Opdyke 1976. Oxygen isotope and paleomagnetic stratigraphy of Pacific core V28-239, Late Pliocene to latest Pleistocene. *Geol. Soc. Am. Mem.* **145,** 449–64.

Sturm, M. 1979. Origin and composition of clastic varves. In *Moraines and varves, origin, genesis, classification,* C. Schüchter (ed.), 281–5. Rotterdam: A. A. Balkema.

Thompson, D. M. and E. A. Bettis III 1980. Archeology and Holocene landscape evolution in the Missouri drainage of Iowa. *J. Iowa Arch. Soc.* **27,** 1–60.

VanZant, K. 1979. Late glacial and postglacial pollen and plant macrofossils from Lake West Okoboji, northwestern Iowa. *Quat. Res.* **12,** 358–80.

Walker, P. H. 1966. Postglacial environments in relation to landscape and soils on the Cary drift, Iowa. *Iowa Agric. Exp. Sta. Res. Bull.* **549,** 838–75.

Wendland, W. M. 1982. Geomorphic responses to climatic forcing during the Holocene. In *Space and time in geomorphology,* C. E. Thorn (ed.), 355-71. London: George Allen & Unwin.

Wendland, W. M. and R. A. Bryson 1974. Dating climatic episodes of the Holocene. *Quat. Res.* **4,** 9–24.

11

Variability of rainwash erosion within small sample areas

Shiu-hung Luk

INTRODUCTION

Spatial variations of geomorphic processes exist at two different levels, those among dissimilar environments – *between-site* variations – and those within nominally 'homogeneous' environments – *within-site* variations. Among investigations of water erosion rates, most were concerned with *between-site* rather than *within-site* variations (Soons 1971, Wischmeier 1976) and, in many cases, it was implicitly assumed that where major environmental factors such as rainfall, vegetative cover, slope and soil are held constant, soil loss test results are quantitatively reproducible.

In laboratory studies of soil loss using the Edmonton rainfall simulator, it was found that the test results were far from replicable, even if the major as well as the subsidiary factors (including slope shape and moisture content) were kept perfectly constant. Rainwash (material entrained by direct drag forces, plus material entrained by splash and transported by wash) pertaining to an alpine podzolic soil from the Canadian Rockies was found to range from 5.62 to 10.78 g m^{-2} min^{-1}, with a coefficient of variation *(CV)* of 18.4% (Luk 1975). Bryan (1979) tested the Devon and Orchard soils from Alberta in the same simulator. For the Devon sample, rainwash ranged from 1.50 to 3.17 g m^{-2} min^{-1}, with $CV = 25.0\%$. For the Orchard soil, the equivalent figures were 15.50 to 64.17 g m^{-2} min^{-1} and 39.9%. It is quite apparent from these results that the measured soil loss variabilities (as represented by the coefficients of variation) substantially exceed the expected margin of measurement error and therefore must be attributed to (a) uncontrolled factors and/or (b) random variation. The known uncontrolled factors include the size of both individual particles and aggregates on the surface, microrelief and various soil characteristics. Random variation may exist as a result of the interactions among soil property variations at the microscale, widely variable depths of sheetflow, and raindrops varying greatly in size and in kinetic energy generated. However, the influence of random variation can only be established after the significance of measurable factors and sources of measurement error are determined.

The implications of these results and hypotheses are considerable in the field situation. Considering the fact that there are factors of soil moisture content, infiltration rate, slope shape and surface microrelief which cannot be precisely controlled in the field, it would seem likely that field-determined soil loss variabilities should be considerably higher than that in the laboratory. The significant laboratory results therefore suggest that even higher variabilities are to be expected in the field. If this is substantiated, then possibly a large number of replicate runs will be required to provide a reasonable estimate of the soil erodibility at a given site. In practice, replicate tests may prove to be too expensive for most soil management work and it would be more desirable to predict soil loss variability by using commonly available soil information. If prediction of soil loss variability can be achieved, then the accuracy of soil erodibility estimates and possibly some soil loss predictions can be quantitatively assessed.

The purpose of this paper is to report and synthesize results obtained from several laboratory and field projects that were completed recently. The magnitude of soil loss variability, its probable causes and the potentials of predicting the variabilities are discussed.

EXPERIMENTAL PROCEDURES

The Edmonton rainfall simulator previously described (Bryan 1970) was used in the laboratory experiments (Bryan & Luk 1981). Minor design modifications were incorporated to improve the separation of materials transported by rainwash and splash-saltation. Rainfall intensity was adjusted to an average of 63.5 mm hr^{-1} (range = 61.6 to 68.6 mm hr^{-1}) with drop size varying from 0.2 to 5.8 mm. Kinetic energy applied was calculated to be 0.227 J m^{-2} s^{-1}, approximately 61% of the value for equivalent natural storms (Wischmeier & Smith 1958). Samples were air-dried, sieved through an 8-mm square-hole sieve, and split into homogeneous subsamples. A 4-cm depth of subsample was placed in the sample pan (30.5 x 30.5 cm) and the soil was levelled to the top of the pan after a 100-g sample was removed for aggregate analysis. Prior to testing, each subsample was wetted by a fine spray 60 cm above the soil surface. A Pentax 6x7 camera (f = 105 mm) was used to take stereo pairs of black and white photographs before and after each experiment. In total, samples of three soils from southern Ontario – Lockport clay, Pontypool sand and Milliken loam (Fig. 11.1) – were tested, each by twenty 60-minute runs at a slope of 12.5° in the simulator (Fig. 11.2).

The Toronto simulator (Morgan 1979) was used in the field pilot study (Luk & Morgan 1981). The simulted rainfall was sprayed from nozzles located on two radial arms which were rotated atop two sections of TV antenna towering. Each radial arm was 2.45 m in length, the speed of rotation was 4 rpm, and there were 28 nozzles on each arm. With an apical fall height of 6.55 m above plot surfaces, drop sizes ranging from 0.2 to 5.8 mm in diameter, and a rainfall intensity of 50 mm hr^{-1}, the simulator achieved 88% of the kinetic energy of equivalent natural storms. In the annular wetted by the simulator, the surface was uniformly scalped to a depth of 35–40 mm by using a 'sod-cutter' (Fig. 11.3). Twenty 30.5 x 30.5-cm plots were then established in the annular by inserting 15 cm high

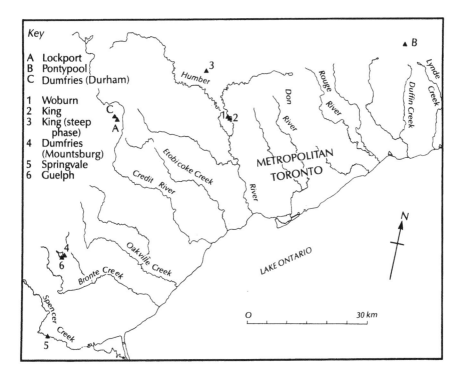

FIGURE 11.1
Location of experimental sites.

boards to a depth of approximately 15 mm on the upslope edge and sides of the plot. At the downslope edge of each plot, a simple backless 'dust-pan'-type collector was installed and securely attached to a 5-l capacity heavy polyethylene bag. The lip of each collector was inserted 15 mm into the soil, 5 mm below the initial surface. Three soils from southern Ontario – Lockport clay, Pontypool sand and Dumfries loam – were tested and, in each case, the rainfall experiment lasted 30 minutes.

To extend the results obtained from the pilot study, six additional soils from southern Ontario (Fig. 11.1) were tested. The Toronto simulator was modified by increasing the number of radial arms from two to six, thus increasing the wetting frequency from 7.5 to 2.5 sec (Fig. 11.4). Calibration data collected under calm atmospheric conditions show that both the original and the modified Toronto simulator are capable of producing rainfall which has no significant areal variation. However, when winds $\geqslant 8$ km hr^{-1} were present, it was found that the wetted annular was disturbed, the finest drops were removed from the area subjected to the simulated rainfall, and the areal uniformity of rainfall intensity was also distorted. To minimize these variations when conducting the field exper-

FIGURE 11.2
A subsample of Pontypool sand under laboratory-simulated rainfall. The dark stains on the sample surface are due to methyl blue solution injected to determine sheetflow velocity.

iments, days with high winds were avoided. Rainfall data obtained were examined and plots which had not received an amount of rainfall within ±10% of the target (25.4 mm in 30 minutes) were discarded. In addition, a correction factor [(target rainfall) ÷ (actual rainfall)] was applied to the accepted plot data.

Various field and laboratory soil characteristics were determined at the plot level. To measure surface microtopography, a microrelief gauge was designed and constructed (Campbell 1970). It is made up of an aluminum frame (30.5 × 30.5 cm) with four slanting and adjustable legs (Fig. 11.5). The gauge is equipped with two sets of detachable protractors and pivoting levels, one mounted on the downslope and the other on the cross-slope edge of the frame. There are five regularly perforated cross pieces in the frame, and twenty-five aluminum rods of 1 cm in diameter and 68.6 cm in length can be lowered through the perforations on to the soil surface. When properly levelled, the length of the rods extending above the top edge of the frame, as measured by a vernier depth gauge which is accurate to ±0.02 mm, indicates the relative heights of the 25 sample points on the ground surface. Resistance of surface soil to torsional shear stress (torsional shear strength) was determined by using a hand-held Pitcon torvane. Bearing capacity, which is closely related to unconfined

FIGURE 11.3
The manual 'sod-cutter' in action.

compressive strength, was measured by a Vicksburg-type proving ring penetrometer. Soil bulk density and soil moisture content were determined by extracting 5.4-cm cores 1–4 cm below the soil surface. In the laboratory, aggregate size distribution was measured by using the standard wet-sieving technique (Yoder 1936). Additional samples were tested for aggregate stability by the drop-test method (McCalla 1944). Soil texture was analyzed by pipetting and wet-sieving after dispersion.

MAGNITUDE OF SOIL LOSS VARIABILITY

The coefficient of variation *(CV)* was used to represent soil loss variability. It is defined as *(s/x̄)* × 100%, where s is standard deviation of sample, and x̄ is sample mean. All the available data on rainwash variability are

FIGURE 11.4
Modified Toronto rainfall simulator used in field experiments. The scalped plots were covered with polyethylene sheets to prevent excessive evaporation overnight.

summarized in Table 11.1 and the range of *CV* is 13–37% for the nine field sites and 13–40% if all other data are considered. By statistical inference, the extreme values would be ±39% to ±120% (or 3 standard deviations) from the sample means. Clearly, the margin of sampling error is substantial and replicate samples are required to provide a reasonably accurate estimate of the population mean. For instance, results from the laboratory experiments suggest that 25, 6 and 29 replicate tests will be required for Lockport, Pontypool and Milliken respectively, to achieve a minimum of ±10% accuracy at the 95% level.

Samples from two soils – Lockport and Pontypool – had been tested in the field and in the laboratory by using essentially identical techniques. The difference in rainfall period was removed by using the laboratory data pertaining to 30-minute rainfall. The only other difference was that Lockport was a coherent sample in the field but it was air-dried, pre-sieved and pre-wetted before it was tested in the laboratory. The measured erodibilities were found to be quite different, largely because of variations in soil moisture content (Table 11.1). However, the measured variabilities are very similar, with a somewhat higher field *CV* for Lockport. Field runoff variability is substantially higher for both Lockport and Pontypool.

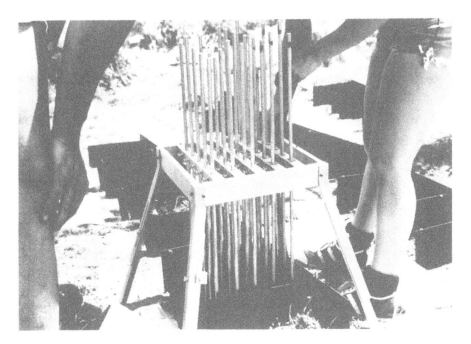

FIGURE 11.5
The frame of the microrelief gauge was 'levelled' so that it orientated parallel to the plot surface. The 1-cm diameter rods were being lowered to the soil surface.

TABLE 11.1
Available data on rainwash variability.

Author	Location and soil series	Plot size (m²)	Rainfall intensity (mm hr⁻¹)	Slope angle (°)	Cover condition	$\bar{x}$ (g m⁻² min⁻¹)	s (g m⁻² min⁻¹)	CV (%)	n	Remarks
Luk (1975)	Canadian Rockies	0.093	93–110	30	bare	8.15	1.85	22.7	9	lab. data
Singer et al. (1977)	California, Auburn	0.372	71–81	5	bare	13.56	2.02	14.9	6	lab. data
					46–55% covered	8.87	2.39	26.9	6	
Bryan (1979)	Alberta, Devon	0.093	102	15	bare	2.11	0.53	25.0	8	lab. data
	Alberta, Orchard					44.30	17.69	39.9	9	
Bryan and Luk (1981)	Ontario, Lockport	0.093	62–69	12.5	bare	6.92	2.12	30.6	10 + 10	lab. data
	Ontario, Pontypool					19.38	4.30	22.2	10 + 10	
	Ontario, Milliken					2.28	0.57	25.2	16	

TABLE 11.1 (continued)
Available data on rainwash variability.

Author	Location and soil series	Plot size (m²)	Rainfall intensity (mm hr⁻¹)	Slope angle (°)	Cover condition	$\bar{x}$ (g m⁻² min⁻¹)	s	CV (%)	n	Remarks
	Ontario, Lockport					23.4	8.28	35.4	8+9	
Luk and Morgan (1981)	Ontario, Pontypool	0.093	51	9.5–10.5	bare	13.7	2.82	20.6	10+10	field data
	Ontario, Dumfries (Durham)					37.0	8.40	22.7	10+10	
	Ontario, Woburn					3.27	1.12	34.3	17	
	Ontario, King					8.22	2.20	26.8	15	
	Ontario, King (steep phase)					6.37	1.46	22.9	16	
Luk (unpublished)	Ontario, Dumfries (Mountsburg)	0.093	51	6.6–8.4	bare	10.93	2.36	21.6	18	field data
	Ontario, Springvale					13.78	5.03	36.5	19	
	Ontario, Guelph					7.66	0.97	12.6	10	

$\bar{x}$ = sample mean; s = sample standard deviation; CV = coefficient of variation; n = sample size.

Whereas the difference in field and laboratory runoff variabilities was expected, the similarity in the rainwash variabilities was not. The higher *CV* for field runoff certainly demonstrates the greater variability of field soil moisture content and infiltration rate, but this was only translated into a moderate increase in rainwash variability for Lockport, and no change at all for Pontypool. The explanation of these results appears to be the mechanics of interrill erosion. Here, rainwash is transported by a non-capacity sheetflow. As long as the flow depth is less than the diameter of erosive raindrops, which is likely to be the case in these experiments, flow depth and presumably runoff volume have only a limited influence on rainwash transport rate (Kilinc & Richardson 1973). In the case of Ponty-pool, the increased variability in runoff had little effect on rainwash be-cause entrainment by splash-saltation was dominant. Where the domi-nance of splash-saltation in causing entrainment was partially eclipsed by turbulence in the raindrop-disturbed flow (Bryan 1976), as in the case of Lockport, some increases in rainwash variability were observed.

CAUSES OF WITHIN-SITE SOIL LOSS VARIABILITY

Runoff rates

Field soils, as opposed to pre-sieved soil samples in the laboratory, have large variations in their distribution of microporosity, the size distri-bution of both unaggregated and aggregated surficial particles, soil mois-ture content (Hills & Reynolds 1969), and therefore they show wide vari-ations in infiltration rates (Hills 1970) and runoff rates (Weyman 1974) within small areas. In the field tests conducted so far, runoff variability ranged from 19% for Guelph loam to 75% for Pontypool sand. However, these high *CV* values for runoff did not translate into a proportionately high rainfall variability, which ranged from 13% for Guelph loam to 35% for Lockport clay. Correlation results show, however, that there are sig-nificant relationships between rainwash and runoff at the 95% level for four sites, and at the 90% level for another two sites (Table 11.2). These results could be interpreted in two ways: (i) rainwash-transported sedi-ments are effectively entrained by surface wash, or (ii) the significant correlations are only co-variations, that is, both variables are affected by similar independent variables. In view of the overall differences in the range of rainwash and runoff variabilities, the second interpretation is preferred, as is shown by the similarity of their respective contributory factors (Tables 11.3 and 11.4). At the same time, the first interpretation cannot be ruled out entirely. The variations in the significance of the relationship between rainwash and runoff seem to suggest that the role of sheetwash entrainment varies with soil type. Data pertaining to Lockport clay and Pontypool sand obtained in the pilot study demonstrate this difference more succinctly (Fig. 11.6). Pontypool sand, which was marginally influenced by wash entrainment, shows no rainwash–runoff correlation. Lockport clay, being partially affected by wash entrainment, registers a strong rainwash–runoff relationship.

Soil properties

In this study, sand content, clay content, organic content, soil bulk density, soil moisture content, aggregate size and stability, as well as the compressive and shear strength of samples at the plot level, were studied. Of these soil properties, organic content, bulk density, soil moisture content and sand content show low variabilities, generally with *CV* values of less than 10% (Table 11.2). Thus, they are unlikely to be significant contributory factors to soil loss variability. This is supported by correlation analysis (Table 11.3), except for bulk density and moisture content. Significant relationships for bulk density were found at the 95% level for Springvale sandy loam and Guelph loam. The positive relationship observed for Springvale is frequently reported (Meeuwig 1965, Luk 1975). For Guelph loam, the relationship is a negative one, suggesting that higher

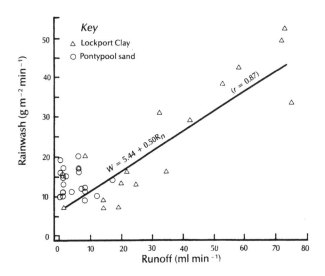

FIGURE 11.6
*Relationship between rainwash (W)
and runoff (R$_n$), field data.*

TABLE 11.2
Soil characteristics.

	Site 1 Woburn loam	Site 2 King clay loam	Site 3 King clay loam (steep phase)	Site 4 Dumfries (Mountsburg) loam	Site 5 Springvale sandy loam	Site 6 Guelph loam
Rainwash (g)	9.1[a] (34.3)[b]	22.9 (26.8)	17.8 (22.9)	30.5 (21.6)	38.4 (36.5)	21.4 (12.6)
Runoff (ml)	153.7 (63.0)	499.3 (43.7)	891.9 (48.1)	1143.3 (50.6)	781.4 (54.8)	242.2 (18.6)
Organic content (%)	6.1 (9.7)	4.3 (12.6)	5.9 (13.2)	2.9 (23.4)	3.7 (11.1)	6.4 (12.0)
Bulk density (g cm^{-3})	1.09 (5.8)	1.33 (4.7)	1.35 (5.4)	1.62 (7.0)	1.39 (3.4)	1.08 (4.5)
Soil moisture content (%)	31.0 (5.8)	25.0 (6.6)	25.0 (12.1)	16.8 (6.8)	18.4 (6.4)	35.9 (10.8)
Sand content (%)	38.1 (5.1)	27.9 (4.2)	18.8 (14.7)	47.5 (6.4)	47.4 (6.2)	41.4 (6.7)
Clay content (%)	13.7 (28.3)	24.3 (8.8)	36.8 (17.5)	16.1 (16.3)	15.9 (18.6)	6.8 (23.4)
WSA[c] > 4 mm (%)	13.6 (30.6)	34.6 (20.6)	27.2 (45.1)	20.0 (42.4)	20.5 (39.0)	12.0 (32.3)
Aggregate stability (median drop number)	NA	NA	192.4 (40.7)	27.5 (31.4)	NA	NA
Torsional shear strength (KPa)	25.0 (19.6)	44.9 (7.9)	57.7 (11.4)	42.4 (12.9)	29.7 (11.1)	18.4 (30.1)

TABLE 11.2 (continued)
Soil characteristics.

	Site 1 Woburn loam	Site 2 King clay loam	Site 3 King clay loam (steep phase)	Site 4 Dumfries (Mountsburg) loam	Site 5 Springvale sandy loam	Site 6 Guelph loam
Bearing capacity (KPa)	366.0 (18.0)	575.9 (12.0)	655.9 (20.7)	556.3 (13.3)	580.6 (8.8)	224.0 (21.0)
Maximum slope (°)	7.5 (29.3)	7.1 (17.4)	8.4 (18.5)	6.6 (16.6)	7.1 (26.4)	7.3 (36.8)
Downslope micro-relief (mm)	4.84 (32.7)	2.31 (31.3)	2.36 (29.1)	2.06 (20.1)	1.79 (36.8)	1.41 (22.9)
Cross-slope microrelief (mm)	4.44 (41.2)	2.64 (20.4)	2.38 (33.3)	2.15 (26.8)	1.95 (27.9)	2.09 (23.0)
Sample size	17	15	16	18	19	10

[a]Sample mean.
[b]Coefficient of variation (%).
[c]Water-stable aggregates.

TABLE 11.3
Correlation coefficients: rainwash vs. other soil characteristics.

Rainwash vs.	Site 1	Site 2	Site 3	Site 4	Site 5	Site 6
Runoff	0.46*	0.56**	0.74**	0.55**	0.89**	0.59*
Organic content	−0.08	−0.18	0.10	−0.06	−0.10	0.49
Bulk density	−0.02	0.16	−0.08	−0.11	0.79**	−0.67**
Soil moisture content	0.08	−0.18	0.17	0.16	0.11	0.59*
Sand content	0.11	−0.13	0.44†	−0.11	−0.28	−0.30
Clay content	0.38†	0.38	0.08	−0.01	−0.04	0.75**
WSA > 4 mm	0.48*	0.28	−0.25	0.28	−0.25	0.22
WSA > 1 mm	0.49*	0.33	−0.30	0.14	−0.07	0.15
WSA > 0.25 mm	0.28	0.40	−0.04	0.05	−0.02	0.53†
Torsional shear strength	0.18	−0.31	−0.35	−0.21	0.21	−0.44
Bearing capacity	0.10	−0.03	−0.01	−0.51**	0.28	−0.40
Maximum slope	0.27	0.64**	0.40†	0.08	0.28	−0.11
Downslope microrelief	0.30	−0.42†	−0.02	0.10	0.10	0.47
Cross-slope microrelief	0.36	−0.22	−0.09	−0.01	0.07	0.04

Significance: ** 95%, * 90% and † 85%.
Site names and locations appear in Fig. 11.1.

TABLE 11.4
Correlation coefficients: runoff vs. other soil characteristics.

Runoff vs.	Site 1	Site 2	Site 3	Site 4	Site 5	Site 6
Bulk density	0.18	0.37	0.09	−0.01	0.69**	−0.82**
Soil moisture content	−0.16	−0.24	0.04	0.16	0.01	0.40
Sand content	−0.19	0.01	0.11	−0.01	−0.16	0.10
Clay content	0.49*	−0.34	0.21	−0.40	0.12	0.34
WSA > 4 mm	0.34	0.19	0.08	−0.01	−0.18	0.25
WSA > 1 mm	0.52**	0.06	0.08	−0.16	−0.04	0.10
WSA > 0.25 mm	0.60**	0.02	0.27	−0.11	0.09	0.48
Torsional shear strength	−0.15	0.24	−0.16	0.24	0.36	−0.33
Bearing capacity	−0.22	0.29	0.07	−0.31	0.25	−0.34
Downslope microrelief	−0.22	−0.07	−0.08	−0.01	0.18	0.28
Cross-slope microrelief	−0.04	−0.23	−0.18	−0.26	0.09	0.18

Significance: ** 95%, * 90%.
Site names and locations appear in Fig. 11.1.

bulk density is associated with lower rainwash. This relationship was observed by Adams *et al.* (1958), but it was not explained. The only reasonable explanation is that compaction in some samples imparted a higher degree of coherence and, therefore, a greater resistance to erosion. Soil cores used to determine soil moisture content and bulk density were extracted after the completion of the rainfall experiment in order to avoid disturbing the initial soil surface. Thus, variations of antecedent soil moisture content in the plots are not known. However, the low variability of soil moisture content observed after the experiments (Table 11.2) tends to suggest that antecedent moisture content may be less variable than was initially hypothesized. The positive relationship observed for Guelph loam may simply reflect the influence of bulk density, for soil moisture content is significantly correlated with bulk density ($r = -0.68$; significant at the 95% level) at this site.

For aggregate size, the measured CV values ranged from 12% to 35% for the parameter WSA (water-stable aggregates) >4.0 mm (Table 11.2). Aggregate stability was tested for two selected samples (King clay loam (steep phase) and Dumfries (Mountsburg) loam) and their variabilities were 41% and 31% respectively. The magnitude of all these CV values suggests that aggregation is a significant contributory factor. However, results from the correlation analysis (Table 11.3) indicate inconsistent relationships. Strong positive correlations were found in Woburn loam and only moderately significant results (at the 85% level) were observed for Guelph loam. No significant correlation exists in the other four soils. Comparison with other results suggests that the inconsistent relationships are not without cause. Woburn loam and Guelph loam both have inferior rainwash–runoff correlation coefficients, as compared with the other samples, and they are also weakly aggregated (Table 11.2). Field observations indicate that their surfaces are much less coherent, as is revealed by their lower shear and compressive strengths (Table 11.2). This implies that particle entrainment is likely to be dominated by splash-saltation rather than flow turbulence and therefore explains the weaker rainwash–runoff relationships. Because aggregation parameters tend to be inferior indices of erodibility among coherent soils (Morgan 1979), which are the well-aggregated samples in this study, it is only to be expected that the two least aggregated soils show the best correlation between aggregation parameters and soil loss. The positive correlations, however, are contrary to many previous findings (e.g. Luk 1977). In the laboratory study (Bryan & Luk 1981), a similar positive correlation was found for Pontypool sand in the first 10 minutes of the experiment, which lasted 60 minutes (Table 11.5). It was hypothesized that, initially, runoff was generated only over aggregates and stones where the infiltration capacity was lower. This would account for the positive correlation with runoff and rainwash. But as rainfall continued, the complete surface became saturated and started to contribute to runoff, so that aggregation was no longer a significant factor. In the field, this initial condition might have been maintained throughout the 30-minute long experiments.

The influence of clay content is usually expressed via soil aggregation (Luk 1979). For these samples, although clay content is not correlated with aggregation at the within-site level, the correlation coefficient at the between-site level is 0.83 ($n = 6$) which is significant at the 95% level. Hence, the effective contribution of clay content to rainwash variability for Woburn loam and Guelph loam (Table 11.3) is expected. The positive relationship is again not usually observed, but it has been found for some Prairie-cultivated soils (Luk 1979). The explanation is that clay particles were released through slaking and dispersion of some aggregates which were weakly cemented, causing the choking of surface pore spaces and increases in runoff as well as soil loss rates.

Because many soils exhibit a coherent surface under field conditions, it was thought that soil strength parameters may be superior to aggregate indices as measures of soil erodibility. The strength parameters investigated were torsional shear strength and unconfined compressive strength (bearing capacity). Their ranges in CV are 8–30% and 9–21% respectively (Table 11.2), which are somewhat lower than those for soil loss. Actual correlation analysis shows that only 1 out of the 12 coefficients is significant at the 95% level. In view of the low CV values in some of the strength parameters and the absence of significant correlations with soil loss, it appears that the strength parameters chosen are not important contributory factors. Moreover, the chosen parameters of shear and compressive strength are significantly correlated with aggregation ($r = 0.80$ and 0.78

TABLE 11.5
Coefficients of correlation: rainwash and runoff vs. initial aggregate size parameters, laboratory data on Pontypool[a].

Time periods (minutes)	Rainwash vs.			Runoff vs.		
	WSA% > 4 mm	WSA% > 2 mm	WSA% > 0.5 mm	WSA% > 4 mm	WSA% > 2 mm	WSA% > 0.5 mm
0– 5	0.24	0.23	0.42*	0.39*	0.45**	0.52**
5–10	0.04	0.18	0.45**	0.24	0.23	0.39*
10–20	0.18	0.15	0.10	0.20	0.17	0.29
20–30	−0.48**	−0.59**	−0.40	0.08	0.01	0.12
30–40	0.02	−0.10	−0.03	0.05	0.01	0.08
40–50	−0.19	−0.01	0.05	0.06	0.01	0.03
50–60	0.09	0.35	0.37	0.25	0.24	0.17

Significance: ** 95%, * 90%.
[a]Sample size = 20.

respectively at the between-site level; $n = 6$) at the 90% level, thus suggesting that the strength parameters are to some extent duplications of the aggregation indices.

Maximum slope steepness

The sites investigated were selected for their conformity of average slope (7.5 ± 1°), but considerable differences do exist within each site. The calculated maximum slope (θ_x) is based on measured angles that are parallel to two perpendicular edges of each plot:

$$\theta_x = \sin^{-1}\sqrt{\sin^2 \theta_d + \sin^2 \theta_a} \qquad (11.1)$$

where θ_d is downslope angle and θ_a is cross-slope angle. The resulting CV for θ_x of the six sites ranged from 17% to 37% (Table 11.2). Given the average θ_x for all sites is 7.3°, the absolute variation of θ_x would range from 7.3 ± 3.7° to 7.3 ± 8.1°(!). Despite these substantial within-site variations, slope angle is only a significant contributory factor for the two clay loam soils (Table 11.3). At the same time, practically no soil property is a significant contributory factor for these soils (Table 11.3) and, at least for King clay loam, within-site variations of soil properties are limited (Table 11.2). It appears that slope angle is only a subsidiary factor, and its importance is only felt where soil properties are more uniformally distributed.

Surface microtopography

Microtopography is, in fact, a complex variable which includes at least three components: *slope shape*, surface microrelief due to undulations on a coherent soil surface (*skin roughness*) and surface microrelief due to variations of particle size distributions of both unaggregated and aggregated surficial materials (*grain roughness*). Each of these components may exist in isolation, but they may also combine in different ways to influence soil loss variability. In the field tests conducted so far, the soil surfaces were largely coherent ones, thus grain roughness was not significant. In the laboratory experiments, pre-sieved air-dried aggregates (≤8 mm in diameter) were used and, therefore, only grain roughness was important.

Microtopography in the field was measured by the microrelief gauge already described. Initially, cross- and down-slope angles were determined by setting a brunton on a piece of 25 x 25 cm plywood board placed on the plot surface. The accuracy of the slope readings was limited where non-plannar slope shapes were present. Then, the two pivoting levels were adjusted according to the determined slope angles. Thus, when the frame was levelled by adjusting the height of attached slanting legs, it should have been parallel to the average plot surface. After the rods were gently lowered through the perforations on the cross-piece of the frame until they touched the soil surface, they were locked in place by screws, and the whole frame (including the rods) was removed from the plot before the length of the rods extending above the upper edge of the frame was determined with the vernier depth gauge.

Because the height data obtained by the above method do not refer to a fixed datum, the differences in height between consecutive points instead of the actual heights were analyzed. The microrelief index (R_i) employed is the standard deviation of the height difference sample, or:

$$R_i = \sqrt{\frac{\Sigma(d_i - \bar{d})^2}{n-1}} \qquad (11.2)$$

where d is height difference and n is sample size. The use of this index has the additional advantage that measurement errors pertaining to the average slope of the plot surface are also eliminated because sample standard deviations are independent of sample means. Two microrelief indices, corresponding to cross-slope (R_a) and downslope microrelief (R_d), were computed.

Measured CV values for the six selected sites ranged from 20–33% and 20–41% for R_a and R_d respectively (Table 11.2). These relatively high CV values certainly suggest the potential contribution of microtopography to soil loss variability, but correlation analysis reveals only one significant coefficient (at the 85% level) out of a total of twelve computed (Table 11.3). When the data were plotted, it was found that for site 3, a curvilinear relationship of the form $y = a + bx + cx^2$ was a better fit to the data (Fig. 11.7). For R_a and R_d, the regression equations were significant at the 90% and 85% levels, respectively. These two results therefore suggest that the influence of microtopography is variable. Under the experimental conditions at site 3, peak soil loss occurred where R_a and R_d were approximately 2.0 mm.

The soils at these sites are the King clay loams which are the most coherent as shown by their high shear and compressive strength (Table 11.2). Because the microtopographic measurements pertain to initial conditions, it could be argued that the microtopography of these coherent soils offered greater resistance to obliteration during the experiments, rainfall erosivity being essentially constant for all sites investigated. This would also explain the somewhat more significant correlations observed at site 3, which is the most clayey soil tested. It is important to bear in mind that both R_a and R_d represent the combined influence of slope shape and skin roughness. No attempt has been made to isolate their individual influence, although the plots were very carefully scalped in the field so that the dominant slope shape tends to be slight concavities with very low curvatures and thus the influence of slope shape is likely to be very limited.

In the laboratory, the influence of microtopography can be studied in greater detail but it is confined to grain roughness. Replicate Lockport clay samples were tested in the Edmonton simulator by placing pre-screened air-dried aggregates and unaggregated particles (≤ 8 mm in diameter) in the sample pan and levelling each sample to the top of the pan. The surface aggregate size distribution was measured off vertical photographs taken before and after each experiment. Surface microrelief of selected samples was also determined by standard photogrammetric techniques. Because soil loss was monitored at 5-minute intervals, it was possible to assess the grain roughness effect over different time periods

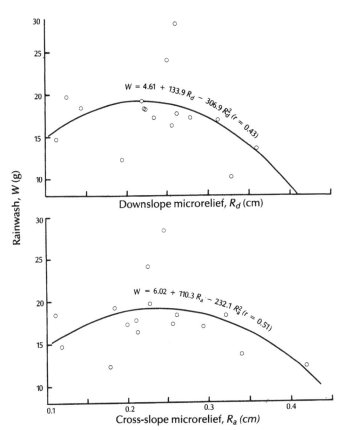

FIGURE 11.7
Relationship between rainwash and microrelief indices, King clay loam (steep phase).

(Table 11.6). It can be seen that significant inverse relationships exist during the initial 10 minutes for coarse aggregates and from 5 to 25 minutes for the finer grains. Large grains in the samples, which are naturally associated with greater grain roughness (Table 11.7), tend to retard surface flow, to increase surface detention and to reduce the transport rate of rainwash. After about 25 minutes, the influence of initial grain roughness became insignificant. This may be the result of several processes. The positive coefficients for D_{15} measured after the experiments imply local deposition of fines which reduced grain roughness. Runoff volume and flow depth increased continuously, negating the initial roughness effects of flow retardation and surface detention. But, on the whole, dispersion, counteracted to some extent by selective entrainment, caused a propor-

TABLE 11.6
Coefficients of correlation: rainwash vs. aggregate size parameters, laboratory photographic data on Lockport[a].

Time period (minutes)	Before experiment			After experiment		
	D_{50}	D_{35}	D_{15}	D_{50}	D_{35}	D_{15}
0– 5	−0.67**	−0.65**	−0.54	−0.55	−0.50	0.02
5–10	−0.57*	−0.56*	−0.67**	−0.61*	−0.38	0.43
10–15	−0.53	−0.52	−0.73**	−0.69**	−0.45	0.38
15–20	−0.41	−0.39	−0.61*	−0.64**	−0.35	0.48
20–25	−0.50	−0.46	−0.62*	−0.55*	−0.26	0.40
25–30	−0.47	−0.41	−0.44	−0.29	−0.07	0.27
30–35	−0.44	−0.39	−0.41	−0.19	0.05	0.31
35–40	−0.42	−0.35	−0.29	−0.13	0.07	0.22
40–45	−0.36	−0.30	−0.34	−0.20	0.04	0.30
45–50	−0.20	−0.16	−0.18	−0.13	0.03	0.02
50–55	−0.24	−0.16	−0.18	0.02	0.25	0.33
55–60	−0.32	−0.23	−0.24	−0.03	0.25	0.63**

Significance = ** 95%, * 90%.
[a]Sample size = 10.

TABLE 11.7
Rainwash, initial aggregate size and initial microrelief, laboratory data for Lockport.

	B13[a]	B11	B16
Rainwash (g min^{-1})	0.50	0.64	0.76
Initial aggregate size (D_{50}, mm)	3.74	3.03	2.88
Initial microrelief (mm)	1.62	1.14	0.97

[a]B13, B11 and B16 are representative, respectively, of low, medium and high rainwash responses. They were selected subjectively from the 20 experimental runs.

tionate reduction in the areal distribution of grains >2 mm in diameter (Table 11.6). Therefore, the grain size distribution (D_{50}) determined after the experiments is also correlated to soil loss during the 5- to 25-minute period.

DISCUSSION AND CONCLUSIONS

In this study, nine field locations were investigated in a regional area where soil types were developed from diverse parent materials and, in each case, it has been demonstrated that substantial soil loss variability exists within the small sample areas ($20m^2$). The measured within-site variability ranged from 13% to 37%. In order to compare within-site and between-site soil loss variability, the latter was computed from the mean rainwash data (Table 11.2). Disregarding the three locations from the pilot project, where a slightly different rainfall simulator was used, between-site variability is 43%. Since site 1 has a substantially higher microrelief (Table 11.2) because control on surface relief was effected by a different operator during plot preparation, a second computation using sites 2–6 was carried out. The resulting CV was 31%. Thus, a reasonable estimate of between-site variability is 31–43%. If this range of values is compared to the 13–37% of within-site variability, it can be seen that the two ranges overlap to some extent, suggesting that in some instances, within-site variability can be as large as between-site variability. This result clearly underscores the significance of soil loss variability studies.

A large number of potentially contributory factors was studied, including runoff rates, soil moisture content, sand content, clay content, organic content, soil bulk density, aggregate size and stability, maximum slope angle, surface microtopography, and soil compressive and shear strength. Results obtained by correlation analysis indicate that very few significant factors could be identified. This is likely to be partially induced by the limitation of the correlation analysis technique itself. The variation of soil characteristics studied in the within-site context tends to be quite limited, as compared to many other studies concerning between-site variations (e.g. Luk 1977). Therefore, significance levels of 90% and 85% were employed, which is not common. Even then, only one or two significant factors per site were observed. The distribution of the significant factors with respect to soil site suggests a possible division of the six locations into three groups: (a) sites 2 and 3 where coherence is high; (b) sites 1 and 6 where the soils are poorly aggregated and coherence is low; and (c) a group with intermediate coherence at sites 4 and 5 (Table 11.2).

For the group (a) sites, initial surface microrelief is more significant because the relief features are more resistant to obliteration due to high soil coherence. Related laboratory data show that, towards the latter part of the experiments, smoothing of the initial microrelief might have occurred (under moderate to high erosivity conditions) due to local deposition of fines. The alteration of the initial microrelief might partially explain the low levels of correlation observed. In addition, the slope steepness factor is also more significant for these soils. How the influence of slope angle is related to microtopography is not known. It might be coincidental because soil property variations are generally small for these soils and, therefore, they did not mask the effects of slope and microtopography.

For the group (b) locations where soil coherence is low, soil properties, especially soil aggregation parameters, tend to be more significant. This is supported by a similar study (Morgan 1979), which showed that a semi-coherent soil surface when subjected to high intensity rainfall might behave more like a layer of aggregates. The positive relationship between aggregate size and rainwash obtained in this study is contrary to results observed in studies involving many different soil types. This relationship was also found in similar laboratory experiments and was explained by the generation of runoff over moist aggregates where the initial infiltration capacity was lower. For these soils, the significance of clay content is related to poor soil aggregation. Choking of pore spaces occurred earlier at plots with higher clay content, resulting in higher runoff rate and higher soil loss. Moreover, the lower rainwash–runoff relationship in these soils implies that splash-saltation was dominant over flow turbulence in the entrainment of surface particles or aggregates.

For the group (c) locations, where soils with intermediate coherence were found, neither the slope–microtopography nor the aggregate–clay factors are significant. Instead, bearing capacity (compressive strength) and bulk density are more important contributory factors at sites 4 and 5, respectively.

The above interpretation of results recognized that the significance of contributory factors to soil loss variability is dependent on soil coherence (or cohesion). Certainly, a considerable amount of work has been done on soil cohesion from the soil mechanics viewpoint, but applications to the study of field soil erodibility have rarely been attempted. The unconfined compressive strength (bearing capacity) and torsional shear strength determined in the field are indices of soil coherence, but they do not appear to be the most appropriate. Equipment that is capable of precisely determining the strength of a very thin layer (<1 cm) of crust-type surface soil might provide better indices. Such equipment is to be designed.

On the whole, the amount of explained within-site variation in soil loss is limited. For instance, the strongest correlation observed is -0.79 for soil bulk density at site 5 (Table 11.3). This yields a r^2 value of 0.62, or 62% of the total variation may be explained. In most other cases, the explained variation is much smaller. Use of multiple regression analysis did not seem to improve the results to any large extent. Therefore, given the fact that all the known contributory factors have been investigated, the unexplained variations must be attributed to measurement error and 'random variation.' Because measurement error tends to be quite small in most cases, it is tentatively proposed here that a substantial amount of within-site soil loss variation is random variation. Here, random variation refers to the interaction of three groups of factors which are spatially and temporally variable – raindrops, soil surface characteristics and moisture-flow depth factors. Raindrops vary from less than 1 mm to over 5 mm in diameter and they deliver vastly different amounts of kinetic energy over soil surfaces that are characterized by large variations in aggregate size, for non-coherent to semi-coherent surfaces, and by a constantly changing surface microtopography for coherent samples. Flow depth is determined by the interactions among raindrops, soil surface characteristics and antecedent moisture content. It affects the relative significance of both splash-saltation and flow-turbulence entrainment. In addition, local increases in flow to depths that are greater than the diameter of raindrops will also impede splash-saltation. Thus, for rainfall extending over a relatively short period, a characteristic of most rainstorms, different rates of soil loss may be observed on diverse surfaces that pertain to the same soil within a small sample area. But even when the *average* surface aggregate and microtopographic characteristics of two given surfaces of the same soil are similar, the precise physical distribution patterns are never the same, which is critical in effecting the number of grains transported across the lip of the collector and thus qualifying these grains as soil loss! All the above-mentioned relationships are individually deterministic. However, their spatial and temporal variations are so complex that the sum of their total effects may be considered essentially random, an idea that has been applied to other geomorphological studies (Melton 1958, Scheidegger & Langbein 1966). The conclusion that random variation constitutes a substantial proportion of within-site rainwash variation is arrived at by the process of elimination and therefore, strictly speaking, is not supported by *direct* evidences. However, it is considered the most likely interpretation of available data because the results were derived from six sets of experiments, which are thus replicable to some extent, and no alternative explanations are available for further investigation at this time.

From the viewpoint of soil conservation, it is unlikely that a sufficiently large number of replicate tests can be performed to determine the erosion variability of given soils, and thus the predictability of erosion variability needs to be investigated. Data available from this study indicate that mean torsional shear strength, microrelief variability, and maximum slope variability are the most closely related to the measured rainwash variability (Fig. 11.8). The direct relationship between rainwash and microrelief variability is self-explanatory. Peak rainwash variability is associated with intermediate values of mean shear strength and maximum slope variability. The physical bases of these two relationships are not entirely known. Given the limited number of data available, they are best considered to be tentative empirical relationships. Among the three variables presented, torsional shear strength is preferred because it can be readily determined in the field. Earlier discussion of soil coherence in this paper further supports the significance of shear strength parameters. However, more data are required to determine the utility of the presented variables for prediction. Finally, it should be emphasized that the measured CV values and observed relationships are specific to the testing conditions employed. For instance, the laboratory and field data reported pertain to a testing period of 30 minutes. Other laboratory data (Bryan & Luk 1981) show a progressive reduction in rainwash variability as rainfall duration is increased. Similarly, within-site soil loss variability may be sensitive to

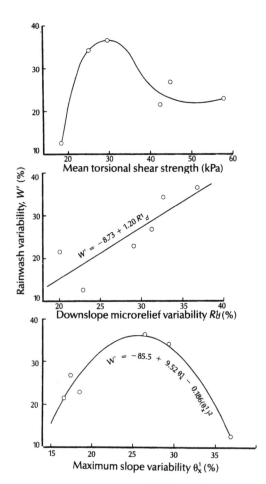

FIGURE 11.8
Relationship between rainwash variability and mean torsional shear strength, downslope microrelief variability, and maximum slope variability.

differences in plot size, rainfall intensity and other factors. Hence, like soil erodibility, soil loss variability should be treated as a relative concept as well.

ACKNOWLEDGEMENTS

The author is grateful to Professor Rorke Bryan (University of Toronto) and Dr Christopher Morgan (Brock University) for their participation in and contribution to the laboratory and the pilot field project, respectively, which laid the foundation for the more detailed investigation presented in this paper. The research projects were financially supported by the Natural Sciences and Engineering Research Council of Canada and the University of Toronto.

REFERENCES

Adams, J. E., D. Kirkham and W. H. Scholtes 1958. Soil erodibility and other physical properties of some Iowa soils. *Iowa State J. Sci.* **32**, 485–540.
Bryan, R. B. 1970. An improved rainfall simulator for use in erosion research. *Can J. Earth Sci.* **7**, 1552–61.
Bryan, R. B. 1976. Considerations on soil erodibility indices and sheetwash. *Catena* **3**, 99–112.
Bryan, R. B. 1979. The influence of slope angle on soil entrainment by sheetwash and rainsplash. *Earth Surf. Proc.* **4**, 43–58.
Bryan, R. B. and S. H. Luk 1981 Variation of soil erosion under laboratory simulated rainfall. Submitted to *Geoderma*. (in press).
Campbell, I. A. 1970. Microrelief measurements on unvegetated shale slopes. *Prof. Geog.* **22**, 215–20.
Hills, R. C. 1970. *The determination of the infiltration capacity of field soils using the cylinder infiltrometer.* Br. Geomorph. Res. Group Tech. Bull. 3.
Hills, R. C. and S. G. Reynolds 1969. Illustrations of soil moisture variability in selected areas and plots of different sizes. *J. Hydrol.* **8**, 27–47.
Kilinc, M. and E. V. Richardson 1973. *Mechanics of soil erosion from overland flow generated by simulated rainfall.* Colorado State University Hydrol. Pap. 63.
Luk, S. H. 1975. *Soil erodibility and erosion in part of the Bow River Basin, Alberta, Canada.* Unpublished PhD dissertation. University of Alberta.
Luk, S. H. 1977. Rainfall erosion of some Alberta soils, a laboratory simulation study. *Catena* **3**, 295–309.
Luk, S. H. 1979. Effect of soil properties on erosion by wash and splash. *Earth Surf. Proc.* **4**, 241–55.
Luk, S. H. and C. Morgan 1981. Spatial variations of rainwash and runoff within small homogeneous areas. *Catena* (in press).
McCalla, T. M. 1944. Water drop method of determining the stability of soil structure. *Soil Sci. Soc. Am. Proc.* **7**, 209–14.
Meeuwig, R. O. 1965. Effects of seeding and grazing on infiltration capacity and soil stability of a subalpine range in Central Utah. *J. Range Mgmt.* **18**, 173–80.

Melton, M. A. 1958. Correlation structure of morphometric properties of drainage systems and their controlling agents. *J. Geol.* **66**, 442–60.

Morgan, C. 1979. *Field and laboratory examination of soil erosion as a function of erosivity and erodibility for selected hillslope soils from southern Ontario.* Unpublished PhD dissertation. University of Toronto.

Scheidegger, A. E. and W. B. Langbein 1966. *Probability concepts in geomorphology.* U.S. Geol. Surv. Prof. Paper 500-C.

Singer, M. J., G. L. Huntington and H. R. Skatchley, 1977. Erosion prediction on California rangeland: research developments and needs. In *Soil erosion: prediction and control,* 143–51. Soil Conserv. Soc. Am. Sp. Publ. 21.

Soons, J. M. 1971. Factors involved in soil erosion in the southern Alps, New Zealand. *Z. Geomorph.* **15**, 460–70.

Weyman, D. R. 1974. *Runoff processes, contributing area and streamflow in a small upland catchment.* Inst. Br. Geog. Spec. Publ. 6, 33–43.

Wischmeier, W. H. 1976. Use and misuse of the Universal Soil Loss Equation. *J. Soil Wat. Conserv.* **31**, 5–9.

Wischmeier, W. H. and D. D. Smith 1958. Rainfall energy and its relationship to soil loss. *Trans Am. Geophys. Union* **39**, 285–91.

Yoder, R. E. 1936. A direct method of aggregate analysis of soils, and a study of the physical nature of erosion losses. *J. Am. Soc. Agron.* **28**, 337–51.

12

The influence of topography on the spatial variability of soils in Mediterranean climates

Daniel R. Muhs

INTRODUCTION

The purpose of this paper is to provide new data on the effects of slope position on soil formation in an arid Mediterranean climate in southern California and to show how processes active in that region are active in other areas with Mediterranean climates. Milne (1936a) coined the term **catena** to describe the lateral or spatial variability of soils on a hillslope. Since Milne's (1936a, 1936b) early studies, other workers have considered the catena in terms of both static and dynamic causes of differentiation (Young 1972) and open *vs.* closed systems (Ruhe & Walker 1968, Walker & Ruhe 1968). The catena concept is important in geomorphological studies because it demonstrates spatial variation in soil properties that can exist on a single landform or geomorphic surface.

STUDY AREA

Soil studies were conducted on San Clemente Island, California, an uplifted fault block about 100 km southwest of Los Angeles (Fig. 12.1). Smith (1898) and later Olmsted (1958) mapped the geology of the island; it is composed mainly of Miocene andesite with smaller areas of dacite, rhyolite and marine sedimentary rocks. Quaternary deposits consist

269

mainly of marine terrace deposits, calcareous eolianite, dune sand and alluvial fans, which have been recently mapped on the northern half of the island by Muhs (1980). Above the highest marine terraces (elevations of ⩾ 300 m), Quaternary deposits are largely absent and the landscape is characterized by gently undulating volcanic bedrock.

The present climate of San Clemente Island is arid Mediterranean (Csb in the Köppen system). Mean annual temperature is 16°C, with very little variation during the year; mean summer temperatures are about 18°C and mean winter temperatures are about 14°C (Naval Weather Service Environmental Detachment 1973). Mean annual precipitation is 165 mm/yr, virtually all of which falls as rain during the winter months. The vegetation in the central part of the island where the present study was conducted is mainly Mediterranean annual grasses (Raven 1963).

A reconnaissance soil survey was made on the bedrock surface above the highest terraces on the central part of the island (Fig. 12.1) and, after numerous exploratory pits had been dug and examined, a representative catena was sampled for detailed laboratory studies. Soils were described and sampled by horizon from the crest, shoulder and toe positions on a 15% andesite bedrock slope extending over a horizontal distance of about 50 m (Fig. 12.2).

LABORATORY METHODS

Particle size distribution was determined by the pipette method (Day 1965) and bulk density by the clod method (Blake 1965). Iron and aluminum were extracted by the dithionite-citrate-bicarbonate method (Mehra & Jackson 1960); quantities were determined by atomic absorption spectrometry. Soil pH was determined on 1 : 1 soil–water pastes with a glass electrode. Clays were separated by sedimentation, mounted on warm ceramic tiles and X-rayed after air drying, glycolation and heat treatment (550°C).

MORPHOLOGY AND PHYSICAL PROPERTIES

Field observations indicate that there are significant differences between the three soils. The crest and shoulder soils are both Lithic Haploxeralfs with thin (c. 20 cm) sola, 5YR or 7.5YR hues in the B horizon and do not exhibit pedogenic structure (Table 12.1). In contrast, the soil at the base of the slope, although it also has a lithic contact, has a thicker (46 cm) solum with 10YR hues, well-developed angular blocky structure and distinct slickensides in the lower part of the profile. Cracks were also evident at the time of sampling (August, 1978) and were 4 cm wide at the surface and 1 cm wide at a depth of 38 cm. These characteristics, along with the laboratory data (Table 12.2), indicate that this soil approaches the central concept of the Typic Pelloxererts. Technically, Typic Pelloxererts must have cracks extending to 50 cm and are not allowed to have a lithic contact within 50 cm of the surface (Soil Survey staff 1975). However, cracks probably extend to depths greater than 38 cm during the peak of the dry season in the fall. In addition, although the toeslope soil in this study does have a lithic contact at 46 cm, its true classification as a Lithic

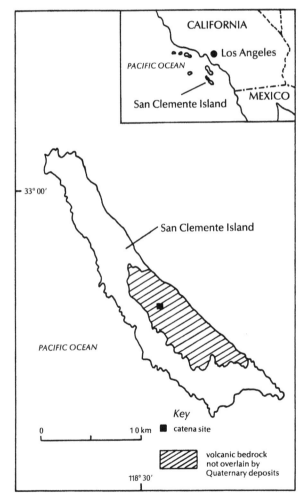

FIGURE 12.1
Location of study area.

Xerorthent does not reveal its actual vertic character. Hence, since it nearly meets the requirements for a Vertisol, it will be referred to as a 'proto-Vertisol' or simply as a Pelloxerert.

Laboratory data support the differences between the soils noted in the field (Table 12.2). There is a significant increase in clay content from the A1 to the B2 horizons of both the crest and shoulder soils (Fig. 12.3), suggesting that illuviation has taken place to form argillic horizons. Field observations, however, did not reveal the presence of clay films in either of these two soils and thin sections did not indicate the presence of argillans. It is possible, therefore, that the increase in clay content with depth is due to parent material layering and the A1 horizons of these two soils are equivalent to the silty eolian mantle found on soils farther north on the island. Muhs (1980) found that this layer was derived from the Mojave Desert under Santa Ana wind conditions and is present on other islands off the California coast. The silt content of the A1 horizons here, however, is only about 40% (Table 12.2), as opposed to the eolian mantle farther north, which usually has around 60% silt (Muhs 1980). It may be that the A1 horizons on bedrock in this part of the island are composite features formed in part *in situ* and in part from the eolian influx.

The lack of evidence for clay movement in the form of clay films and argillans may be due to shrink–swell activity. Thin sections revealed skelsepic fabric in the B2 horizons of both soils, indicative of intense swelling

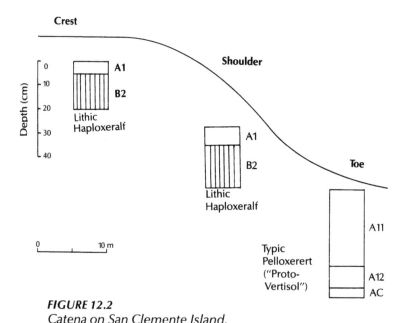

FIGURE 12.2
Catena on San Clemente Island.

pressures (Brewer 1976). Illuviation may have taken place, therefore, (and may be taking place at the present) but shrinking and swelling may have destroyed all evidence of it.

The crest soil appears to be on a slightly more stable position. This is suggested by its clay content in the B2 horizon (Fig. 12.3, Table 12.2), which is higher than the clay content of the B2 horizon of the shoulder soil. Clearly, however, both of these soils have experienced some erosion as indicated by their thin sola, whereas the toeslope soil, with its thicker solum, has experienced some accumulation of material.

The proto-Vertisol at the toeslope position has nearly uniform properties with depth (Table 12.2). Clay content is high (47% to 51%), distinct horizons are not readily apparent in the field (Table 12.1) and the AC horizon is designated by a slight decrease in the grade of structure and the presence of weathered bedrock fragments. Bulk density is also high (1.79–1.91 gm/cm³) and is probably related to the high clay content.

The thicker solum and higher clay content of the toeslope soil, by comparison with the crest and shoulder soils, suggests that water movement (overland flow) has entrained and transported particles downslope.

TABLE 12.1
Field data for catena on San Clemente Island.

Profile	Horizon	Depth (cm)	Color, dry[a]	Textural class	Structure[b]	Boundary
crest	A1	0–4	10 YR 4/4	sandy loam	m	clear smooth
	B2	4–20	7.5 YR 3/4	clay	1cabk	abrupt smooth
	R	20+	bedrock (andesite)			
shoulder	A1	0–7	10 YR 2/1, 4/3	sandy loam	sg	clear smooth
	B2	7–23	5 YR 3/3	sandy clay	1cabk	abrupt smooth
	R	23+	bedrock (andesite)			
toe	A11	0–33	10 YR 3/1	clay	3m,cabk	clear smooth
	A12	33–41	10 YR 3/1	clay	3cabk	gradual wavy
	AC	41–46	10 YR 3/1	clay	1cabk	abrupt smooth
	R	46+	bedrock (andesite)			

[a]According to the Munsell notation.
[b]Symbols from Soil Survey Staff (1951).

This process is common on slopes in humid regions and it appears that it is important in arid regions as well, because here it may explain much of the difference in profile morphology between the upper and lower parts of the slope.

CHEMICAL PROPERTIES

Chemical data also indicate important differences between the crest and shoulder soils and the toeslope soil that are probably related to water movement. Soil pH increases with depth in all three soils, but also increases significantly from the crest and shoulder soils to the toeslope soil (Table 12.2). The more alkaline conditions in the toeslope soil may be due to movement of bases (Ca^{++}, Mg^{++}, Na^+) downslope and accumulation at the toe position. This could occur not only from overland flow, but also from throughflow above the less permeable B2 horizons of the crest and shoulder soils and the bedrock surface.

Greater oxidation of iron-bearing minerals (augite, hypersthene and magnetite) from the andesite parent material upslope is suggested by the relatively high values for dithionite-extractable iron in the crest and shoulder soils (Table 12.2, Fig. 12.3). The increase in Fe content from the A1 to the B2 horizons in the upper two soils also suggests that Fe in these two profiles has begun to move downward (Fig. 12.3); this process is not apparent in the toeslope soil. Aluminum shows less of a trend with slope position (Table 12.2); this may be related to the fact that even though pHs are significantly different between the three soils, all pH values (5.7 to 7.3) are in the range where A1 is essentially insoluble (Black 1967). Hence, only small amounts have been released by weathering and there is no marked differentiation on the slope. Fe, on the other hand, can be soluble at pH values just under 6 (Black 1967); these values are present in the A1 horizons

TABLE 12.2
Laboratory data for catena on San Clemente Island.

Profile	Horizon	Depth (cm)	Sand (%)	Silt (%)	Clay (%)	Bulk density (g/cm³)	pH	Fe (%)	Al (%)
crest	A1	0–4	47.2	39.3	13.5	1.79	5.8	1.16	0.16
	B2	4–20	20.7	26.2	53.1	1.82	6.2	1.19	0.15
shoulder	A1	0–7	45.3	40.3	14.4	1.19	5.7	1.16	0.14
	B2	7–23	34.2	34.4	31.4	1.41	6.2	1.64	0.21
toe	A11	0–33	14.9	33.6	51.5	1.90	6.3	1.00	0.20
	A12	33–41	17.6	34.4	48.0	1.91	7.2	0.90	0.17
	AC	41–46	17.8	34.7	47.5	1.79	7.3	0.84	0.13

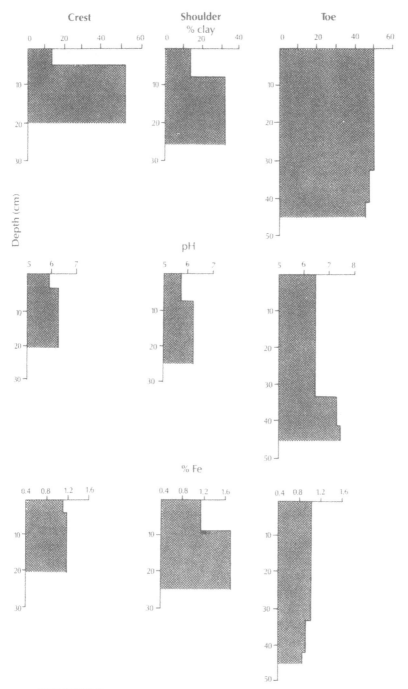

FIGURE 12.3
Laboratory data for the catena studied on San Clemente Island.

of the crest and shoulder soils, but not in the proto-Vertisol at the base of the slope. Greater water movement through the crest and shoulder soils (especially through the shoulder, where Fe has clearly begun to migrate down through the profile) could be responsible for the higher Fe values. The higher Fe values in these two soils probably explain the redder hues (5YR and 7.5YR *vs.* 10YR in the toeslope soil) visible in the field.

CLAY MINERALOGY AND WEATHERING

Clay mineralogy

Clay mineralogy generally supports the observations on the dominant processes discussed above. In the A1 horizons of the crest and shoulder soils, mica and kaolinite are present (Fig. 12.4), but the dominant minerals are quartz and plagioclase. The kaolinite in these horizons is indicated by a broad, low (001) peak at 7.1 Å, which disappears after heating to 550°C. The mica is indicated by a peak at 10.0 Å and is present in the B2 horizons of the crest and shoulder soils as well. Its presence here, along with the abundance of quartz and the overall similarity of the A1 horizon clay mineralogy with the eolian mantle clay mineralogy farther north on the island (Muhs 1980), lends strength to the idea that the A1 horizons here are at least partly eolian. Kaolinite is present in the silty eolian layers of the soils on marine terraces farther north, and it may be of eolian origin here.

The clay mineralogy of the B2 horizons of the crest and shoulder soils differ markedly from one another. The crest soil exhibits a relatively sharp 17 Å peak upon glycolation (Fig. 12.4), which indicates the presence of smectite (montmorillonite). This peak collapses to about 10 Å after heating to 550°C. Poorly crystalline kaolinite is indicated by a broad peak centering on about 7Å; chlorite may be contributing to this peak, because a small shoulder peak is present at 14 Å and remains after 550°C heat treatment. However, heat treatment greatly reduces the intensity of the peak at 7 Å, suggesting that the main cause of the 7 Å peak is either kaolinite or Fe-rich chlorite. The B2 horizon of the shoulder soil has a somewhat sharper (001) kaolinite peak at 7 Å, suggesting somewhat better crystallinity. However, the smectite peak at 17 Å in this horizon is much broader and lower, indicating smaller quantities of this mineral, poorer crystallinity, finer particle size or some combination of these factors.

The proto-Vertisol at the base of the slope exhibits a clay mineral assemblage somewhat different from that of the crest and shoulder soils (Fig. 12.4). Glycolated samples show strong peaks at 17 Å in all horizons and some expand to 18–20 Å. Johnson *et al.* (1962) made similar observa-

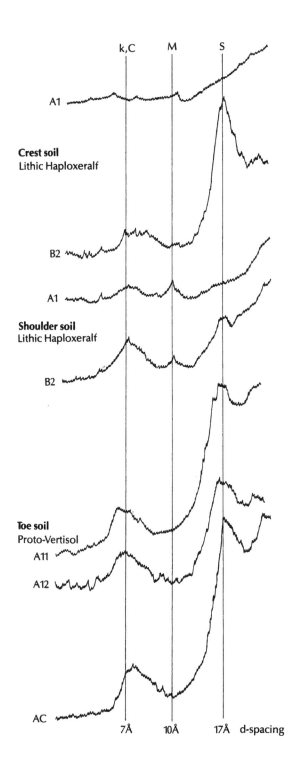

k,C   M   S

A1

Crest soil
Lithic Haploxeralf

B2

A1

Shoulder soil
Lithic Haploxeralf

B2

Toe soil
Proto-Vertisol

A11

A12

AC

7Å   10Å   17Å   d-spacing

FIGURE 12.4
X-ray diffractograms
for glycolated samples
from the catena
studied on San
Clemente Island.
S = smectite,
M = mica,
K = kaolinite,
C = chlorite.

tions of smectites in basalt-derived Vertisols in Arizona, and attributed the greater expansion to complexing with organic matter. This kind of complexing in Vertisols has been reported from other regions and may account for the relatively dark color of Vertisols despite their low organic matter contents (Mohr *et al.* 1972). Poorly crystalline kaolinite is again suggested by the broad peak centering on 7 Å, but the presence of chlorite is also suggested by a very small peak which remains at 7 Å after heat treatment; the lack of a (001) chlorite peak at about 14 Å suggests that this chlorite is Fe rich. Only trace amounts of mica are present in the lower two horizons of the toeslope soil (Fig. 12.4). Abundant quartz and plagioclase and smaller amounts of K-feldspar are present in all three horizons.

Weathering processes

The data presented above suggest that the removal of bases from the upper horizons of the crest and shoulder soils and their movement downslope, as indicated by the pH data, have strongly influenced the direction of clay mineral formation. There are apparently two sources for clay formation, and both of these are influenced by water movement downslope.

Some clay forms *in situ* by weathering of primary minerals (plagioclase and pyroxenes) in the parent andesite. Thin sections reveal both plagioclase and augite being replaced by clay in the crest and shoulder soils. The evidence for grain replacement is not apparent in the toeslope soil, but this may be due to destruction of evidence by shrinking and swelling or during transport to the toe position by overland flow. Another source of clay is eolian material; mica and K-feldspar, which are not present in the bedrock, are almost certainly of this origin, and some plagioclase may be of eolian origin as well. All three of these minerals can alter to clay.

Removal of bases from the surface horizons of the crest and shoulder soils to positions at depth in the profiles and downslope has created a base-rich environment favorable for smectite formation. These conditions are best displayed in all horizons of the toeslope soil and in the B2 horizon of the crest soil, to a lesser degree in the B2 horizon of the shoulder soil, and not at all in the A1 horizons of the crest and shoulder soils.

The lower amounts of mica in the proto-Vertisol suggest that it may be altering directly to smectite, with this alteration keeping ahead of eolian influx of mica. Transformation of mica to smectite may take place readily in the presence of high amounts of Ca^{++} and Mg^{++}, with these elements exchanging for K^+ in the interlayer position of mica. Because of the higher pHs in the toeslope soil discussed earlier, this seems to be a likely process. This direct transformation is also suggested by the absence of intermediate weathering products such as hydrobiotite (randomly interstratified mica-vermiculite) and mica-smectite, which are present in soils on marine terraces on the northern half of the island (Muhs 1980).

CATENAS IN OTHER AREAS WITH MEDITERRANEAN CLIMATES

Other investigators have shown similar patterns in the spatial variation of soils due to topography in Mediterranean climates elsewhere. Interestingly, similar patterns occur regardless of the amount of precipitation or the type of parent material. This suggests similar processes in all areas, with the amount of water movement passing through the profiles probably being the most important.

San Diego County, California

On the California mainland, Nettleton *et al.* (1970) have described a catena developed on tonalite (quartz diorite) in an area receiving about 360 mm/yr of precipitation. Hilltop and backslope (crest and shoulder) positions are occupied by Xerochrepts and Haploxeralfs, whereas Natrixeralfs occupy the footslope positions and areas of restricted drainage have Pelloxererts. The Natrixeralfs and Pelloxererts have higher clay contents, higher pHs, higher base saturations and more smectite clays than the Xerochrepts and Haploxeralfs. No data were given on the soil color or Fe and Al content.

Coastal plain of Israel

A well-studied soil catena on the Sharon Plain of Israel has been described by Dan and Yaalon (1964) and Dan *et al.* (1968). The soils have developed from cemented calcareous dunes (eolianite) in a climate which receives about 520 mm/yr of precipitation. Soils on the crests of these dunes are Rhodoxeralfs (terra rossa), lower shoulder position soils are either Typic or Aeric Albaqualfs or Glossaqualfs and soils in the toe position are Typic Pelloxererts. The Pelloxererts have much higher clay contents and the most poorly drained of these have higher pHs than soils on the upper part of the slope. Smectite is also the dominant clay mineral in the Pelloxererts. The Rhodoxeralfs on the crest and shoulder positions, however, have thinner sola, lower clay contents, redder hues, higher free Fe in the clay fraction and kaolinite as the dominant clay mineral. The Albaqualfs and Glossaqualfs appear to have properties intermediate between the Rhodoxeralfs and the Pelloxererts. The pattern of soils described in this study is very similar to that described for San Clemente Island; in addition, Dan *et al.* (1968) suggest the importance of eolian dust as a source of clay in these soils. A later study by Yaalon and Ganor (1979) documented the importance of eolian dust additions to soils in Israel from the Saharan, Libyan, Sinai and Negev deserts.

Golan Heights

Dan and Singer (1973) examined a toposequence on the basalt plateau of the Golan in an area receiving about 750–900 mm/yr precipitation. Haploxeralfs are present on the crest of the hillslope, Xerorthents are present on the shoulder position and Chromoxererts are in the toe position. The Pelloxererts have thicker sola, higher clay content, lower free Fe content, higher pHs and more smectite than the Haploxeralfs and Xerorthents higher on the slope; these latter two soil types have more kaolinite. Although free Fe values are higher for the Haploxeralfs than for the Pelloxererts, the differences are not so great as to have caused a hue change observable in the field.

Southern Lebanon

Verheye (1973) mapped the soils of southern Lebanon and provided detailed laboratory analyses of representative profiles. Parent materials for these soils are mainly limestones and sandstones; precipitation is 650–1400 mm/yr, depending on altitude. Well-drained sites are occupied by Haploxeralfs or Rhodoxeralfs (terra rossa) on hard carbonatic rocks and ferriferous sandstones; Rendollic Xerochrepts are present on soft limestones. Poorly drained sites have Vertic Rendollic Xerochrepts. Clay content and solum thickness are actually higher in the Alfisols than in the Vertic Rendollic Xerochrepts and pHs are not much different. However, free Fe content is considerably higher in both of the Alfisols, and kaolinite is the main clay mineral in the Rhodoxeralfs; the Vertic Rendollic Xerochrepts are dominated by smectite. In this study, it is clear that there is something of a departure from the other areas already discussed, in that well-drained sites have thicker sola, suggesting that there has been less erosion by overland flow. However, the direction of chemical weathering and clay mineral formation appears to be similar to other regions.

Andalusia, Spain

Taboadela (1953) described clay mineral suites of various soils from Spain and included a discussion of both well-drained soils (terra rossa, probably Rhodoxeralfs) and poorly drained soils (Andalusian black soils, probably Xererts). The parent materials are calcareous and, at least in the case of the Andalusian black soils, are calcareous sandstones; precipitation is 300–700 mm/yr. The Andalusian black soils generally have a higher clay content than the terra rossa soils and are characterized by smectite with smaller amounts of illite. The terra rossa soils, on the other hand, are dominated by illite with smaller amounts of kaolinite. Unfortunately, not many field data are given in this study, so it is not possible to compare all of the characteristics of these soils with those already discussed. However, even with the few data that are given, it seems clear that a similar pattern of soil differentiation along hillslopes has developed here as well.

CONCLUSIONS

Soil development on hillslopes in Mediterranean climates appears to take a similar direction, regardless of the amount of precipitation or the nature of the parent material. On well-drained sites, thin, red or reddish soils with argillic B horizons develop. The pH may be relatively low, and there is a tendency toward kaolinite formation. On poorly drained down-slope sites, the soils that form are Vertisols, or at least soils with vertic characteristics. These soils are generally thick, have high clay contents, high pH and smectite as the dominant clay mineral. Illuviation processes do not dominate in Vertisols, but pedoturbation processes are active and the result is a relatively uniform profile with no evidence of argillic B horizon development.

The observations made in the present study on San Clemente Island and in the other studies cited above suggest that water movement, both as overland flow and throughflow, is the dominant process in variability of soil type along a hillslope in a Mediterranean climate. A complicating effect is the influx of eolian materials. This process appears to be common in areas with Mediterranean climates, probably because most of them are adjacent to deserts, which are often sources for airborne materials. However, subsequent water movement downslope appears to decrease the importance of this process; regardless of the mineralogy of the eolian materials, those that are reactive seem to 'adjust' to good drainage and removal of basic cations on the crest and shoulder positions and poor drainage and abundance of basic cations in the toe position. A summary of the processes active on each portion of the hillslope is shown in Figure 12.5.

Catenas in Mediterranean climates appear to be good examples of the concept of 'thresholds' in geomorphology (Chorley & Kennedy 1971, Schumm 1977). In the case of a catena, however, it is a pedologic thresh-old. Dynamic metastable equilibrium, as discussed by Chorley and Kennedy (1971) and Schumm (1977), represents variations about a changing average condition, with episodic change as thresholds are exceeded. As conceptualized, this describes possible conditions in fluvial systems along a temporal dimension, but the idea may be applied to soil systems along a spatial dimension. In the Mediterranean catena case, soils on the crest and shoulder positions have similar dominant processes (Fig. 12.5) and their profiles represent the average condition resulting from this multi-plicity of processes. Specifically, the soil profiles with their argillic B horizons represent changing average conditions resulting from inputs of weathering products from the bedrock and eolian influx, throughputs of clay from illuviation and outputs of soluble constituents and solid particles lost by throughflow and overland flow. The vertisols in the toeslope positions may be regarded as being located on the other side of a pedologic threshold. This is because clay buildup is sufficient to make pedoturbation dominant; conversely, illuviation is no longer dominant and argillic B horizons are absent. Hence, the spatial variability *between* thresholds along such a catena is represented by the variability between the crest and shoulder soils.

The threshold concept can be applied to soils along a temporal dimension as well. For example, soils on the youngest marine terraces on the northern half of San Clemente Island are Typic Natrixeralfs, but soils on older terraces are Typic Chromoxererts (Muhs 1980), indicating that a temporal pedologic threshold has been crossed. The application of the threshold idea in pedology as well as in geomorphology promises to be a fruitful conceptual framework for process-oriented studies.

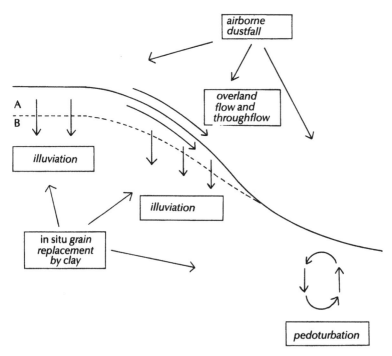

FIGURE 12.5
Sketch summarizing the dominant processes on each portion of a hillslope in a Mediterranean climate.

ACKNOWLEDGEMENTS

This study was supported in part by grants from the US Geological Survey (Earthquake Hazards Reduction Program) and the Geological Society of America through its Penrose and J. Hoover Mackin Awards. I thank Tracy Rowland (University of Wisconsin) for help in the field and in the laboratory and Scott Burns (Lincoln College, New Zealand) for help in the laboratory. The Department of Geological Sciences, University of Colorado, provided laboratory facilities. Peter Birkeland (University of Colorado) and Nel Caine (University of Colorado) made helpful comments on an earlier draft of the paper. Peggy Lathrup typed the manuscript.

REFERENCES

Black, A. 1967. Applications: electrokinetic characteristics of hydrous oxides of aluminum and iron. In *Principles and applications of water chemistry*, S. D. Faust & J. V. Hundter (eds), 274–300. New York: Wiley.

Blake, G. R. 1965. Bulk density. In *Methods of soil analysis, Part I*, C. A. Black (ed.), 274–90. Am. Soc. Agron. Monogr. 9.

Brewer, R. 1976. *Fabric and mineral analysis of soils*. Huntington: Robert E. Krieger.

Chorley, R. J. and B. A. Kennedy 1971. *Physical geography: a systems approach*. London: Prentice-Hall.

Dan, J. and A. Singer 1973. Soil evolution on basalt and basic pyroclastic materials in the Golan Heights. *Geoderma* **9**, 165–92.

Dan, J. and D. H. Yaalon 1964. The application of catena concept in studies of pedogenesis in Mediterranean and desert fringe areas. *Int. Congr. Soil Sci. Trans*, 8th, Bucharest, 751–8.

Dan, J., D. H. Yaalon, and H. Koyumdjisky 1968. Catenary soil relationships in Israel, 1. The Netanya catena on coastal dunes of the Sharon. *Geoderma* **2**, 95–120.

Day, P. R. 1965. Particle fractionation and particle-size analysis. In *Methods of soil analysis, Part I*, C. A. Black (ed.), 545–67. Am. Soc. Agron. Monogr. 9.

Johnson, W. M., J. G. Cady and M. S. James 1962. Characteristics of some brown Grumusols of Arizona. *Soil Sci. Soc. Am. Proc.* **26**, 389–93.

Mehra, O. P. and M. L. Jackson 1960. Iron oxide removal from soils and clays by a dithionite–citrate system buffered with sodium bicarbonate. *Clays and Clay Minerals* **5**, 317-27.

Milne, G. 1936a. *A provisional soil map of East Africa*. East African Agric. Res. Sta. Amani Memoir.

Milne, G. 1936b. Normal erosion as a factor in soil profile development. *Nature* **138**, 548–9.

Mohr, E. C. J., F. A. van Baren and J. van Schuylenborgh 1972. *Tropical soils*, 3rd edn. The Hague: Mouton.

Muhs, D. R. 1980. *Quaternary stratigraphy and soil development, San Clemente Island, California*. Unpublished PhD dissertation. Boulder: University of Colorado.

Naval Weather Service Environmental Detachment 1973. *Station climatic summary: San Clemente Island, California*.

Nettleton, W. D., K. W. Flach and R. E. Nelson 1970. Pedogenic weathering of tonalite in southern California. *Geoderma* **4**, 387–402.

Olmsted, F. H. 1958. *Geologic reconnaissance of San Clemente Island, California.* U.S. Geol. Survey Bull. 1071-B.

Raven, P. H. 1963. A flora of San Clemente Island, California. *Aliso* **5**, 289–347.

Ruhe, R. V. and P. H. Walker 1968. Hillslope models and soil formation I. Open systems. *Int. Congr. Soil Sci. Trans,* 9th, Adelaide, 551–9.

Schumm, S. A. 1977. *The fluvial system.* New York: Wiley.

Smith, W. S. T. 1898. A geological sketch of San Clemente Island. In *18th Annual Report,* Part II. U.S. Geol. Survey.

Soil Survey staff 1951. *Soil survey manual.* U.S.D.A. Agric. Handbook 18.

Soil Survey staff 1975. *Soil taxonomy: a basic system of soil classification for making and interpreting soil surveys.* U.S.D.A. Agric. Handbook 436.

Taboadela, M. M. 1953. The clay mineralogy of some soils from Spain and Rio Muni (West Africa). *J. Soil Sci.* **4**, 49–55.

Verheye, E. 1973. *Formation, classification and land evaluation of soils in Mediterranean areas, with special reference to the southern Lebanon.* Ghent.

Walker, P. H. and R. V. Ruhe 1968. Hillslope models and soil formation II. Closed systems. *Int. Congr. Soil Sci. Trans,* 9th, Adelaide, 561–7.

Yaalon, D. H. and E. Ganor 1979. East Mediterranean trajectories of dust-carrying storms from the Sahara and Sinai. In *Saharan dust: mobilization, transport, deposition,* C. Morales (ed.), 187–93. New York: Wiley.

Young, A. 1972. The soil catena: a systematic approach. In *International geography* (Vol. 1), W. P. Adams & F. M. Helleiner (eds), 287–9. Toronto: University of Toronto.

13

Temporal variability of a summer shorezone

Antony R. Orme

INTRODUCTION

Coastal change occurs in many forms, at rates varying from imperceptible to instantaneous, and over timescales ranging from a few seconds to millions of years. Understanding of these changes is important to society, especially today when industrial, residential and recreational development of the coastal zone may proceed at a scale and rate that far outpace effective coastal planning. Fortunately, over the past 40 yr, coastal geomorphology has developed rapidly, aided by advances in shorezone monitoring systems and dating techniques. Various conceptual models have been invoked to explain observed phenomena, for example the process–response and morpho–dynamic models which seek to show how, either directly or through feedback mechanisms, coastal forms interact with the physical processes and materials of their environment. Nevertheless, because the subject matter is so dynamic in time and space, and often difficult to study in the field, explanations are frequently complex and sometimes elusive, as this study of changes on a summer shore in California will show.

DURATION, RATE AND NATURE OF COASTAL CHANGE

Coastal change occurs over a wide range of different timescales nested one within another. At one extreme are the slow changes in relative land and sea levels related to the differential motion of continental and oceanic lithospheric plates. Such changes are often of long duration, measurable over timescales of 10^5 or 10^6 yr or more, but rates of change may vary from negligible to quite rapid, as with the lateral motion of the outer Baja California coast which, over the past 5×10^6 yr, has been moving north-westwards at a rate of 0.06 m/yr on the western limb of a spreading sea floor associated with the East Pacific Rise beneath the Gulf of California (Orme 1980b). Related vertical motions may also be quite rapid, as with the uplift of 0.01 m/yr observed along the coast of the Transverse Ranges near Ventura, California (Wehmiller *et al.* 1977, Yeats 1978).

Changes related to glacioeustatic forces cause further adjustments in land and sea levels on timescales of 10^3 or 10^4 yr, most recently in the Flandrian transgression which raised the sea to roughly its present level between 15 000 and 5000 yr BP. For example, once the sea rose through the Golden Gate some 10 000 yr ago, sea level in southern San Francisco Bay rose about 0.02 m/yr until 8000 BP, after which it declined and then averaged 0.001 – 0.002 m/yr from 6000 BP to the present (Atwater *et al.* 1977). Before 8000 BP, the shoreline moved horizontally inland as rapidly as 30 m/yr.

Within episodes of relatively stable sea level, seacliff erosion generally proceeds within timescales of 10^1 to 10^3 yr, but rates are highly variable. For example, since the Coast Survey was conducted in 1859, Bolinas Point and Duxbury Point, northwest of San Francisco, have retreated at mean rates of 0.5 m/yr. South of San Francisco, cliff retreat over the past 100 yr has varied from 0 – 0.1 m/yr, depending mainly on lithology.

Although the above average rates may conceal some rapid movements, as in earthquake-related uplift or sudden cliff collapse, most such changes are very slow and become significant only over longish periods of time. Changes that are significant over shorter periods are more often associated with unconsolidated sediments than with hard rock. For example, major changes in beach volume may be linked to sediment influxes from nearby watersheds, reflecting the frequency of flood events which may range from several times a year to once every 10 yr or more. The longer the interval between sediment influxes, the more prolonged the changes in the shorezone, and the more difficult it is to predict the nature of long-term coastal change from short-term records. Then there are the seasonal changes, between stormy winters and tranquil summers. Lastly, within a season, changes may last several days, a few hours, even a few minutes or less. These within-season changes form the focus of this paper.

The nature of temporal coastal change assumes many forms, for example linear, non-linear, cyclic and instantaneous. As will be shown, both linear and non-linear trends are found with progressive shorezone erosion or deposition within a season. Over longer periods, many non-linear changes occur, such as the negative exponential curve associated with cliff retreat, initially rapid but decelerating as the widening shore platform absorbs more and more wave energy. Cyclic changes range from predictable tidal cycles to quasi-cyclic Quaternary sea-level oscillations. The dangers inherent in interpreting temporal changes are well exemplified by such sea-level curves which, though related to cyclic sequences, may exhibit positive exponential, linear, and negative exponential trends at different times, each part of an S-shaped curve of changing sea level.

Instantaneous changes often disturb the above trends, some as perturbations from which a system soon recovers, others as thresholds from which a system may or may not recover. A threshold is an abrupt change in time or space when or where stress suddenly overcomes a system's resistance. Stress may accumulate within a system until an intrinsic threshold is reached and rapid change occurs, as when marine sediments accumulate on a slope to a critical mass sufficient to generate turbidity currents, or when stresses accumulating within a seacliff suddenly exceed the cliff's resistance to failure and the cliff collapses. Extrinsic thresholds occur when a system suddenly responds to external forces, as when an equilibrium shorezone succumbs to changes in the wave and current regime.

The relatively rapid changes suffered by the California coast in spring and autumn are forms of extrinsic threshold that separate distinctive summer and winter regimes. To evaluate what happens between the spring and autumn thresholds, several time series of shorezone data collected near an experimental groin over 128 consecutive summer days are analyzed and interpreted. Those aspects of the data pertinent to the impact of the groin have been discussed elsewhere (Orme 1978, 1980a). This paper focuses on the temporal variability of the data. It first describes the research facility and design, then the process variables, then analyzes the geometric and textural data, and concludes by evaluating the observed changes in the context of various explanatory models.

RESEARCH FACILITY AND DESIGN

This research was conducted around an experimental groin built by the Coastal Engineering Research Center between Point Mugu and Port Hueneme, 90 km west of Los Angeles, California (Fig. 13.1). The shoreline trends N 52° W at the site and describes a broad arc, convex seaward, at the southwest edge of the Oxnard alluvial plain. The steel groin extends 150 m seawards perpendicular to the shoreline and ranges from 3 m above mean lower low water at its landward crest to about mean lower low water at its seaward crest. A timber pier, 5 m wide and 208 m long, provides a platform above the groin from which experiments may be conducted and from which a movable boom operates (Fig. 13.2). The pier extends from the groin's landward end to a point 58 m beyond the groin's seaward end.

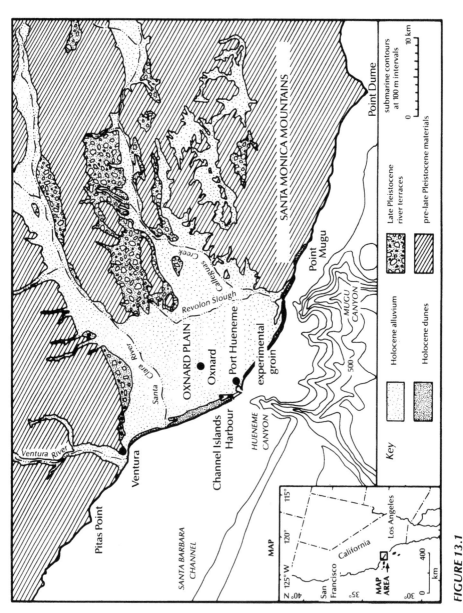

FIGURE 13.1
Location of experimental groin, between Point Mugu and Port Hueneme, California.

Predominant littoral drift along this shore during any year is from northwest to southeast, producing a net movement of about 750 000 m³ of sediment per year in a southeasterly direction. Accordingly, the northwest face of the groin is termed the upcoast side, whereas the southeast face is the downcoast side. Littoral drift reaching the site comes mainly from fluvial sediments introduced to the shorezone during the winter months by the Santa Clara River, some 17 km to the northwest. Hueneme sub-

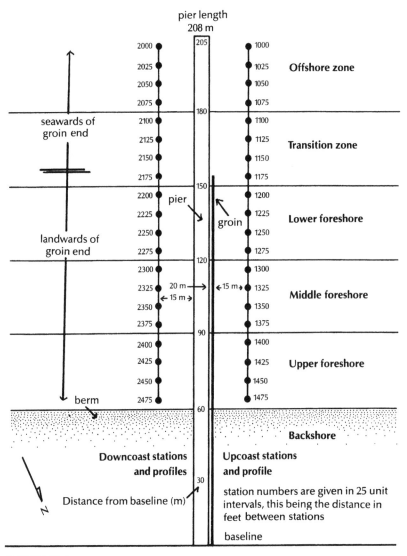

FIGURE 13.2
Experimental groin and pier facility, showing survey stations and profiles, and shore zonation.

marine canyon, which heads close to shore 6 km northwest of the site, siphons off part of this drift into deeper water, whence it is not normally returned shoreward. Jetties confining the entrances to Channel Islands Harbor and Port Hueneme also impose significant barriers to littoral drift, barriers that are overcome by periodic mechanical bypassing operations. Periodic reversals in the direction of littoral drift, notably in summer, involve relatively small volumes of sediment. Sediment discharged by Calleguas Creek southeast of the site is mainly trapped in Mugu Lagoon, while Mugu submarine canyon and Point Mugu present significant barriers to the small volume of sediment moving from farther to the southeast.

For this study, environmental, geometrical and textural data were obtained daily over 128 consecutive summer days, from May 11 to September 15, 1972. The environmental data comprised nine variables: wave period, wave height, wave type, wave angle, longshore current velocity and direction, width of surf zone, and wind velocity and direction. Wave period was determined by timing the duration of ten consecutive wave crests passing the outer end of the pier, and then dividing by ten. Wave height at breaker point was estimated visually and recorded to the nearest 0.1 m. Spilling, plunging, surging, and mixed spilling and plunging waves were nominally identified. Wave angle at breaker point was obtained by measuring, with a protractor on the pier rail, the direction from which waves approached the shore, 90° representing a perpendicular approach. Longshore current velocity was estimated by using small dye packets that dispersed upon immersion. The net displacement of the dye-patch centroid in meters parallel with the shore was measured over 1 minute and converted to meters per second (m/s). Current direction was recorded as movement upcoast towards the northwest or downcoast towards the southeast. The width of the surf zone from shore to breakers was measured in meters. Wind velocity was measured using a cup anemometer and converted to km/hr. Wind direction was observed in terms of compass octants.

To further the study, two 145 m long profiles were established parallel to the groin, the upcoast profile 15 m northwest of the groin, the downcoast profile 20 m southeast of the structure, and both extending from the berm to the seaward end of the pier. These profiles thus extended seawards beyond the breakers on most occasions. Sampling stations were established at 7.5 m intervals along these profiles and defined by marker buoys anchored to the sea floor by screw anchors. Upcoast stations were numbered from 1000 at the seaward end of the profile, through 1025, 1050, and so on, at 7.5 m (25 feet) intervals to 1475 near the berm. Downcoast stations were similarly numbered from 2000 to 2475 (Fig. 13.2). Elevations (water depths) for all stations were measured daily relative to the top of the boom on the pier and referenced to mean lower low water, providing an accuracy of ±0.03 m.

Sediment samples were also acquired daily for all stations along the two profiles, generally around the time of lower low water. The 40 sampling stations and 128 d (days) of observation thus yielded 5120 sediment samples, each initially weighing 500–700 g. After preliminary analysis, stations at 30-m intervals were found to be representative of discrete zones along the profiles. Thus, sediment samples from 5 upcoast and 5 downcoast stations at 30-m intervals (stations 1000, 1100, 1200, 1300, 1400, 2000, 2100, 2200, 2300, and 2400) were analyzed for the 128 days, i.e. 1280 samples. These samples were air-dried, presplit, washed twice in warm tap water and thrice in cold distilled water, oven dried for 8 hours at 100°C, sieved to remove particles larger than 1 mm (weighed separately), then microsplit to provide 0.2–0.25 g portions for processing by settling velocity techniques. Particle diameters were calculated using the formula proposed by Gibbs *et al.* (1971) as follows:

$$D = 20 \; \frac{0.55804V^2\rho_f + 0.003114V^4\rho_f^2 + [g(\rho_s - \rho_f)][4.5\eta V + 0.008705V^2\rho_s]}{[g(\rho_s - \rho_f)]}$$

(13.1)

where D = particle diameter (mm), g = acceleration of gravity (cm/s^2), V = velocity (cm/s), η = dynamic viscosity of fluid (poises), ρ_f = density of fluid (g/cm^3), ρ_s = density of a quartz sphere (g/cm^3). These diameters were then converted to phi units (ϕ) where (ϕ) = $-\log_2 D$, and six parameters were computed for each sample, namely phi median diameter, phi mean diameter (mean grain size), phi sorting coefficient, phi skewness, phi second skewness, and phi kurtosis. Of these, the phi mean grain size, $M\phi = (\phi_{84} + \phi_{16})/2$, and phi sorting (or Graphic standard deviation), $\sigma_G = (\phi_{84} - \phi_{16})/2$, are used in this analysis because they are considered most expressive of the dynamic environment.

PROCESS VARIABILITY

Visual inspection of the nine environmental variables presented graphically in Figure 13.3 shows significant deviations from the mean over short time intervals and some brief trends lasting from 3 to 8 days but rarely longer. Over the 128-day observation period, trends are insignificant or weak and cyclicity is not readily apparent.

Wave period averages 14.53 s (σ = 2.07 s) and an extreme range from 7.5 to 21.0 s. Least variability occurs during high summer between days 40 and 100. Wave height at breaker point averages 1.12 m (σ = 0.39 m) and an extreme range of 0.46–2.74 m. Waves during high summer tend to be smaller than those of early and later summer. Spilling breakers occur on 63 d and mixed spill-plunge breakers on 43 d, leaving only 19 d for plunging breakers and 3 d for surging breakers. The predominance of spilling and

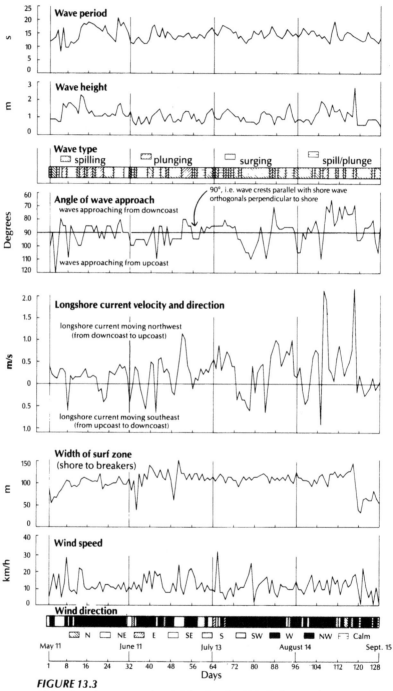

FIGURE 13.3
Littoral environmental variables for the 128-d observation period.

mixed breakers is not conducive to scour at the beach face. Waves are observed approaching perpendicular to the shore on 12 d but only once, between days 29 and 31, does this persist for more than one day, although on several additional occasions wave angle passes through 90° en route to or from an oblique upcoast or downcoast approach. Commonly, waves approach from a more southerly or downcoast direction on 50 d and from a more westerly or upcoast direction on 66 d. Persistence of wave approach is noteworthy, especially during later summer when, for example, waves approach from downcoast between days 61 and 72, 86 and 85, and 106 and 118. The above wave properties are typical of summer conditions at the site and correlations between individual parameters are not strong. It may be noted, however, that relatively small waves of spilling type and shortish period often approach from upcoast, whereas higher waves are associated with more southerly swells approaching from downcoast, reflecting late summer storm activity off western Mexico (Fig. 13.3).

During the 128-d observation period, the longshore current flows upcoast towards the northwest for 93 d, downcoast towards the southeast for 33 d, and is not measurable on 2 d during current reversals. This predominance of a current flowing upcoast is typical of summer conditions at the site but untypical of winter conditions or the year as a whole, when the current normally flows downcoast. Furthermore, it is noted that during the summer months the angle of wave approach does not have a precise effect on longshore current direction; specifically, the current flows upcoast for 93 d, but favorable waves approach from downcoast on only 50 d. This is because such currents are normally set up by forces beyond the immediate study area, and are often sufficiently strong to persist against relatively weak waves approaching slightly obliquely from upcoast. Thus, although waves approach from upcoast over 66 d, the current flows downcoast for only 33 d, and then rather weakly. Towards the end of summer, from day 106 to 118, powerful currents flow upcoast at velocities > 2 m/s in sympathy with strongly oblique waves of above average height approaching from the south, associated with the aforementioned storminess off western Mexico. This episode terminates abruptly after day 118 with wave height decreasing from 2.74 to 0.61 m, wave angle changing from 68° to 95°, and the longshore current reversing from flowing upcoast at 2.19 m/s to flowing weakly downcoast at 0.18 m/s, all more or less overnight. Similarly, surf-zone width diminishes from 146 m to 33.5 m over 4 d. This seems to mark the end of summer and the onset of change towards winter conditions. Before day 118, the surf zone is generally over 100 m wide, averaging 105 m ($\sigma = 20$ m) for the entire period.

Local winds have little impact on surf-zone activity. Winds flow onshore from the northwest for 44 d, from the west for 46 d, from the southwest for 22 d, and from the south for 9 d. Offshore winds occur on only 4 d and calms on only 3 d. Wind velocity averages 12.09 km/hr (3.36 m/s), a gentle breeze, and only once freshens to over 30 km/hr (>8.3 m/s).

RESPONSE VARIABILITY – SHOREZONE GEOMETRY AND TEXTURE

A basic tenet of the process–response model is that shorezone geometry and texture respond to changes in the magnitude and direction of energy variables, notably waves and related currents, to which they are exposed. By feedback, however, shorezone properties may in turn affect the operation of these external variables, leading to conditions wherein geometry, texture and energy become interdependent, to a greater or lesser extent. The nature of this interdependence is exceedingly complex and by no means well understood. Furthermore, in a littoral environment where a groin partially interrupts the alongshore components of the energy and sediment flux, interrelationships between geometry, texture and energy may vary substantially from one side of the structure to the other.

To explore these interrelationships further, the 128-d time series of elevation data and textural parameters are examined by several statistical procedures, specifically correlation, factor, trend, and spectral analyses. Initially, the mean elevation of each station over 128 d is correlated with similar data from every other station on the same side of the groin, yielding one correlation matrix for the upcoast side, another for the downcoast side. Predictably, most stations correlate strongly with neighboring stations, and correlations weaken progressively with distance. Negative correlations between small groups of stations indicate that net scour occurs around one locality while net fill occurs elsewhere. In particular, the correlation matrices suggest certain behavioral zones of stations in terms of erosion and deposition. To distinguish these zones of similar geometric behavior further, a factor analysis was performed, first on the upcoast data, second for the downcoast data, and finally for the combined upcoast and downcoast data. A principal axis solution was computed, and squared multiple correlations were used to estimate communality values. The program iterated on the communality values until maximum change in communality was less than 0.001. The factor matrix was rotated orthogonally using the varimax method. Of the eight factors extracted from the combined upcoast and downcoast data, none was combined, thereby suggesting that process–response relationships across the groin were independent of one another. For example, there is no reason to suppose that fill on one side of the groin will be accompanied by scour on the other side, thus casting doubt on one of the basic assumptions in the design and use of groins. That apart, the four factors each extracted from the separate upcoast and downcoast data, explaining 76% and 86% of the common variance respectively, yield a useful zonal framework for further analysis.

In essence, the correlation and factor analyses reveal five zones of similar geometric response – an offshore zone; a transitional zone immediately seaward of the groin; a lower foreshore zone inside the groin's seaward end; a middle foreshore zone; and an upper foreshore zone leading to the berm. These zones are shown schematically in Figure 13.2, but it should be noted that their boundaries are imprecise and rather mobile. Data concerning elevation, mean grain size, and sorting characteristics over 128 d are presented graphically for each zone in Figures 13.4, 13.5, 13.6, 13.7 and 13.8. Upcoast stations, northwest of the groin, are shown on the left side of each diagram, and the downcoast stations on the right side. Elevations are related to a zero datum of mean lower low water.

Visual inspection of these plots reveals several interesting points common to all stations. First, over short periods of time, elevation, grain size and sorting each deviate significantly from the mean for the observation period as a whole, and more so for stations seaward of the groin than for those to landward. Second, despite the short-term deviations, medium-term and long-term trends are observable at most stations, although these trends are usually weak over the total observation period. Third, with the longshore current flowing predominantly upcoast over the observation period, earlier studies of groin impacts suggest that net fill should occur against the downcoast side of the groin, while net scour occurs on the upcoast side. In reality, the observed response of the shorezone to these energy conditions is less consistent and more complex than anticipated. Let us now examine the behavior of each zone.

The offshore zone

As a minimum, the offshore zone comprises stations 1000–1075 and 2000–2075, but often extends into the adjacent transition zone. This zone experiences scour and fill of greater magnitude than other zones closer to shore, mainly because higher energy conditions promote the growth and migration or collapse of successive offshore bars and troughs. The upcoast stations, in particular, reveal frequent scour and fill as high waves and swift currents move upcoast. With smaller waves and currents, modest offshore bars form and, when the current flows downcoast, widespread fill may occur. The nearby downcoast stations exhibit a similar but more muted response. In terms of trends over the entire observation period, upcoast stations reveal a negative linear trend towards net scour that

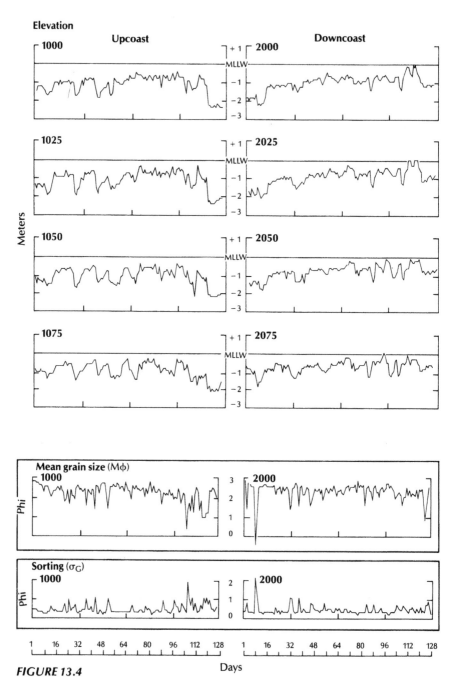

FIGURE 13.4

Offshore zone – 128-d time series of elevation, Mφ and σ$_G$. (MLLW = mean lower low water.)

increases seawards. Within this trend, the bars and troughs often reveal asymmetric profiles, reflecting rapid initial scour followed by slower more erratic fill, a feature that we will return to later. In contrast, downcoast stations show a weak positive linear trend towards net fill over the period, a trend that again increases seawards. The trends at both upcoast and downcoast stations, however, are somewhat colored by high-energy events during late summer, when significant scour upcoast and apprecia-ble fill downcoast tend to accentuate differences between the two sets of stations.

For the entire period, $M\phi$ at station 1000 is 2.32, while at station 2000 it is 2.39, or slightly finer. σ_G averages 0.50ϕ at station 1000, and 0.46 ϕ at station 2000. Like the elevation data, the textural data reveal some large short-term fluctuations, but such long-term trends as can be discerned are very weak. To explore these trends further, the textural data were subjected to tests for goodness of fit for linear, quadratic and cubic trends. Station 1000 shows an increase in $M\phi$ over the observation period that is weakly expressed in the linear fit and explained only slightly better in the quadratic and cubic trends. Sorting at station 1000 shows a significant decrease over the period, showing that as $M\phi$ increases so sorting wors-ens, a common occurrence. On the downcoast side, trends in $M\phi$ at station 1000 are so weak as to need higher degree polynomials for expres-sions, while sorting shows no trend.

The transition zone

This zone represents a transition between the high-energy and bar-and-trough topography of the offshore zone to seaward, and the lesser energy and ridge-and-runnel topography of the lower foreshore. At times it is clearly transitional, at other times it assumes the properties of one of its neighbors. This zone, arbitrarily defined by stations 1100–1175 and 2100–2175, witnesses the creation or passage of scour troughs and inter-vening bars enroute towards the shore. These features are particularly evident at the upcoast stations, less so downcoast. Responses at the outer edge of the zone are similar in magnitude and persistence to those off-shore, whereas responses at stations 1175 and 2175 close to the groin end are very muted.

For the entire period, $M\phi$ at station 1100 is 2.18, while at station 2100 $M\phi$ is 2.32, or significantly finer. σ_G values average 0.50 ϕ at station 1100 and 0.44 ϕ at station 2100. No trends are measurable in the textural data from station 1100, but station 2100 yields a very weak decrease in $M\phi$ and a weak increase in σ_G over time.

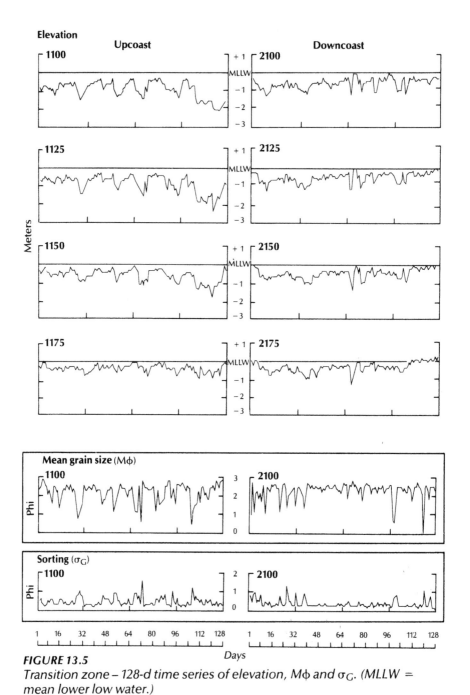

FIGURE 13.5
Transition zone – 128-d time series of elevation, Mφ and σ_G. (MLLW = mean lower low water.)

The lower foreshore

Probably because of the groin's protection, the lower foreshore is the most strongly identifiable zone on both profiles and is well emphasized by factor analysis. It embraces stations 1200–1275 and 2200–2275, but commonly extends seawards to the groin end and landwards well up the middle foreshore. It is a zone of muted but fairly continuous responses to energy pulses arriving from deeper water but individual scour or fill epi-

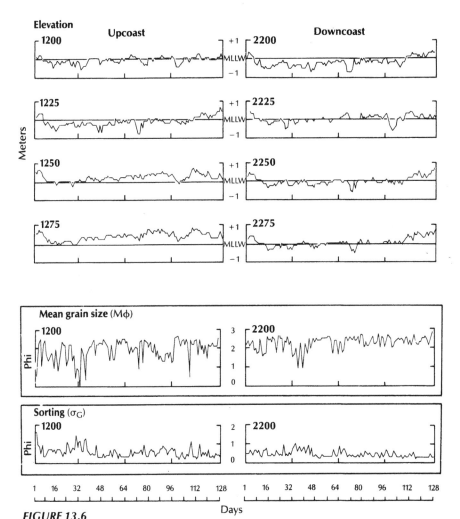

FIGURE 13.6
Lower foreshore – 128-d time series of elevation, Mφ and σ_G. (MLLW = mean lower low water.)

sodes rarely >0.5 m. Upcoast stations 1200–1325 often witness small scour troughs migrating landward under high energy conditions, with modest fill at other times, both regardless of longshore current direction. Downcoast stations 2175–2350 are also typified by alternating scour and fill that diminish in magnitude landward. A modest trough ≤1 m deep may sometimes form during moderate wave action, but more commonly this is the zone of ridges and runnels found on so many lower foreshores. Most stations show no significant trend throughout the period. Scour occurs in the first 8 to 16 d at several stations, and modest fill characterizes the last 3 weeks of summer as sediments entrained offshore are moved more vigorously towards the beach. During high summer, little change occurs.

For the entire period, $M\phi$ is 1.94 at station 1200 and 2.30 at station 2200. σ_G averages 0.60 ϕ at station 1200 and 0.46 ϕ at station 2200. Station 1200 shows a weakly significant decrease in $M\phi$ expressed in a linear fit, while higher-degree polynomials yield an even weaker decrease for station 2200. Stations 1200 and 2200 both exhibit weak linear increases in sorting over the period, again reflecting that as sediments become finer so sorting improves.

The middle foreshore

This poorly distinguished zone, embracing stations 1300–1375 and 2300–2375, occupies a transition between the ridge-and-runnel topography of the lower foreshore and the cusp terrain of the upper foreshore. It lacks much intrinsic character. Modified bars and troughs, ridges and runnels, pass through the zone towards the beach face and impart modest variability (Fig. 13.7). Troughs ≤ 1 m deep occasionally develop but runnels of 0.1–0.5 m are more common. These small perturbations apart, stationarity persists for long periods and no trends are observable. For the entire period, $M\phi$ is 1.97 at station 1300 and 2.27 at station 2300. σ_G averages 0.47 ϕ at station 1300 and 0.45 ϕ at station 2300. Station 1300 shows a weak linear increase in $M\phi$ over the study period but no other textural trends were observable from these data.

The upper foreshore

This innermost zone, landwards of stations 1400 and 2400, is typified by steep slopes and by quite different responses between the upcoast and downcoast stations. The upcoast stations 1400–1475 are characterized by net fill during the period, interrupted by prolonged stationarity when the berm lies seaward of one or more stations in the zone. The stations correlate negatively with stations 1100–1175, suggesting that fill on the

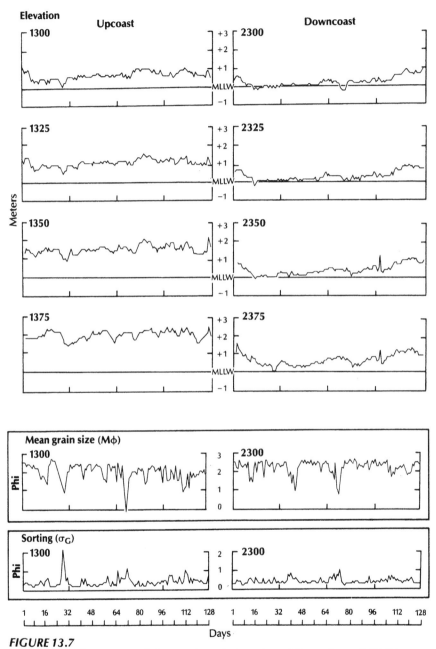

FIGURE 13.7
Middle foreshore – 128-d time series of elevation, $M\phi$ and σ_G. (MLLW = mean lower low water.)

upper foreshore corresponds with scour just beyond the groin. The down-coast stations, lying seawards of the berm for most of the period, and thus exposed to waves and currents moving upcoast, show much more variety,

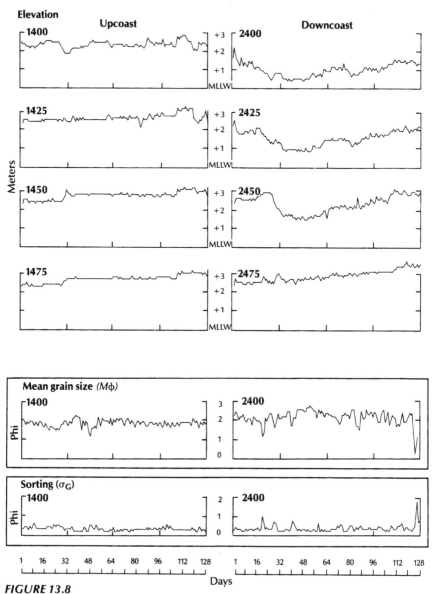

FIGURE 13.8
Upper foreshore – 128-d time series of elevation, Mϕ and σ_G. (MLLW = mean lower low water.)

including substantial scour in early summer, followed by a significant depositional trend that outlasts the study period. Cusps also developed well on the beach face in this zone.

For the entire period, $M\phi$ 1.90 is at station 1400 and 2.20 at station 2400. σ_G averages 0.36 ϕ at station 1400 and 0.40 ϕ at station 2400. Station 2400 shows a weakly significant increase in $M\phi$ and a responsive decrease in sorting over the period, but no trends are observed for station 1400. The sediment pattern thus shows a landward coarsening of $M\phi$ and generally much coarser sediments on the upcoast side of the groin, where winter lag deposits survived, than on the downcoast side where there is a paucity of coarse source materials.

Spectral analysis

The time series represented by the elevation and textural data are also amenable to spectral and cross-spectral analysis. Spectral analysis decomposes the time series into a set of frequency components, allowing those components that contribute significantly to variability in the data to be identified and interpreted. A least-squares fitting method was used to detrend the observed series. Autocovariances were computed for 64 lags in order to establish a frequency interval of 2 d. Raw estimates of the spectrum were obtained by a Fourier transform of the autocovariances. The raw data were then smoothed by hamming and plotted.

The power spectra for elevation show similar characteristics for all stations, namely high power at very low frequencies, moderate power at frequencies between 0.04 and 0.06 d^{-1} (cycles/d) and no significant peaks at higher frequencies. Station 1000 illustrates these relationships found throughout the upcoast profile (Fig. 13.9). In addition to high power at very low frequencies, the power spectrum for downcoast station 2000 also reveals a secondary peak at 0.15 d^{-1}. Moving landwards, this secondary peak shifts towards the higher frequencies until, from station 2350 to the berm, there is only one solitary peak at very low frequencies.

The power spectra for $M\phi$ and σ_G also exhibit high power at very low frequencies, but there are several additional peaks in the middle to high frequency parts of most spectra. $M\phi$ at station 1000, for example, exhibits additional peaks at frequencies of 0.070, 0.234, and 0.344 d^{-1} (Fig. 13.10). The sorting spectrum for the same station shows several peaks at higher frequencies, the more important of which occur at frequencies of 0.078, 0.234, 0.344, and 0.406 d^{-1} (Fig. 13.11). At station 2000, the power spectrum for $M\phi$ shows sharp narrow peaks at frequencies of 0.047, 0.133, 0.164, and 0.211 d^{-1}. The sorting spectrum at station 2000 shows additional peaks at frequencies of 0.078, 0.125, 0.164, and 0.195 d^{-1}.

The peaks consistently found at very low frequencies result from cycles with periodicities greater than the observation period and, as such, are not readily interpreted. Tests on the shorter periodicities reveal no statistical significance for the cycles involved. However, because these periodicities are often consistent between stations, notably with respect to periodicities of 6–8 and 12–14 d, it is possible that some periodic or rhythmic changes do occur, especially in texture. This suggestion cannot be explored further in this context because the observation period, though 128 d long, is too short for additional periodicities to be verified.

Cross-spectral techniques were also applied to $M\phi$ and σ_G data in order to identify the degree of association and amount of time separation, if any, between the two series. The degree of association and the amount of lead or lag time of one series over the other is estimated by the coherence and the phase angle respectively. In this instance, the technique yields high values for coherence and constant values for phase. Coherence values are generally greater than 0.6, indicating a strong relationship between the power spectra of $M\phi$ and σ_G. The phase values lie generally between 170° and 190°, indicating that $M\phi$ and σ_G are almost exactly out of phase. Thus, as $M\phi$ increases, σ_G worsens, a common attribute of sedimentary assemblages.

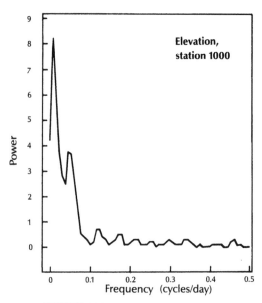

FIGURE 13.9
Spectral analysis – relation between power and frequency for elevations at station 1000.

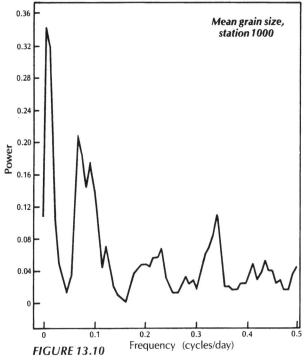

FIGURE 13.10
Spectral analysis – relation between
power and frequency for Mφ at station 1000.

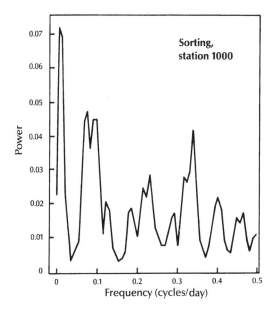

FIGURE 13.11
Spectral analysis –
relation between
power and frequency
for σ_G at station 1000.

Changes in time and space

In order to explore temporal changes further, daily station elevations along the two profiles are plotted against time and then linked by contours at 0.5 m intervals (Fig. 13.12). On these plots, fill is shown by contours that trend upwards to the right, scour by contours that trend downwards to the right. Scour troughs are shown by inverted contour crenulations pointing towards the shore, bars by crenulations pointing seawards. Comparing these plots with the process variables (Fig. 13.3) shows that short-lived energy changes generate little if any response, whereas more persistent conditions may produce significant mobility within the shorezone.

On the upcoast side, when high waves and swift currents move northwest, substantial scour occurs seawards of the groin but less scour typifies the lower foreshore while, landward of station 1300, there is a delay of several days before scour develops. On the plots, therefore, the off-shore–onshore axis of the scour phase trends oblique to the shore, tailing off to the right as it nears the berm. Such conditions occur for days 5–13, 26–32, 42–46, 66–72, 105–112 and 115–118. With smaller waves and modest currents moving northwest, a bar often develops offshore, while fill also occurs in the transition zone and lower foreshore, for example during days 59–65. The middle and upper foreshore show little change under these conditions. When waves approach the shore from the west and the longshore current reverses direction to flow southeast, fill occurs throughout the upcoast profile, for example on days 20–23, 32–38, 46–50, 55–57, 75–78, and 118–126. A small trough sometimes develops on the lower foreshore when the swifter currents entrain sediments landward from the groin end.

On the downcoast side, changes in shorezone geometry are less dramatic and the trend towards net fill is shown by contours moving upwards to the right. With high waves and swift currents moving northwest, there is less change in the downcoast profile than in the upcoast profile, as shown by only modest scour for days 26–32, 42–46, 51–54, and 115–118. Under these conditions, waves expend most of their energy on the offshore bar, well seen from day 115–117, whereas water on the upcoast side is sufficiently deep for waves to penetrate closer to shore. More dramatic changes occur downcoast on days 5–13 and 66–72 when significant scour shows a lag in the landward deepening of the profile similar to that found upcoast. On days 105–112, scour occurs only on the first 2 d, followed by fill. With short-period waves and longshore currents moving southeast, scour occurs offshore and may extend onto the lower foreshore, as on days 32–38, 73–78, 82–84 and 118–126. On days 20–23, 46–50, and 55–57, longer-period waves were not associated with scour. With small waves and slow currents from any direction, the downcoast profile generally fills along its entire length, with accumulation across the foreshore and a bar forming offshore.

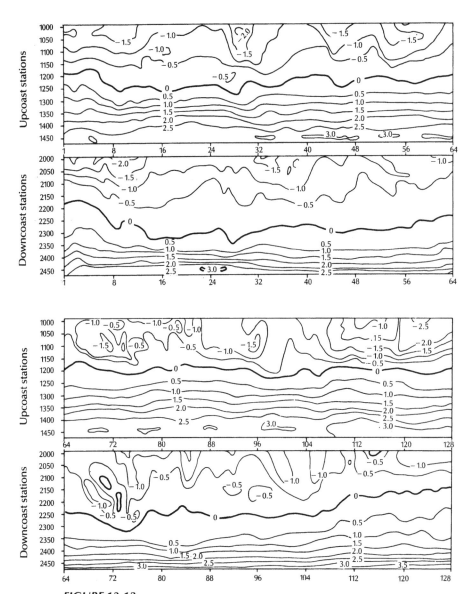

FIGURE 13.12
Relation between profile elevations and time over 128 d. Contour interval is 0.5 m above and below mean lower low water.

Bar and trough behavior

Closer inspection of the time series for the outer stations reveals more about the behavior of bars and troughs. In the offshore and transition zones lying 90–150 m seawards of the berm, both stationary and migratory bedforms may develop at one time or another. During earlier summer, in particular, the time-series profiles of elevation are asymmetric, most notably at upcoast stations 1000 through 1150 (Figs. 13.4 & 13.5). Evidently, bars accumulate gradually to heights of 1.0–1.5 m above their base over a period of 7–10 d and then either persist at their new elevations for 5–8 d or begin eroding immediately. In either case, when bar destruction begins it is generally accomplished within 1 or 2 d, much more rapidly than the construction phase. In some instances, bar destruction is correlated with changing energy conditions, scour being particularly favored by a change to higher, longer-period waves and swifter currents. In other cases, destruction occurs without significant environmental change, suggesting that some bars reach an intrinsic threshold which renders its sedimentary form unstable under existing energy conditions. Bar magnitude and asymmetry are more pronounced in the deeper offshore zone than in the shallower waters of the transition zone (Fig. 13.13). Conversely, in these profiles, the troughs scour rapidly and fill slowly, and are normally characterized by coarser, less-sorted sediments than are found in the bars.

At other times, notably during later summer, the time-series profiles of elevation are more symmetrical, the bars and troughs smaller, and

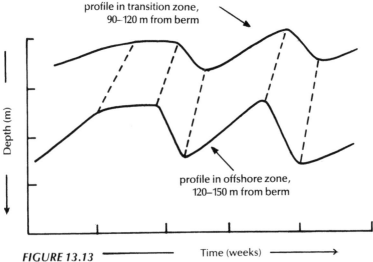

profile in transition zone, 90–120 m from berm

Depth (m)

profile in offshore zone, 120–150 m from berm

FIGURE 13.13 —————— Time (weeks) ———→

Schematic plot of bar and trough profiles against time for the offshore and transition zones.

migratory forms develop which eventually weld onto the beach face. The onshore migration of the bar takes anything from 7 to 15 d, but commonly averages 10 to 11 d (exemplified in Figs. 13.14 and 13.15, which cover days 65–75 of the study period). During this period, wave period increased from 14 to 18 s, and then declined to 13.5 s; wave height rose from 0.9 m to 1.6 m and back; waves approached from SSW; the longshore current flowed northwest at velocities of 0.4–0.9 m/s until day 72, when it reversed to flow southeast in sympathy with waves approaching from the west; the surf zone averaged 110 m wide, and during the earlier part of the period winds from the south accentuated the current flowing upcoast. At the beginning of the period (July 15 or day 1 on Fig. 13.14), the downcoast profile presents a sloping plane but, within 2 d, an offshore bar and trough develop which, over the next several days, migrate towards the berm. By day 10, these forms are largely welded onto the beach face and a deep-water plain lies to seaward. The upcoast profile also begins with a sloping plain which by day 4 has been transformed to a modest bar and deep broad trough. These then migrate shorewards, the bar growing in size while the trough first deepens further and then fills as it nears the berm, until on day 11 both features are close to the beach face. The upcoast

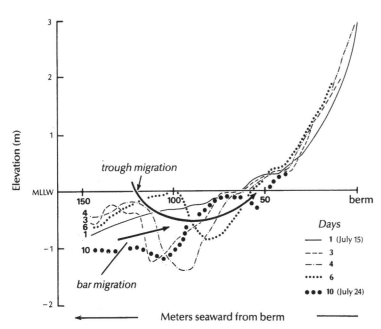

FIGURE 13.14
Bar and trough behavior along the down coast profile over 10 d.
(MLLW = mean lower low water.)

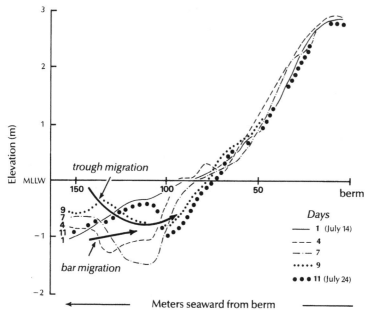

FIGURE 13.15
Bar and trough behavior along the upcoast profile over 11 d.
(MLLW = mean lower low water.)

profile lags slightly behind the downcoast profile because the energy regime was first directed against the downcoast side.

Whether bars grow and collapse *in situ* or whether they migrate towards the shore, the period of 7–15 d involved in both instances invokes comparison with the non-significant periodicities noted in the spectral analysis.

CONCLUSION

By any criterion, the southern California coast is unstable. Over the longer term near the study site, late Cenozoic compression of the Transverse Ranges has caused tectonic uplift of the Ventura Hills and Santa Monica Mountains, and subsidence of the Oxnard Plain, at rates of ≥0.01 m/yr. Continuation of these movements into the Holocene beyond the assumed close of the Flandrian transgression confounds the explanation of shoreline change. Over the shorter term, further instability is imposed on the shorezone by the erratic influx of sediments from nearby watersheds. Nevertheless, it is usual to distinguish between a winter shorezone of net scour and a summer shorezone of net fill, these seasons being separated by transitional conditions, often as extrinsic thresholds, in spring and autumn. During the relatively dry summers, the coast is rarely replenished with river sediments, and waves and currents must necessarily fashion the shorezone from whatever is available, namely the sediments remaining from the last winter influx. Thus, changes in the summer shorezone commonly occur against a background of a relatively stable or gently diminishing sediment budget.

In the foregoing analysis, we have seen how changes take many forms. Some are instantaneous perturbations, often random in time and space, that increase in magnitude and frequency seawards. Ignoring these perturbations, the data obtained over 128 consecutive summer days sometimes reveal no trend whatsoever, or weak linear trends that are only slightly better expressed by quadratic and cubic fits. Summarizing the elevation trends for the period, upcoast stations seaward of the groin experience net scour that increases seawards, notably when high waves and swift currents move northwest. Upcoast stations on the lower and middle foreshore show little trend, while the upper foreshore shows fairly persistent fill. Thus, the pattern of persistent scour on the sheltered side of the groin, which might have been predicted from other studies because the longshore current moved northwest for 93 out of 128 d, is not revealed. Downcoast stations seaward of the groin experience net fill over the study period, lower foreshore stations reveal little trend, while middle and upper foreshore stations witness initial scour followed by prolonged fill. Similarly, the textural data show either no trends or trends that are often so weak as to be revealed only by quadratic and cubic fits. However, where trends do occur, the reciprocal relationship between mean grain size and sorting is usually revealed. The search for cyclicity in the data was not very rewarding, but it is possible that some periodicities of 6–8 and 12–14 d may occur, perhaps related to bar and trough behavior. In this latter context, the recognition of both stationary and migratory bars offers some interesting prospects for further elucidation of the threshold hypothesis.

Finally, to what extent do the changes observed in the foregoing analysis fit one or more of the many schemes that have been invoked to explain geomorphic change? Geomorphologists often assume that, over specific periods of time, landforms tend towards some type of equilibrium with the energy and materials of a particular system. Many such assumptions have characterized the search for explanation in fluvial geomorphology. Drawing on earlier work, for example, Schumm (1977) illustrates five types of equilibria with respect to changes in valley-floor elevation with time—namely static equilibrium, steady-state equilibrium, dynamic equilibrium, metastable equilibrium, and dynamic metastable equilibrium. How appropriate are these models to the foregoing observations? Clearly static equilibrium, where no change occurs with time, is quite unsuited to the frequently changing shorezone. The other four models, however, each finds some expression in the observed data. Examination of Figures 13.4 to 13.8 shows that many stations exhibit steady-state equilibrium over the observation period, wherein fluctuations in elevation, $M\phi$ and σ_G occur about a constant no-trend condition. Dynamic equilibrium is expressed by those stations whose data show fluctuations about a changing condition whose trend may be linear or non-linear. Metastable equilibrium is demonstrated by elevation data from station 1475, which show prolonged stationarity for backshore elevation separated by episodic pulses of sediment which raise beach level across depositional thresholds. The reverse effect occurs in winter as episodic storms lower the backshore across erosional thresholds. Dynamic metastable equilibrium sees episodic change occur as thresholds are exceeded in a changing system already experiencing variations. This condition is probably exemplified by the episodic collapse or migration of offshore bars from stations already experiencing temporal variability around a changing state.

There is an assumption in fluvial geomorphology that the above types of equilibria either do not occur simultaneously or, if they do, they are nested within one another. Schumm (1977) proposes that channel gradient and landscape change reveal static equilibrium over a brief timespan, perhaps a day, that steady-state equilibrium occurs over a span of 10^2 to 10^3 yr, that dynamic metastable equilibrium spans perhaps 10^6 yr and embraces episodic erosion as thresholds are exceeded, and that progressive landscape reduction may take 10^7 yr or more. This model invites comparison with the timescales involved with the long-term modification of the coastal zone ranging from 10^1 to 10^6 yr discussed earlier in this paper.

In contrast, in the soft sedimentary environments afforded by sandy shorezones, different types of equilibria are neither mutually exclusive nor necessarily nested within one another. Change is rapid and variability is ubiquitous, but equilibrium of one type or another may often be found more or less simultaneously at different stations within the shorezone. Clearly, the contrast with Schumm's fluvial landscape is an expression of response times, but the simultaneity of different equilibria of short duration within the shorezone cautions against the tacit transfer of theoretical constructs from one category of environment to another.

The present study has sought to explain the temporal variability to be found in a summer shorezone along the southern California coast. The data characterize the short-term behavior of a zone around a long, low, impermeable groin and are thus important to coastal engineering practice and shorezone management. Over the medium term, the study is framed within the cyclic sequence of summer fill and winter scour experienced by southern California beaches. Over the longer term, say 10^2 to 10^3 yr, one scenario for the study site sees the maintenance of a steady-state equilibrium as intermittent inflows of fluvial sediment offset the subsidence imposed on the local coast by tectonic forces. However, with the sediment inflow now partially restricted by dam construction in the Ventura and Santa Clara watersheds, an alternative scenario sees the coast crossing a threshold into an erosional phase that will persist until a new equilibrium is reached. In this sense, the coast may be viewed as in dynamic metastable equilibrium, poised on the threshold of change that may already have been signalled by the significant erosion of beaches between Port Hueneme and Ventura over the past 30 yr.

ACKNOWLEDGEMENTS

This study was funded, in part, by the Coastal Engineering Research Center, U.S. Army Corps of Engineers, Contract DACW72-73-C-007, A. R. Orme, Principal Investigator. The author is also grateful for the assistance of the late Professor W. C. Krumbein and of Dr. W. R. James and D. Christinaz.

REFERENCES

Atwater, B. F., C. W. Hedel and E. J. Helley 1977. *Late Quaternary depositional history, Holocene sea-level changes, and vertical crustal movement, southern San Francisco Bay, California*. U.S. Geol. Survey, Prof. Paper 1014.
Gibbs, R. J., M. D. Matthews and D. A. Link 1971. The relationship between sphere size and settling velocity. *J. Sed. Petrol.* **41**, 7–18.
Orme, A. R. 1978. Impact of a low impermeable groin on shorezone geometry. *Geosci. and Man* **18**, 81–95.
Orme, A. R. 1980a. Energy–sediment interaction around a groin. *Z. Geomorph.* **34**, 111–28.
Orme, A. R. 1980b. Marine terraces and Quaternary tectonism, northwest Baja California, Mexico. *Phys. Geog.* **1**, 138–61.
Schumm, S. A. 1977. *The fluvial system*. New York: Wiley.
Wehmiller, J. F., K. R. Lajoie, K. A. Kvenvolden, E. Peterson, D. F. Belknap, G. L. Kennedy, W. O. Addicott, J. G. Vedder and R. W. Wright 1977. *Correlation and chronology of Pacific coast marine terrace deposits of continental United States by fossil amino acid stereochemistry – technique evaluation, relative ages, kinetic model ages, and geological implications*. U.S. Geol. Survey, Open-File Rept 77–680.
Yeats, R. S. 1978. Neogene acceleration of subsidence rates in southern California. *Geology* **6**, 456–60.

14

Using the normal generated distribution to analyze spatial and temporal variability in geomorphic processes

H. Charles Romesburg and Jerome V. DeGraff

INTRODUCTION

Increasing recognition is being given to the significance of thresholds in geomorphic processes and landform evolution (Coates & Vitek 1980). While research frequently focuses on conditions across a threshold, examination of variability among thresholds may prove equally important. This seems for research on landform evolution. Many processes responsible for landscape development feature extrinsic and intrinsic thresholds as significant elements in their operation (Gardner 1980, Howard 1980, King 1980). Schumm (1977) states that, where thresholds are a significant factor, it is possible to modify the classic landform evolution model of progressive erosion to include the concept of dynamic metastable equilibrium. This conceptual modification allows the model to accommodate diverse aspects of landscape development by emphasizing episodic erosion in response to exceeding threshold conditions (Schumm 1976, 1977). One of the problems in using such models is discovering the temporal and spatial variability of the processes (Thornes & Brunsden 1977). Quantitative analytical techniques are frequently required to provide estimates of these variabilities.

ANALYZING PROPORTIONAL GEOMORPHIC DATA

Data relating to thresholds can frequently take the form of a ratio (Bull 1980) or a proportion, p. Looking at proportional data and writing $p = r/k$, there are two interpretations possible for the variables r and k. First, r and k may be counts in a sample, i.e. r is the number of times the object or event of interest occurs in the sample and k is the total number of all objects or events. An example is the proportion composed of counts of a specific landslide type among all landslides sampled. In turn, two different conceptual views are commonly taken for the underlying probability model for count-based proportions. If the assumption that the 'true' p is spatially invariant within the sampled body is valid, then r can be considered a random variable, binominally distributed over k trials in any given sampling unit. Alternatively, if spatial heterogeneity is assumed to exist, then p can be considered as the parameter of the binominal distribution for a given sample, with p in turn assumed to have a beta distribution over the whole population. The compound distribution formed in this latter case was first derived by Skellam (1948) and is termed the beta-binominal distribution (BBD). It is also referred to in the literature as the negative hypergeometric distribution, Polya distribution, and compound binominal distribution. Chatfield and Goodhardt (1970) discuss the fitting of the BBD to data. Maximum likelihood estimates of the two BBD parameters are shown to be computationally difficult to find, and they discuss two simpler methods—one the method of moments and the other the 'method of mean and zeros'.

A second interpretation is to treat r and k not as counts but as measurements on an interval or ratio scale, e.g. r might be the weight of an item, with k corresponding to the total weight of the sample. The proportion of suspended sediment within a total sediment load, proportion of mapped areas covered by alluvium, and proportion of sand-size particles in a sediment sample are examples. In practice, the measured values of p are found to vary among samples, suggesting a spatially dependent model. For the distribution of p, it seems natural to prescribe the beta distribution because its domain is $0 < p < 1$, and because it is flexible enough to allow a 'satisfying' variety of fitted shapes. However, there are attendant disadvantages. The two beta distribution parameters must be estimated from sample moments—maximum likelihood estimates are biased in non-consistent fashion (Gnanadesikan *et al.* 1967) and their sampling distributions are not readily tractable—and probability statements about the distribution of p require tables of the beta density function or computer integration. Moreover, in performing hypothesis tests concerning p, the beta distribution is usually abandoned. Davis (1973), for example, tests the proportion of porosity in Tensleep Sandstone by assuming the observed n-sample $p_1, p_2 \ldots p_n$ as realizations from a normal population, i.e. by invoking the central limit theorem. Some of the difficulties involved with using normal distribution theory to analyze proportion data are the somewhat arbitrary decisions of when the analysis should be carried out on transformed data, e.g. using an arcsin transform of the observed p_1, and dealing with cases where the normal sampling distribution 'spills over' the domain $0 < p < 1$.

The normal generated distribution (ngd) as the probability density functions of proportion p is proposed as an alternative to the beta distribution. Its use is intended whenever p is measured on interval or ratio scales, or for the case when p is composed of counts and k is 'large' and approximately invariant among samples, so that p can be regarded as a continuous metric. The ngd surmounts several difficulties surrounding the beta distribution. First, the two ngd parameters, ξ and τ^2, can be estimated by maximum likelihood, are unbiased, their distributions derived, and hypothesis tests involving p translated to tests of ξ and τ^2. Second, the need to transform the observed p_i is eliminated.

THE NORMAL GENERATED DISTRIBUTION

The ngd, invented by Chiu (1974) and later used by Romesburg (1976) for estimating biological population survival, has its probability density function $q(p)$ written as follows:

$$q(p) = \phi\left[1/\tau[\Phi^{-1}(p)-\xi]\right]/\tau\phi[\Phi^{-1}(p)] \tag{14.1}$$

where ξ and τ^2 are its parameters and $\phi(\cdot)$ and $\Phi(\cdot)$ are the PDF and cumulative distribution function (CDF) of the standardized normal distribution. The ngd takes its name from the fact that it is obtained by a transform of the normal distribution. The characteristic curves of the ngd as a function of its parameters are shown in Figure 14.1. These curves

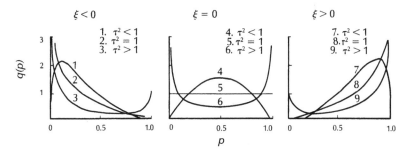

FIGURE 14.1
Typical curves for the normal generated distribution
PDF $q(p)$ as a function of the parameter space: $\xi <, =,$
> 0; $\tau^2 <, =, > 1$.

closely follow those of the beta distribution and are sufficiently general to fit most unimodal and many bimodal cases arising in practice.

The mean, $E(p)$, and variance, $Var(p)$, of $q(p)$ are given by

$$E(p) = \Phi(p) \tag{14.2}$$

and

$$Var(p) = Pr[y \leqslant c, z \leqslant; w] - E^{2(p)} \tag{14.3}$$

where the new parameters $c = \xi/(1+\tau^2)^{1/2}$ and $w = \tau^2/(1+\tau^2)$ are used to curb cluttered notation; y and z have a joint bivariate normal distribution with zero means, unit variances, and correlation w, and Pr $[\cdot]$ denotes their joint bivariate probability.

The CDF, $Q(p)$, is obtained by integrating Equation 14.1 and is:

$$Q(p) = \Phi[1/\tau[\Phi^{-1}(p) - \xi]] \tag{14.4}$$

The problem of estimating either $q(p)$ or $Q(p)$ comes down to the problem of estimating its parameters. This can be done in either of two ways. If a random sample of size n-sample proportions is available, then the maximum likelihood estimates (MLEs) are:

$$\hat{\xi} = \sum_{i=1}^{n} \Phi^{-1}(p_i)/n \tag{14.5}$$

$$\hat{\tau}^2 = \sum_{i=1}^{n} [\Phi^{-1}(p_i) - \hat{\xi}]^2/(n-1) \tag{14.6}$$

That is, $\hat{\xi}$ and $\hat{\tau}^2$ turn out to be the mean and sample variance of the transformed proportions $\Phi^{-1}(p_i)$. Chiu (1974) described how estimates of ξ and τ^2 can be found using the method of moments, but for most applications the MLE estimates seem preferable.

The second method of estimating ξ and τ^2 is suitable for subjective estimates of $q(p)$ and $Q(p)$. All that is required are two estimates of $Q(p)$ at any two different values of p. Denoting their particular values of p as p_1 and p_2 and the corresponding values of $Q(p)$ as $Q(p_1)$ and $Q(p_2)$, then the procedure goes as follows. Suppose a researcher wants to estimate the CDF of the proportion of area covered by landslide deposits but he lacks data. He turns then to his experience and subjectively estimates, say, that the probability that 10% or less area is in landslides is 30%, whereas the probability that 25% or less area is in landslides is 50% or less–that is, he has obtained $p_1 = 0.1$ and $Q(p_1) = 0.30$, and $p_2 = 0.25$ and $Q(p_2) = 0.50$. Next, Equation 14.4 is written in its inverse form and algebraically rearranged to:

$$\xi + \tau\Phi^{-1}(Q(p)) = \Phi^{-1}(p) \tag{14.7}$$

This equation is linear in ξ and τ. Using the above data to obtain two simultaneous equations in the form of 14.7, we can easily solve for $\xi = -0.67$, $\tau = 1.17$, and $\tau^2 = 1.38$. In turn, these values of ξ and τ^2, when placed into Equation 14.4, allow estimates of $Q(p)$ for any value of p. For example, the question, 'What is the probability that 15% or less area is in landslides?' is:

$$Q(p=0.15) = \Phi[1/1.17[\Phi^{-1}(p=0.15) - (-0.67)]] = 0.38 \qquad (14.8)$$

or 38%. The approach has merit whenever 'quick-and-dirty' estimates are in order, or when it is infeasible to obtain data.

Although the ngd can be fitted by hand calculation, the tedium and chance for error make a computer program desirable. Such a computer program is available (see Romesburg et al. (1980)); it will produce numerical values for Equations 14.1 through 14.6.

Landsliding is a process commonly recognized as a threshold-dependent phenomena (DeGraff & Romesburg 1980). It can play a major role in landform evolution under certain conditions. The following two examples examine the temporal and spatial variability of landsliding to illustrate the usefulness of ngd analysis (Fig. 14.2).

FIGURE 14.2
A map showing part of the western United States. The letters indicate the location of study areas referenced in the temporal and spatial examples. A, Russian River Valley near Cloverdale, California and B, Mt Terrill on Fishlake Plateau, central Utah.

TEMPORAL VARIABILITY

Time is a pervasive aspect of geomorphic research (Thornes & Bruns-den 1977). Clearly, the variability of threshold events over time is central to landform evolution. It is often difficult to obtain reliable data on the temporal distribution of past threshold events. In some instances, tree-ring data can provide the required record (Agard 1979, Shroder 1980).

The northern margin of the Fishlake Plateau in central Utah is greatly modified by widespread mass movement. Wetter, colder climatic condi-tions in the late Pleistocene enhanced mass movement activity, shaping the margin slopes of the plateau. Rock glaciers, solifluction lobes, rock streams and related features indicate a periglacial environment with wide-spread gelifluction and geliturbation. Mass movement activity may still play a significant role in landscape development on the high-elevation slopes of this area.

Mass movement activity was investigated on the northern slope of the Fishlake Plateau near Mt Terrell. This northeast-facing slope ranges from 3130 m to 3250 m ASL. The slope hosts sag ponds, lobate, bouldery masses and hummocky relief indicative of mass movement activity. Throughout this forested slope, ground cracks, scarps, exposed soil or other indicators of active movement are absent. However, many trees display recurve growth oriented downslope, which implies that active mass movement has occurred in recent time.

Twenty trees were sampled in a zone across the lower part of the slope. Sample density was 1 tree/0.35 ha over the area of approximately 6.5 ha (DeGraff 1980). An increment borer was used to extract cores at breast height and near the base of each tree. Microscope-aided measurement of annual growth rings identified the years in which eccentric growth was great and represented response to mass movement rather than common environmental fluctuations. Only events large enough to produce this effect in both breast-height and base-level cores for the same year are considered significant mass movements. Activity was detected as long ago as 1915 and as recently as 1979. A 34-year period, 1946–1979, was examined using the ngd technique. The proportional data are the number of trees among the 20 sampled showing mass movement disturbance in a particular year. The time period represents the years common to all 20 trees. Figure 14.3 shows the PDF and CDF curve generated from this proportional data.

The PDF $q(p)$ curve is like Case 1 in Figure 14.1. Its mean $E(p) = 0.12$, variance $\text{Var}(p) = 0.004$. From the fitted CDF $Q(p)$ we can say, for example, that the probability of $p \leq 0.1$ is $Q(p = 0.1) = 0.42$, or 42%; the probability of $p \leq 0.2$ is $Q(p = 0.2) = 0.89$, or 89%.

These values can be restated to provide insight on mass movement activity during the last 34 years. Each tree can be assumed to represent 0.35 ha, or about 5% of the study area. The fitted CDF $Q(p)$ yields an estimated percentage of 42 that active movement affected 10% or less of the area. It yields an estimated percentage of 89 that active movement affected 20% or less of the area. The degree to which active movement has contributed to landform evolution in this area can be estimated using the fitted CDF $Q(p)$ curve for the sampled data. Comparing curves developed from neighboring slopes could lead to deductions on how episodic erosion (mass movement) is contributing to local slope evolution.

SPATIAL VARIABILITY

Location can be considered a property of the landscape. Maps are measures and displays of this property (Lewis 1977). The spatial distribution of threshold events and related, or controlling, landscape characteristics may be important for understanding how a process is affecting evolving landforms. One question that can be asked concerns the degree of covariance of the threshold-dependent process with a controlling landscape characteristic. Varnes (1974) examined covariance among map units for engineering geology maps. Many of the analytical techniques used

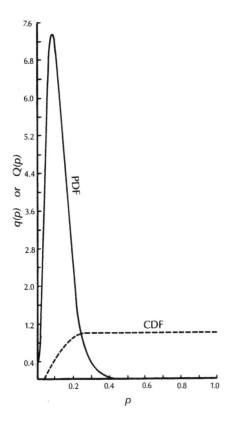

FIGURE 14.3
Graph showing PDF q(p) and CDF Q(p) curves generated from proportional tree-ring data (n = 34). The PDF curve is a solid line; the CDF curve is a dashed line. Maximum likelihood estimators are: ξ, τ^2.

were graphical or tabular methods. The normal generated distribution offers a statistical technique for examining variability among map units.

Radbruch-Hall (1976) investigated geologic factors affecting landsliding in the northern Coast Ranges. An area in the Russian River Valley near Cloverdale, California, was studied. Four maps were generated for this area. Landsliding was found to be controlled by bedrock lithology and physical state, i.e. sheared or crushed. Relatively soft rocks that are sheared are most prone to massive landsliding. One of the three bedrock units–the Franciscan melange–generally is composed of sheared, soft lithologies.

Map variability of the threshold-dependent process, landsliding, and its controlling landscape characteristic–the Franciscan melange–can be examined for expected covariance. Both the landslide, geologic, and land-slide/geologic maps have the General Land Office grid showing sections of one-mile square areas (Radbruch-Hall 1976). Within each of these mapped areas, on the landslide map, the proportion of the square mile covered by landslide deposits was determined. The proportion of the square mile underlain by Franciscan melange was determined from the geologic map. On the superimposed geologic/landslide map, the proportion of the square mile where landslide deposits and Franciscan melange coincided was determined. In all cases, the amount of landslide deposits, underlying Franciscan melange, or landslide deposits coincident with Franciscan melange measured by an electronic planimeter was divided by the total square mile area to produce the proportion. Square mile areas, on the map, which yielded values of either 0 or 1 were excluded from the ngd analyses. This was done to emphasize the process variability where landsliding, Franciscan melange, or covariance of landsliding deposits and Franciscan melange was neither absent nor totally dominant.

For comparative purposes, Figure 14.4 shows the ngd curves of $q(p)$ computed using the landslide (a), bedrock (b), and covariance (c) data. Table 14.1 shows the related parameter values for landslide, bedrock and covariance data. Referring to Figure 14.1, the landslide curve's shape corresponds to Case 1, the bedrock corresponds to Case 3, and the covariance curve to Case 1. The curves have two uses. First, as already illustrated, the fitted CDFs of the curves allow probability statements to be made about the variate (p). Second, hypotheses tests regarding the true parameters ξ and τ^2 are possible. To illustrate this, suppose that we want to see if the landslide curve, which is highly skewed, could have come from a process characterized by the uniform distributions PDF. The uniform distribution, Case 5 of Figure 14.1, is characterized by $\xi = 0$ and $\tau^2 = 1$. Therefore, using the fact that $\hat{\xi}$ and $\eta\hat{\tau}^2/(n\text{-}1)$ are normally and chi-square distributed (n-1 df) respectively when $\Phi^{-1}(p_i)$ is a normally distributed variate (Romesburg 1976), the following tests of hypotheses can be carried out: $H_0 : \xi = 0.0$ vs. $H_a : \xi \neq 0.0$ and $H_0: \tau^2 = 1$ vs. $H_a : \tau^2 \neq 1$.

Carrying this out, we reject H_0: $\xi = 0.0$ ($P < 0.00001$) while failing to reject H_0: $\tau^2 = 1$ ($P = 0.39$). Thus, we reject the hypothesis that the uniform distribution is the true underlying process that gives rise to the data. Note that the transform $\Phi^{-1}(p_i)$ has the characteristic shape of the arcsin transform which is required when making hypothesis tests about the parameters of a binominal population (for example, see Ostle 1963, p. 340). The $\Phi^{-1}(p_i)$ arise naturally out of the derivation of the ngd; they therefore achieve naturally what the arcsin transform is contrived to do, i.e. make the transform of the variate p_i approach a normal distribution so that parametric statistical tests can be used. We could also test whether a curve is symmetrical in much the same way as testing for the hypothesis of uniform distribution. In this case, the test would disregard τ^2 values and simply be carried out as H_{-0} : $\xi = 0.0$ vs. H_a : $\xi = 0.0$.

In Figure 14.4, it is clear from the shape of the curves that areal distribution of landslide deposits (a) and Franciscan melange (b) is quite different. The differing processes responsible for their respective areal distributions is the probable reason for these differences. However, the shape of the curves and their attendant parameter values are quite similar for landslide deposits (a) and covariance of landslide deposits and Franciscan melange (c). From our earlier test, we concluded that the landslide data did not arise from a uniform distribution. Instead, landslide deposits are distributed over the area in a way that indicates a relationship to the Franciscan melange. If no previous investigation had examined this inter-

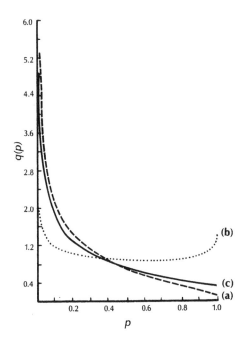

FIGURE 14.4
Graph showing the PDF q(p) curves for (a) landslide, (b) Franciscan melange and (c) covariance of landslide and Franciscan melange. Sample and ngd parameter values are in Table 14.1.

TABLE 14.1
Sample and ngd parameter values for landslide, Franciscan melange, and covariance of landslide and Franciscan melange proportional data.

Proportional data	n	$\bar{p}$	S^2	ξ	τ^2	$E(p)$	$Var(p)$
Landslide	29	0.269	0.078	−0.887	1.016	0.266	0.061
Franciscan melange	21	0.468	0.107	−0.074	1.285	0.481	0.095
Covariance	21	0.311	0.085	−0.749	1.197	0.307	0.075

relationship, the ngd curves would serve to develop hypotheses for future research. In this example, a detailed investigation showed that the lithology and physical state of this bedrock unit is a principal factor controlling landsliding in this area (Radbruch-Hall 1976).

DISCUSSION

In sampling, the observed proportions usually exhibit variability. It is natural, therefore, to look for a theoretical model that could reasonably give rise to the observed proportions. Because they can assume a wide variety of unimodal and bimodal shapes, both the beta distribution and ngd are candidates for theoretical models. However, the beta distribution, when estimated from the data, is biased, whereas the ngd is not. Moreover, estimating the ngd using subjective estimates is easier, as is making hypothesis tests about the true values of the parameters. Of the models available for fitting, the ngd has good overall features.

Threshold-dependent processes causing episodic erosion are important elements requiring study to refine our understanding of landform evolution. The ngd is clearly useful in examining variability between threshold events. Additionally, it can be the basis of an axiomatic model for examining change across a threshold. This approach has been successfully applied in biologic studies (Romesburg 1976). For landform evolution studies, the ngd can provide deductive insight to the process being investigated and generate hypotheses directing future research.

REFERENCES

Agard, S. S. 1979. *Investigation of recent mass movements near Telluride, Colorado, using the growth and form of trees.* Unpublished MS thesis. Boulder: University of Colorado.
Bull, W. B. 1980. Geomorphic thresholds as defined by ratios. In *Thresholds in*

geomorphology, D. R. Coates & J. D. Vitek (eds), 259–63. London: George Allen & Unwin.

Chatfield, C. and G. J. Goodhardt 1970. The beta-binominal model for consumer purchasing behavior. *Appl. Stat.* **19**, 240–51.

Chiu, W. K. 1974. A new prior distribution for attributes sampling. *Technometrics* **16**, 73–102.

Coates, D. R. and J. D. Vitek 1980. Perspectives on geomorphic thresholds. In *Thresholds in Geomorphology*, D. R. Coates & J. D. Vitek (eds), 3–23. London: George Allen & Unwin.

Davis, J. C. 1973. *Statistics and data analysis in geology*. New York: Wiley.

DeGraff, J. V. and H. C. Romesburg 1980. Regional landslide-susceptibility assessment for wildland management: a matrix approach. In *Thresholds in geomorphology*, D. R. Coates & J. D. Vitek (eds), 401–14. London: George Allen & Unwin.

DeGraff, J. V. 1980. *Geologic survey of landslide activity adjacent to the proposed Mt. Terrill park development*. Open-file Report. Geologic Services, Fishlake National Forest, USDA Forest Service.

Gardner, J. S. 1980. Frequency, magnitude, and spatial distribution of mountain rockfalls and rockslides in the Highwood Pass Area, Alberta, Canada. In *Thresholds in geomorphology*. D. R. Coates & J. D. Vitek (eds), 267–95. London: George Allen & Unwin.

Gnanadesikan, R., R. S. Pinkham and L. P. Hughes 1967. Maximum likelihood estimation of the parameters of the beta distribution from smallest order statistics. *Technometrics* **9**, 607–20.

Howard, A. D. 1980. Thresholds in river regimes. In *Thresholds in geomorphology*, D. R. Coates & J. D. Vitek (eds), 227–58. London: George Allen & Unwin.

King, C. A. M. 1980. Thresholds in glacial geomorphology. In *Thresholds in geomorphology*, D. R. Coates & J. D. Vitek (eds), 297–321. London: George Allen & Unwin.

Lewis, P. 1977. *Maps and statistics*. New York: Wiley.

Ostle, B. 1963. *Statistics in research*, 2nd edn. Ames: Iowa State University Press.

Radbruch-Hall, D. H. 1976. *Maps showing areal slope stability in parts of the northern Coast Ranges, California*. Misc. Invest. Map, U.S. Geol. Survey., I-982.

Romesburg, H. C. 1976. Use of the normal generated distribution for estimating population survival. *J. Theor. Biol.* **61**, 447–57.

Romesburg, H. C., K. Marshall and J. V. DeGraff 1980. NGD–a FORTRAN IV program for fitting the normal generated distribution. *Behavior Res. Methods and Instrument.* **12**, 385–6.

Schumm, S. A. 1976. Episodic erosion: a modification of the geomorphic cycle. In *Theories of landform development*, W. Melhorn & R. Flemal, (eds), 69–85. London: George Allen & Unwin.

Schumm, S. A. 1977. *The fluvial system*. New York: Wiley.

Shroder, J. F. Jr 1980. Dendrogeomorphology: review and new techniques of tree-ring dating. *Prog. Phys. Geog.* **4**, 161–88.

Skellam, J. G. 1948. A probability distribution derived from the binomial distribution by regarding the probability of success as variable between sets of trials. *J. R. Statist. Soc.* (Series B) **10**, 257–61.

Thornes, J. B. and D. Brunsden 1977. *Geomorphology and time*. New York: Wiley.

Varnes, D. J. 1974. *The logic of geologic maps, with reference to their interpretation and use for engineering purposes*. U.S. Geol. Survey Prof. Paper 837.

15

Problems in the identification of stability and structure from temporal data series

John Thornes

INTRODUCTION

Chorley and Kennedy (1971) provide a useful classification of systems equilibrium behavior in terms of the perceived response of the system through time by observation of a state variable such as channel width, discharge, vegetation or temperature. Static equilibrium is said to exist when the value of the state variable is unchanging through time (Fig. 15.1a). This is characteristic only of deterministic systems with completely fixed control on inputs and system behavior and may be expected to occur only rarely in natural systems. Steady state equilibrium is identified as having a constant mean value with respect to a given timescale, but fluctuations occur about this mean due to the presence of interacting variables such as the imperfect buffering systems of water chemistry or the competition between basins for space for growth (Fig. 15.1b). Dynamic equilibrium is defined as having 'balanced fluctuations about a constantly changing systems condition which has a trajectory of unrepeated "average" states through time.' It is apparently assumed here that the mean and variance of the fluctuations remain constant (Fig. 15.1c). We might add to these, three further classes which are of particular interest to this paper – pure oscillating systems (Fig. 15.1d), steady oscillating systems (Fig. 15.1e), and dynamically oscillating systems (Fig. 15.1f). These last two might be represented, for example, by the passage of dunes past a point in time or the progressive trend of a seasonally dominated but variance-stationary series such as change in the BOD (biochemical oxygen demand) content of a stream with a regularly increasing pollutant load.

This classification is predicated by a systems analysis context in which the objectives are usually couched in terms of the general model

$$Y_t = SX_t \text{ for } K \geqslant 0 \qquad (15.1)$$

in which S is an operator called the system transfer function. The effort is then applied to identifying the transfer function (i.e. Y_t and X_t known, find S) or identifying the input Y_t and S known, find X_t). The first is used to infer the internal dynamics of the system, such as in the relationship between rainfall and overland flow quantities. The second is a refined version of inferring past events (e.g. climatic controls) from response characteristics (such as form or sediment properties).

This general strategy imposes a number of well-known limitations (Bennett 1978). From a systems analysis point of view, these limitations arise mainly from the fact that extraneous 'noise' is mixed with the input, arises from the transfer function model of the processes and is present in the measured output. Input noise may be generated either from errors in measuring the input or from the presence of ignored variables. Model

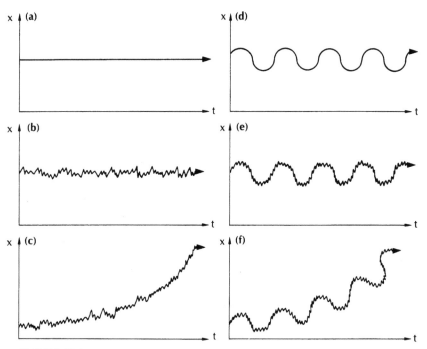

FIGURE 15.1
Types of equilibrium in dynamical systems. (Based on a diagram by Chorley & Kennedy 1971.)

noise arises from procedures such as linearization and mis-specification, whereas output noise occurs as a result of data manipulation such as aggregation, smoothing and sampling problems.

There are three other rather severe limitations to the use of this strategy in a geomorphological context which form the subject of this paper. The first is that they reveal little of the nature of the equilibrium and have little to say about its sensitivity to change, other than by reference to the amplitude of environmental fluctuations in relation to empirically defined thresholds. Typically, for example, it might be argued that a point lying closer to the discriminant line of meandering-braiding (Leopold & Wolman 1957) is more likely to be carried over the boundary than one lying further away. This tends to lead to an emphasis on extreme fluctuations as a controlling mechanism rather than emphasizing the fundamental stability properties of the system.

The second issue is that, given that certain patterns of equilibrium behavior may be established, the systems analysis approach can provide little help in identifying the several different ways in which such patterns may arise. Several workers (e.g. Chorley 1962) have drawn attention to this problem, highlighting the point that time-independent behavior appears to make geomorphological history rather irrelevant to an understanding of contemporary process–form relationships. However, even if it were to be argued that 'characteristic' (i.e. equilibrium) forms and behavior do exist, this does not obviate the need to understand the different ways in which they are generated and give rise to similar, if not identical, patterns of observed temporal response. Periodic behavior, for example, may arise in several different and sometimes rather surprising ways.

The third limitation of inductive systems analysis is that, in geomorphology, the available data for reconstructing such patterns are usually short in length and often imperfect with respect to frequency and techniques of collection. Given the recognized deficiencies of the procedure even when a complete and nearly perfect series is available (points 1 and 2 above), one might legitimately enquire what, if anything, is to be gained from the spotty and imperfect information which is generally available. This is the subject of the third part of the paper.

The concept of a stable, dynamic equilibrium has had a long history in geomorphology. Attention has focused on two particular themes. The first of these is the observation of stable forms such as the hydraulic geometry relations of Leopold and Maddock (1953) or the network characteristics of Horton (1945). The second is the determination of equilibrium forms directly from mechanical considerations (Lane 1955), or from maximizing assumptions (Langbein & Leopold 1966), or from both (Kirkby 1977). Neither of these approaches say anything about the trajectory along which equilibrium is reached or the stability which remains once the equilibrium is reached. Recent interest in rapid shifts between equilibria (Schumm 1979) and the formal application of stability analysis to systems of geomorphological interest (notably by Parker 1976) invites a discussion of the wider concept of stability. Such a discussion follows.

STABILITY OF EQUILIBRIA

The characteristic form proposition argues that, under a fixed and constant set of controls, processes and resistances, a characteristic form will emerge which represents a balance between forces and resistances (Hack 1960). Two examples are soil-cover thickness and sediment load. In the first, the thickness is a function of the rate of erosion and the weathering of bedrock. In the second, the load is a function of the rate of supply of material into the reach, the rate of entrainment of material and the rate of settling. Both are dynamical systems in that they can be represented by differential equations, where time is the independent variable of the general form:

$$\frac{dx_i}{dt} = f_i(x_1, x_2...x_m) \tag{15.2}$$

The condition that equilibrium exists is the condition that $dx_i/dt = 0$ for all i, or in other words the changes in the state variables of the system are zero over time.

Stability usually refers to neighborhood stability. This is defined to mean that if a system is perturbed by a small amount around its equilibrium, then the perturbation dies away. In an unstable system the perturbation grows in time. This idea has often been expressed qualitatively in the geomorphological literature (see Brunsden & Thornes 1979). The basic concept is shown in Figure 15.2a where the perturbation $x = e^{\alpha t}$ grows if α is positive and decreases if α is negative. In the case of the soil cover, if the system is stable and the erosion rate is perturbed (increases, say) then weathering will increase to cause damping of the perturbation. Notice that if $\alpha = 0$, then the perturbation neither grows nor decreases. It represents a knife-edge between the two behaviors and hence is itself very unstable. Bifurcation is said to occur in the behavior at $\alpha = 0$, the equilibrium being stable to one side and unstable to the other. In the geomorphological literature this point is a stability threshold (Kirkby 1980).

In formal terms, neighborhood stability analysis amounts to determining whether the change in the magnitude of the perturbation about equilibrium is positive or negative. This is achieved by obtaining an approximate differential equation for the perturbation using a Taylor expansion, neglecting the non-linear terms and then solving to find the behavior of the perturbation through time. These methods can be extended to a general linear system of the type described above, provided that a solution exists to the perturbed equations. The equations are usually expressed in the general form

$$(\mathbf{x})' = \mathbf{A}\mathbf{x} \tag{15.3}$$

where $\mathbf{x}'$ is the derivative with respect to time and $\mathbf{A}$ the matrix of coefficients in the equations defining the perturbations. The general properties of homogeneous differential equations can then be used (Hirsch & Smale 1974) to determine the stability or otherwise of the system, provided that the coefficients are constant. The key lies in the eigenvalues of the matrix

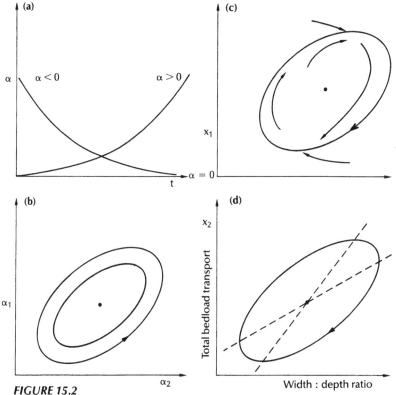

FIGURE 15.2
Types of stability situation: (a) bifurcation; (b) neutral stability; (c) stable limit cycle behaviour; (d) hypothetical model of stable limit cycle in a channel system.

A. These eigenvalues are complex numbers, i.e., they are the sum of real and imaginary parts. The real parts determine the sign of the change, whereas the imaginary part determines whether the growth or decay will be oscillatory. For stability, all the real parts of the eigenvalues must be negative; for instability, all the eigenvalues must have positive real parts. If the real parts are all zero (i.e., the eigenvalues are purely imaginary) then the system is neutrally stable.

For a simple example of the application of stability analysis, consider the equation:

$$\frac{dQ_s}{dt} = \alpha(K - Q_s) - \beta B_s Q_s \tag{15.4}$$

where Q_s is the total transported sediment, B_s is the availability of material from the bed and K is the capacity of total transport. α is a rate coefficient related to the speed with which material of a constant size is entrained. The term $\beta B_s Q_s$ represents an interaction between the bed material availability and the transported bed material. The greater is this reaction, the less the uptake into the flow. β governs this interaction rate. Then, for equilibrium transport:

$$Q^*_s = \frac{\alpha k}{(\alpha + \beta B_s)} \tag{15.5}$$

Obtaining the Taylor expansion for a perturbation of the sediment load x_0 at Q^*_s, neglecting the higher terms and solving the differential equation for x at time t gives

$$x(t) = x_0 e^{at} \tag{15.5}$$

where

$$a = -(\alpha + \beta B_s) \tag{15.6}$$

Since this can only be unstable for β negative and $\beta B_s > \alpha$, the system is likely to remain stable for perturbations about the equilibrium value.

A common feature in geomorphological systems is stable oscillatory behavior. This is evident, for example, in the pool and riffle cycle occurring in perennial channels. I have already indicated one particular kind of oscillatory equilibrium (Fig. 15.2b, neutral stability), but it can be shown that this form is an extremely special case which usually degenerates into a source or a sink, as I implied earlier. Moreover, the oscillations are entirely dependent on the initial values of the perturbation for their amplitude. The effect would be to produce an infinite set of possible oscillations which would appear in the response variable (for example, channel depth) as a random series with a mean about the center. Actual channel series tend to show much greater regularity, though they may be super-imposed on trends due to other causes, such as downvalley increase in discharge. This dynamic behavior implies that perturbations around the stable oscillations are 'caught' by the system and suppressed into the stable oscillations. This kind of behavior is illustrated in Figure 15.2c. It is well known from electrical circuit theory as the Van der Pol equation and from population biology (May 1974) as a stable limit cycle.

One way in which such a stable limit cycle may occur is if the feedback of the system is delayed. The situation may be represented by posing the oscillating channel width problem in the context of the density-dependent growth equation in biology. Assume that stream dynamics are such that the rate of increase of channel width in a downstream direction is a function of the width, by assuming that the increased sediment there-by produced requires increased efficiency and hence an increased width : depth ratio. Moreover, assume that there is a limiting upper width (L) because roughness effects dominate. The behavior of the channel under this simple model can be described by:

$$\frac{dw}{dt} = rw(1-w/L) \tag{15.7}$$

in which r is a growth rate coefficient. In spatial terms:

$$\frac{dw}{dx} = rw(1-w/L) \tag{15.8}$$

If the regulatory effects operate with a lag T in the feedback, the equation can be written as:

$$\frac{dw}{dx} = rw(1-w(x-T)/L) \tag{15.9}$$

This equation has a monotonically damped stable point if the feedback delay is of the same order as the natural response time, an oscillatory damped stable point for longer delays and for long delays ($rT > \frac{1}{2}\pi$) the perturbations exhibit stable limit cycles (May 1976). The period of the cycles generated by this general form of the equation is of the order of $4T$. This implies, for natural channels, that T is of the order of πw, on the basis of empirical behavior. Allen (1974) has shown that time lags resulting from different systems reaction times can generate periodicity in plan, depth, width and roughness. The time lags reflect the speed with which sediment can adjust to changes in the flow conditions and the properties on which the transport depends.

A more common way in which limit cycles occur is in the solution of the paired equations:

$$\frac{\partial x}{\partial t} = F(x,y) \tag{15.10}$$

and

$$\frac{\partial y}{\partial t} = G(x,y) \tag{15.11}$$

and hence they arise in systems where the response variables are each functions of themselves and of each other. In biology, this is typically represented by the predator–prey type of relationships. Intuitively, in the channel problem outlined above we might expect to find a stable limit cycle solution governing the relationship between width and sediment load. This is sketched intuitively in Figure 15.2d. Here, the behavior of the channel with respect to sediment load and width to depth ratio is viewed as a cycle in which the dominant increase in load occurs by relative increase in the width. Eventually, the limit described above comes into play causing sedimentation, a decrease in width : depth ratio and a fall in the total sediment transport. Notice that the maximum sediment transport occurs prior to the maximum width, when sedimentation is under way. The governing equations of such a model have rather strict requirements, but it seems intuitively reasonable to pose the problem in these terms. A significant requirement would be that sediment from the bed is uniformly available, i.e., no armouring is taking place (Thornes 1980), otherwise structural instability might be inherent in the model. In either of these two simple models, an overall increase in width in a downstream direction coupled with a stable limit cycle-type oscillation might be expected to induce the behavior described by Figure 15.1f.

The third way in which steady cyclical behavior can be obtained in the dynamic system is that most familiar to geomorphologists, namely to have a cyclical input to the system driven, for example, by a seasonal climatic component. Indeed, it is most often assumed in geomorphological literature that a cyclical response *necessarily* results from a cyclical input, which is why some time has been taken to show that this need not be the case. However, there is a large class of periodic inputs to geomorphological systems reflecting seasonal, daily and other cyclic components which in linear systems invoke sinusoidal responses. Some of these and the techniques for their analysis are discussed in Bennett (1978).

The general behavior of the equilibrium systems is determined by their relative stability, which is determined by the coefficients of the interaction matrices, especially the magnitude of the negative real parts which determine the magnitude and speed of damping. In the real world, these effects

are influenced by the variability of the parameters involved in the differential equations of the process, forcing us to consider the equilibrium of
stochastic differential equations. Consider, for example, the simple model
in which the rate of change of soil thickness is assumed to be a function
of the actual thickness at time t. This might be because residence time
and weathering increase up to a point D when the increased residence
time is offset by the decreased throughput of minerals (Kirkby 1980). It
might be expected, then, that D is a stochastic variable in (long) time
affected by average moisture conditions. This is then expressed as:

$$\frac{dz}{dt} = z(t)\,(D - z(t)) \tag{15.12}$$

where D is a stochastic variable. May (1974) shows that the distribution
function of the variable z corresponds to a Pearson Type III Gamma
distribution subject to the conditions that (1) the variation of D is purely
random in time and (2) that the mean value of D is greater than half of the
variance of D. This is required because D is the stabilizing element of the
system.

It is no accident that all the examples in this section have involved
some element of positive feedback. This is because it provides the de-
stabilizing element in the differential equations. Many geomorphological
systems are relatively stable, so that the role of fluctuations in shifting
these equilibria is particularly important. The problem is to identify in a
noisy output which is the signal and which is the noise, and having identified the components, to estimate them.

IDENTIFICATION OF STRUCTURE

Given observed data, part of the problem is to infer from what *was*
observed what *might* have been observed. In other words, one suggests
a probability model for the processes (signal and noise) generating the
data in which some parameters are unknown and therefore to be inferred
from the data. The statistical analysis is then concerned with parameter
identification, i.e., determination of the parameter values by estimation
or hypothesis testing. A model is called **structural** if its parameters have a
natural or structural interpretation. Such models provide explanation and
control of the processes generating the model. When no models are
available from theory or from experience, it is still possible to find models
for prediction (from what has been observed to what will be observed)
and simulation (from what has been observed, generate more data with
similar characteristics). The second type of model is called **synthetic** (Parzen 1974). This distinction is somewhat artificial and it is not uncommon

to find descriptions of structural components defined by models couched in synthetic terms.

A core feature of the steady state concept as described by Figure 15.1b is that the mean and variance of the series remain constant through time. Additionally, the perturbations, the time-varying elements in this simple case, are pure white noise, in other words, they are normally distributed, independent random variates. In geomorphological terms, the steady state so defined usually implies either that the fluctuations are randomly generated (and hence of little interest) or that they are the result of a large number of interacting and complex forces which lead to the near balance implied by the steady state (and hence incapable of resolution). In statistical terms, white noise represents the end point of the identification and estimation process. If after extracting the model of the process from the original data the resulting series is a white noise series, then in a certain respect the model is considered adequate. The white noise property is indicated by a flat power spectrum in the frequency domain and a set of insignificant autocorrelation coefficients in the discrete time domain.

Recently, it has been demonstrated that under certain circumstances difference equation versions of the density-dependent type of relationship explored in the first part of the paper can give rise to an extraordinary range of dynamic behavior. In particular, it is found that the difference equations yield switching behavior analogous to the stable limit cycle of the continuous delayed differential equations for constrained density dependence at certain values of the growth parameter. When the latter is greater than two, a two-point stable cycle is produced. This means there are two stable equilibria which the system may occupy. As the parameter increases, the system enters a chaotic regime in which, for any parameter value in this domain, there exists an infinite number of different periodic orbits (Li & Yorke 1975, May 1976). Because this would be characterized by a flat spectrum, the result might be indistinguishable from white noise. Situations thought to be of great complexity might reflect, therefore, rather simple macroscopic behavior.

Where the systems response pattern is more complex, there are usually two approaches. The first is to attempt to identify the deterministic components and model them, subtracting each to leave a stochastic element to be modelled. The second is to assume from the outset that the system is to be represented by an *undefined* 'deterministic' parameter which describes the general equilibrium or drift and an undefined stochastic term which is also to be paramaterized. The first type is most commonly used when a strong trend or periodic component is present, the second when the response pattern is less clearly partitioned. Identification commences with inspection of the autocovariance or its transform – the power spectrum.

If strong trend and periodic components are revealed, then the general structure identified usually takes the classical form:

$$X_t = T_t + P_t + S_t + R_t \tag{15.13}$$

in which X_t is the value of the response variable at time t and it is assumed to incorporate trend (T_t), periodic (P_t), persistence (S_t) and pure random terms (R_t). These may be considered additive (linear) as above or multiplicative. The persistence term is a neighborhood effect wherein the value of the variable at t is affected by that at $t - k$, where k is small.

The trend and periodic components are usually estimated by ordinary least-squares regression techniques. A simple example is shown in Figure 15.3. The upper graph shows the electrical conductivity of the river Trent at Lea Marston (UK) observed at 15-minute intervals. Removal of the diurnal harmonic and the persistence effects fitted by an autoregressive-moving average model produces the series shown in the middle and lower graphs respectively. Where the periodic component is clear in the data and easily accounted for, this approach yields highly satisfactory results. Often, however, this is not the case. Data processing may induce apparent cyclicity and the use of conventional Fourier series may create more seasonal residuals because the periodic component is asymmetrical. Nonetheless, spectral identification and Fourier estimation tend to yield much better results than techniques such as seasonal differencing.

The second approach to the identification and estimation of a univariate series is to assume a simple linear system in which the process is represented by a model of the general type:

$$\Phi(B) X_t = \theta(B) a_t \tag{15.14}$$

if the series is assumed stationary and non-periodic. This is the Box–Jenkins (1970) mixed autoregressive-moving average model. B is the backward shift operator defined by $X_t B = X_{t-1}$ and ϕ and Θ are coefficients of the autoregressive and moving average components respectively. a_t is a series of white noise. In this procedure it is usual to assume that the process is linear and driven by the stationary white noise. It provides an essentially black-box approach to the identification and estimation of the parameters and hence has particular value, mainly in forecasting, unless the model can be specifically linked to governing continuous differential equations of the system, which is sometimes the case. Another difficulty of this approach is that the original model identification is a rather subjective procedure based on visual inspection of the correlograms and partial correlograms and the estimation is carried out over the entire series. Some of these objections are overcome by adaptive estimation of the parameters. In the forecasting mode, this means that new data can be added and

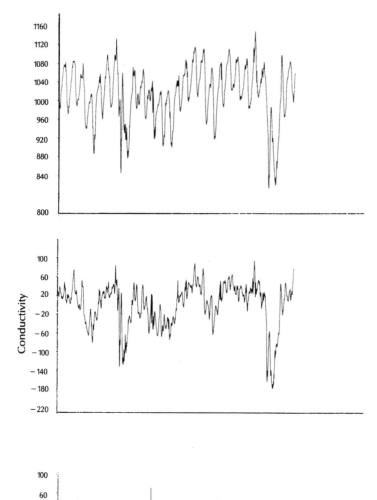

FIGURE 15.3
Progressive variance reduction by harmonic removal and autoregressive-moving average modelling of Lea Marston conductance series.

immediately influence the parameter values and hence the future forecast. In another sense, if the parameter values of an autoregressive-moving average type model can be assumed to reflect other (unsteady) controls on the system, identification of the time or space shifts in the parameter values may provide information on the system operation. Unlike the Box–Jenkins technique, it focuses on the underlying system rather than the input–output formulation (Culling in press). Although Bennett (1976) has used the technique to identify the changing parameter values in a channel height series, and although the technique has been widely used in hydrology and water quality studies, it has hitherto attracted little attention from geomorphologists.

Attempts to identify causal relationships between structurally identified variables are at the core of most time-based geomorphological studies. Conventional (non-temporal) bivariate and multivariate regression frequently proves unsuccessful, not only because the statistical requirements of regression analysis, particularly independence, are not met, but also because the relationships are often lagged. The failure to find consistent relationships between sediment yield, grazing densities and precipitation may reflect lags of 10 to 50 years and non-linearities due to feedback. Establishing the transfer function relationships between several series may be carried out in the time or the frequency domain. In both cases, the data have to be discretized. In deterministic systems, causality is defined by the equations. When the problem is approached from the data, series X_t is said to cause Y_t if one is better able to predict Y_t when X_t is used than if it is not used. In other words, the series, X_t is said to cause or partially cause Y_t if it reduces the mean error variance of Y_t.

In simple cases of cause and effect, the use of cross-spectral analysis enables one to identify the frequency at which the correlation (coherence) between the two series is greatest and to identify the lag between the variables at that frequency. Edwards and Thornes (1973), for example, evaluated the pattern of lags between peak annual discharge and various water quality determinands for the River Stour (UK), so developing a lead–lag matrix for the systems behavior. Where such systems are affected by feedbacks, the coherence and phase estimates based on usual techniques break down. This is important in the context of equilibrium dynamics. Granger (1969) suggested a theoretical decomposition which circumvents this problem for simple cases. Unfortunately, adequate estimation techniques for the theory have still to be worked out.

Again the Box–Jenkins techniques provide for the estimation of simple and multiple input transfer functions analogous to and based on the same procedures as the univariate case. For steady systems, this technique provides an adequate representation because the parameters are estimated from the entire series. For non-stationary series, the trend and seasonal components have first to be removed for each series before the transfer function is obtained. Generally speaking, the identification techniques briefly reviewed in this section are adequate for most geomorphic problems. The main difficulty lies in the hard geomorphological fact that the data whereby such procedures could be employed to investigate systems dynamics are sparse and of poor quality.

THE DATA PROBLEM

The extent to which data inadequacy constitutes a problem depends upon the use to be made of the data. In the context of dynamic equilibria, they may be taken to be the description and interpretation of data for identification of the underlying structures, as outlined in the second section. The problem is discussed here in the context of a water quality data collection exercise. Given that data are collected on a regular basis for evaluation of water quality conditions, a key question is whether there are too many or too few data for the purposes in hand. Characteristically, these purposes include a basic 'alarm' function, the identification of a mean threshold exceedance, understanding of the basic operating system by structure identification and extending this for use in forecasting. If there is too much information, one needs to know how much can be thrown away. If there is too little, one may ask whether interpolation of various kinds will help in reconstructing the original series and what effects it is likely to have on other procedures.

Although in geomorphology we are mainly concerned with structural identification, water quality series provide suitable analogs for much geomorphic data because of their variety of responses. Some series, such as dissolved oxygen and conductance, are basically periodic with added noise, whereas others, such as suspended solids concentrations, have a much lower noise to signal ratio and, while including periodic and trend components, are dominated by stochastic elements. Yet others, such as ammonium, can be modelled as simple autoregressive processes, while pH is a nearly constant series with added white noise.

The data series used in this investigation include 15 minute observations obtained by probes from the river Thame and weekly data from the River Stour. The former is a polluted and highly controlled river in central England which is dominated by human activity (Fig. 15.4), whereas the latter is a little-controlled rural catchment in eastern England. In all cases except pH, the series for both rivers can be fitted with models which achieve over 90% 'explanation' of the original data series. These models are essentially of the 'classical decomposition' type mentioned in section two. For pH in both cases the residual variance (white noise under the

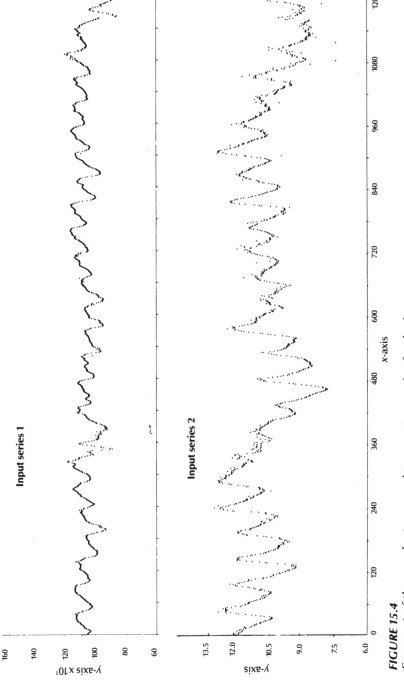

FIGURE 15.4
Segment of the conductance and temperature series for the Lea Marston data.

usual tests) constitutes about 60% of the total variance, the rest being accounted for by autoregressive-moving average models.

Given the general objectives outlined above, there are three possible approaches to the problem. The first is to determine theoretically the effects of decimation (removal of data points) and interpolation on the correlation and partial autocorrelation functions. The extension of this to parameter estimation proves, at this stage, intractable. The second approach is to generate series of known properties synthetically and then to evaluate the effects of decimation and interpolation on identification *and* parameter estimation. The third approach is to take impure series (i.e., actual data) and again to simulate decimation and patching and evaluate the success of these processes for different determinands.

The criteria used to evaluate success depend in part on the uses to be made of the data series. The appropriate criteria here are listed below.

(a) Reproduction of the basic statistics, particularly the first four moments. These are required for comparison with predetermined water quality standards, for evaluation of treatment effects and for simple comparison between different periods.

(b) Preservation of the tails of the distribution of the raw data. The water quality manager needs to be particularly aware of the upper end of the distribution. If the density of data loss is such that this is not well preserved, or if the method of interpolation tends to obscure it, this raises difficulty in treatment policy.

(c) Maintenance of a low mean square error between the patched series and the original data. Besides being an overall check of goodness of fit, this measure is fairly sensitive to extreme values.

(d) Retention of the same basic autocorrelation structure as the original series and values close to the original. This property is critical in identification and forecasting of the underlying models; it may also be used to determine spectral estimates and hence, the frequency characteristics of the series.

(e) Close match between the identified model for the raw data and for the interpolated data; this depends on (d) for Box–Jenkins models.

(f) A good match between the originally estimated parameters of the identified model in the raw data and those estimated from the interpolated data. Such models can be used both for forecasting and for identifying the causal mechanisms as outlined in section two.

It would be impossible, in the space available to outline all the combinations of data treatment, methodological approach and goodness-of-fit criteria. Instead, the effects of missing data on the autocorrelation and partial autocorrelation function are first considered. Then the case of simple linear interpolation is outlined. Finally, the more general results of the exercise, and especially those for related-variable-regression, will be discussed. In what follows, the missing data are assumed to occur at random. For a discussion of systematic decimation see Thornes (1973) and Valdares Tavares (1975). Moreover, the effects of small sample size are not considered here, though they are relevant to many geomorphological problems. Discussion of this problem is to be found in Wallis and O'Connell (1972).

The simplest case is to consider a known stochastic linear model and to ask what effect will be produced on the autocorrelation function (acf) and the partial autocorrelation function (pacf) if the data are decimated to progressively higher densities and if the decimated values are replaced by the expected value of the series as a whole. In general applications, we need to know from the acf and pacf (i) the actual values of the acf and pacf at various lags; (ii) the point at which the function becomes effectively zero; and (iii) the characteristic shape of the various functions. Three situations can then arise: (a) the estimated functions lie within the confidence bands of the original undecimated series; (b) they lie outside these bands but the bias is systematic; (c) they lie outside the bands and the bias is non-systematic.

Consider the first-order autoregressive case as a theoretical example:

$$Y_t = \alpha Y_{t-1} + E_t \tag{15.15}$$

Then, with the usual restrictions:

$$E(y_t^2) = \frac{\sigma_\epsilon^2}{(1-\alpha^2)} \tag{15.16}$$

and

$$E(y_t\, y_{t-r}) = \alpha^r \sigma_\epsilon^2 / (1-\alpha^2) \tag{15.17}$$

where

$$E(\varepsilon_t) = 0;\; E(\varepsilon_{t_1}\, \varepsilon_{t_2}) = 0 \text{ for } t_1 \neq t_2$$

and

$$E(\varepsilon_{t_1}\, \varepsilon_{t_2}) = \sigma_\epsilon^2 \text{ for } t_1 = t_2$$

If m observations are missing and we know which they are, and p pairs are affected, then the natural estimate of $\rho_{r,m}$ is given by:

$$\rho_{r,m} = \frac{\frac{1}{n-r-p} \sum_t y_t y_{t-r}}{\frac{1}{n-m} \sum_t y_t^2} \tag{15.18}$$

A reasonable estimate of p_r is given by $\hat{p}_{r,m}$ if all missing values are set equal to zero and the usual estimate is multiplied by:

$$\frac{n-r}{n}\left[\frac{(n-m)}{(n-r-\rho)}\right] \tag{15.19}$$

where p is the number of cross products $(Y_tY_{t\,+k})$ lost. Assume that we do not know p (i.e. we do not have the information about which cross-product terms involve the missing observations), which is extremely tedious to obtain in a long series. The average value of p must then be obtained combinationally using

$$\bar{p} = (n-r)\left[1 - \frac{(n-m)(n-m-1)}{n(n-1)}\right] \tag{15.20}$$

substituting for p in Equation 15.19 gives the average value for the multiplying factor,

$$\frac{(n-1)}{(n-m-1)} \tag{15.21}$$

which for typically large sample sizes gives $1/(1 - n/m)$ as the expectation of the correction factor for bias in the autocorrelation coefficient in the AR(1) model with m missing data points (and this is true for all other simple models).

For a known α_r we may obtain the effects of missing observations as the reciprocal of Equation (15.21). This is not so good an estimate as that obtained from Equation (15.19), but in real applications p is not so easy to obtain. The expected effects of decimation at various densities on the value of the autocorrelation at lag one are shown in column three of Table 15.1. The next column gives the values obtained by simulation. These were reached by generating a Markov model 1000 observations long, decimating to various densities, replacing the decimated values by the

TABLE 15.1
Results of estimation of r_1 for the Markov model.

n	m	Theoretical	$\hat{p}1,m$	$\hat{p}1,m$ range
1000	0	0.5 ± 0.0529	–	–
	100	0.449	0.465	0.454–0.446
	200	0.401	0.385	0.409–0.389
	300	0.350	0.305	0.359–0.349
	400	0.300	0.287	0.325–0.275

mean and then estimating the acf from the reformed series using the usual equation. The results obtained are close to the expected effects of decimation, except for the value for $m = 300$. Naturally, we expect some variability since we have used the expected value of p. The final column indicates the effects of variance of p. The values were obtained by simulation, decimation and counting the actual deleted cross-product cases. This process was repeated 250 times to obtain the variance of p. The range given relates to two standard deviations about the mean.

This analysis has been extended to show that, by inserting the mean of the remaining series into the gaps and using the usual estimating equation, reasonable acf and pacf may be obtained for simple and mixed (ARMA) discrete stochastic models. According to the model identified at this stage, application of a correction will yield an acf, pacf and parameter estimates close to those for the whole original series. The correction factor, involving p, will itself be subject to variability in p. In integrated (ARIMA) models, two procedures may be followed: (i) difference then fill with the mean value, called post-filling; and (ii) fill and then difference, called pre-filling. Both theory and simulation indicate that the practice of post-filling gives the best results. (For further details see Thornes and Clark 1976.)

These results do not cover the case where there is systematic bias in the observations. Systematic bias is likely to take one of two forms in geomorphological experiments. First, data are missing where a particular value is exceeded and hence, the distribution of the remaining observations is truncated. In these circumstances we may not assume that the statistical characteristics of the missing observations are the same as those of the remaining ones. Second, data are missing on a regular time basis (e.g., weekends). This is not a serious problem, except where it might introduce a truncation of the first type, say due to some periodic phenomenon.

The second illustration refers to simple interpolation. This is achieved by distributing the difference between the last and next observed points over the intervening gaps by adding it successively to the intervening points, as shown in Figure 15.5. Following Friedman (1962), this procedure may be represented by the general transformation:

$$p_i = X_i - [(1 - W_i)x_0 + W_i x_2] \tag{15.22}$$

where p_i is the error in the ith interpolation, X_i is the original series value at the ith gap, x_0 and x_2 are the last and first values about the gap and W_i is the weight attached to the terminal value in computing the straight line between x_0 and that terminal value. In the simplest case:

$$p_1 = x_1 - \tfrac{1}{2}[x_0 - x_2] \tag{15.23}$$

The transformation covers not only the case of more than one intermediate value to be interpolated, but also non-equally spaced intervals. Letting μ and σ represent the mean and standard deviation of the variable designated by the subscript then:

$$\mu_p = \mu_{x_1} - \tfrac{1}{2}(\mu_{x_0} - \mu_{x_2}) \tag{15.24}$$

and for a stationary series $\mu_p = 0$. Similarly, by taking expectations, we can obtain the general error variance equation:

$$\begin{aligned}
\sigma_{p_i}^2 = {} & \sigma_{x_1}^2 + (1 - w_i)^2 \, \sigma_{x_0}^2 \\
& - 2\,(1 - w_i)\, \rho_{x_0 x_1} \, \sigma_{x_0} \, \sigma_{x_1} \\
& - 2\, w_i\, \rho_{x_1 x_2} \, \sigma_{x_1} \, \sigma_{x_2} \\
& + 2\, w_i\,(1 - w_i)\, \rho_{x_0 x_2} \, \sigma_{x_0 x_2}
\end{aligned} \tag{15.25}$$

in which $P_{x_e x_{t+k}}$ is the Kth order correlation coefficient. The magnitude of the error variance thus depends on the assumptions made about W and the basic autocorrelation structure. For example, with $W = \tfrac{1}{2}$, as

$$\rho_{x_0 x_1} = \rho_{x_1 x_2} = \rho_{x_0 x_2} \to 0$$
$$\sigma_{p_i}^2 \to 1.5\, \sigma_x^2 \tag{15.26}$$

which is the white noise case. Similarly, for the first-order Markov process with $\phi = 0.9$ we obtain

$$\sigma_{p_i}^2 = 0.1050\, \sigma_x^2 \tag{15.27}$$

and for the first-order process with $\phi = -0.9$ we obtain

$$\sigma_{p_i}^2 = 3.704\, \sigma_x^2 \tag{15.28}$$

Table 15.2 shows the ratios of $\sigma_{pi}^2 : \sigma_{xi}^2$ for various lengths of gaps, with $\phi = 0.9$. This table gives the variances, for example, of the errors associated with the third element of interpolations across gaps of length five as 0.3076 of the variance of the raw series and, as expected, the function is symmetrical.

Now, in fact, the effect of this on the estimate of the total variance depends on the number of cases of 1, 2, 3... length gaps in the series. The distribution of the number of gaps of length m is determined from the theory of runs. In the case of estimation from an actual geomorphic experiment, we could obtain it directly by counting. It is then possible to multiply the number of cases in each category (e.g. 3rd element in 5-gaps) by the error variance to obtain a correction for the error sums of squares.

The number (n_m) of gap runs of length m in a series of length N with random decimations of density d is given by:

$$n_m = Nd^m(1-d)^2 \qquad (15.29)$$

The distribution of runs of various lengths for different densities of decimation is shown in Figure 15.5b. For example, with a 10% decimation, 81% of the gaps will be length one, 16% will be length two and 3% will be length three.

In summary, to estimate the error variance (which will equal the mean square error if $\mu_{pi} = 0$, combine the expected error variance for a given element in a gap of length m with the number of cases to give the error sum of squares due to interpolation. The former is estimated from Equa-

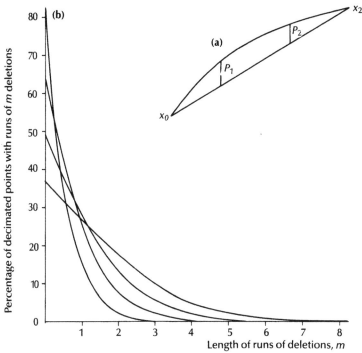

FIGURE 15.5
(a) Method of interpolation. (b) Percentage distributions of runs of various lengths for different percentage decimations.

tion (15.25) knowing the theoretical or empirical autocorrelation function; the latter is obtained from the theoretical decimation density or the actual observed distribution of gaps. This yields a measure of the overall error sum of squares which, when divided by N, gives the approximation variance.

For example, in the first-order Markov process, the effects of decimation of density d will be, for reasonably large N and with only single gaps considered:

$$\rho_k = |\phi|^k/(1-d.c) \tag{15.30}$$

in which ϕ is the first order coefficient in the original model, d is the density and c is the correction to the variance for the error in its estimation. This may be regarded as the 'minimal effects' case and the examples of $\phi = 0.5$ with decimations of 10% and 40% are shown in Table 15.3.

In order to study the effects of interpolation on the variance of the mean and of the mean squared error, I again simulated the decimation–interpolation procedures for artificially and empirically observed series of different lengths and different percentage decimations. For each of the 250 simulations, I have calculated the mean square error between the original and the decimated–interpolated series.

TABLE 15.2
Ratio of error variance for the ith member of a run of m missing data points to true variance for the complete series: for an AR(1) model with $\phi = 0.9$.

M \ i	1	2	3	4	5
1	0.1050				
2	0.1395	0.1395			
3	0.1566	0.2080	0.1566		
4	0.1664	0.2482	0.2482	0.1664	
5	0.1665	0.2740	0.3076	0.2740	0.1665

TABLE 15.3
Effects of simple interpolation (m = 1) for d = 0.1 and 0.4 on the acf of AR(1) with $\phi = 0.5$.

Lag	Original	d = 0.1	d = 0.4
1	0.5	0.5053	0.5129
2	0.25	0.2526	0.2609
3	0.125	0.1263	0.1304
4	0.0625	0.0656	0.0652
5	0.031	0.0313	0.0323

As an example of this, Table 15.4 shows the mean and variance of the mean squared error for an ARMA (1, 0, 1) model. Part (a) gives the raw data. The mean of the mean squared error is identical to that expected from theoretical considerations. The mean of the mean squared error rises steeply in a non-linear fashion, as expected due to the increasing frequency of longer gaps with their associated higher errors of linear interpolation. Over the N-values here considered, there is no identifiable bias in this parameter. The variance of the mean squared error varies as a function of both the length of the record and the degree of decimation in a systematic fashion.

Figure 15.6 shows similar results for simulation of the decimation and simple interpolation of the Stour raw conductance series. The function (b) here is less strongly curved, reflecting the higher degree of autocor-

TABLE 15.4
Mean and variance of the mean squared error for an ARMA (1, 0, 1) model after decimation and simple linear interpolation.

(a) Raw data

Length	Mean	Variance
250	0.2180	3.8157
500	0.0958	4.0080
750	0.0232	3.6807
1000	0.0000	3.4441

(b) Mean of mean squared error

N \ D	40%	30%	20%	10%
250	0.9416	0.5570	0.2882	0.1085
500	1.0020	0.5956	0.3107	0.1197
750	1.0060	0.6055	0.3236	0.1281
1000	0.9318	0.5712	0.2978	0.1151

(c) Variance of mean squared error

N \ D	40%	30%	20%	10%
250	0.0625	0.0304	0.0115	0.0017
500	0.0287	0.0179	0.0062	0.0012
750	0.0176	0.0095	0.0043	0.0011
1000	0.0136	0.0083	0.0023	0.0006

relation, but is found to agree very well with the theoretically expected mean mean square error.

To summarize, if we have missing data, the acf and pacf can be obtained and corrected. With the corrected values, the theory of this last section can be used to determine the likely effects of using simple interpolation. These effects will basically be related to the strength of the autocorrelation in the original series. Knowing the structure of the underlying series, the effects of missing data on the parameters of simple Box–Jenkins models can be determined. Empirical results, not reproduced here, confirm that the procedure is fairly robust. Simple interpolation also appears to be a reasonable procedure for handling randomly occurring gaps, in that the upper tails of the distribution are also quite well preserved.

Finally, Table 15.5 shows the ratio of error variance to total variance due to simple interpolation and interpolation by regression with a related series or pair of series. For example, the gaps in conductance for the River Stour were filled by regression against carbonate hardness and non-carbonate hardness. All the cases are for a 40% loss of data and, in the second

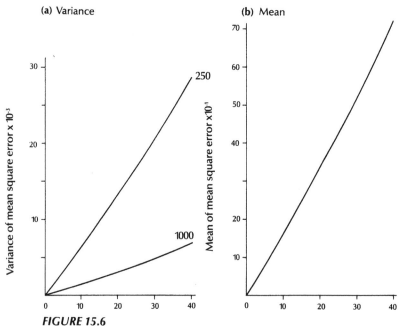

FIGURE 15.6
(a) Variance of mean squared error for 250 iterations of decimation and interpolation of Stour conductance data with different lengths and different decimations. (b) Mean of mean squared error for same data with length of 1121 observations.

set, the regressions used for interpolation were based on the original series. Although the fitness test is a rather general one, three conclusions may be drawn from these results. First, the regression-related interpolation appreciably reduces the error variance (mean squared error) when compared with simple interpolation. Second, there is considerable variation according to the behavior of particular variables. Third, there is a sharp difference between these two sets of series. All these conclusions reiterate the basis of and results from the theoretical work, namely the higher the original autocorrelation the greater the possibilities for identifying the systems behavior. In general, the results seem very encouraging from a geomorphological view. From a water quality manager's point of

CONCLUDING REMARKS

An attempt has been made in this paper to show that the various forms of equilibrium behavior are both complex and interesting. Unless it is assumed that the world is completely chaotic, science has to accept the

TABLE 15.5
The ratio of error variance to total variance with 40% decimation for simple and related-variable regression interpolation. Error variance is the sum of the mean squared difference between the original series and that produced by decimation–interpolation.

	Simple interpolation	Regression interpolation	Variable used
Stour			
1. conductance	15.3	6.4	5 and 6
2. pH	36.0	24.2	7
3. chloride	13.2	9.4	1
4. nitrate	8.7	5.4	5 and 6
5. carb. hardness	25.4	7.1	1 and 6
6. non-carb. hardness	7.1	3.5	4 and 1
7. ammonia	19.8	23.9	2
8. oxygen	42.2	22.4	9
9. discharge	35.3	21.1	8
Thame			
1. discharge	3.6	1.8	2
2. conductance	0.5	1.3	6
3. temp.	0.8	1.9	6
4. suspended solids	2.1	5.3	6
5. ammonium	1.1	2.8	3
6. pH	7.4	17.5	3
7. oxygen	2.2	5.5	6

view, they might be interpreted as indicating that, for most determinands, a few stations with less frequent observations would probably suffice.

overwhelming tendency towards equilibrium, albeit of a dynamic *indeterminate* character. The conceptual models of the first section appear out of the reach of the experimental field geomorphologist. However, these types of behavior may be discovered and identified by the tools described in the second part. The third part of the paper is meant to indicate that, although the data demands for identification and estimation are rather stringent, they may not necessarily be prohibitive. As the data base becomes longer and stronger, the subject will be better able to test the theoretical models. In the meanwhile, geomorphologists should make fuller use of the data already available.

ACKNOWLEDGEMENTS

The work described in this paper was carried out under a grant from the Natural Environment Research Council. Dr M. Knott (LSE) provided discussion of and help with the formulation of the theoretical aspects of the mean-insertion case. Mr Malcolm Clarke worked as research assistant on the project and contributed greatly to the computer simulation and analysis by writing or adapting the appropriate programs. All this help is gratefully acknowledged.

REFERENCES

Allen, J. R. L. 1974. Reaction, relation and lag in natural sedimentary systems: general principles, examples and reasons. *Earth-Science Rev.* **10**, 263–342.
Bennett, R. J. 1976. Adaptive adjustment of channel geometry. *Earth Sur. Proc.* **1**, 131–50.
Bennett, R. J. 1978. *Spatial time series: analysis, forecasting and control.* London: Pion Press.
Box, G. E. P. and G. Jenkins 1970. *Time series analysis: forecasting and control.* San Francisco: Holden-Day.
Brunsden, D. and J. B. Thornes 1979. Landscape sensitivity and change. *Trans Inst. Br. Geogs.* **4**, 485–515.
Chorley, R. J. 1962. *Geomorphology and general systems theory.* U.S. Geol. Survey Prof. Paper 500-B.
Chorley, R. J. and B. A. Kennedy 1971. *Physical geography: a systems approach.* London: Prentice-Hall.
Culling, W. E. H. in press. Atochastic processes in geography. In *Quantitative geography in Britain,* N. Wrigley & R. J. Bennett, (eds). London: Routledge and Kegan Paul.

Edwards, A. M. C. and J. B. Thornes 1973. Annual cycle in river water quality: a time series approach, *Water Resources Res.* **9**, 1286–95.

Friedman, M. 1962. The interpolation of time series by related series, *J. Am. Stat. Assn* **57**, 729–57.

Granger, C. W. J. 1969. Investigating causal relations by econometric models and cross-spectral methods. *Econometrica* **37**, 424–38.

Hack, J. T. 1960. Interpretation of erosional topography in humid temperate regions, *Am. J. Sci.* **258A**, 80–97.

Hirsch, M. W. and S. Smale 1974. *Differential equations, dynamical systems and linear algebra.* New York: Academic Press.

Horton, R. E. 1945. Erosional development of streams and their drainage basins; hydrophysical approach to quantitative morphology, *Geol. Soc. Am. Bull.***56**, 275–370.

Kirkby, M. J. 1977. Maximum sediment efficiency as a criterion for alluvial channels. In *River channel changes*, K. J. Gregory (ed.), 429–42. Chichester: Wiley.

Kirkby, M. J. 1980. The stream head as a significant geomorphic threshold. In *Thresholds in geomorphology*, D. R. Coates & J. D. Vitek (eds), 53–73. London: George Allen & Unwin.

Lane, E. W. 1955. The design of stable channels. *Trans. Am. Soc. Civil Engrs.* **120**, 1234–79.

Langbein, W. B. and L. B. Leopold 1966. *River meanders – theory of minimum variance.* U.S. Geol. Survey Prof. Paper 422-H.

Leopold, L. B. and T. Maddock Jr. 1953. *The hydraulic geometry of stream channels and some physiographic implications.* U.S. Geol. Survey Prof. Paper 252.

Leopold, L. B. and M. G. Wolman 1957. *River channel patterns: braided, meandering and straight.* U.S. Geol. Survey Prof. Paper 282-B.

Li, T. Y. and J. A. Yorke 1975. Period three implies chaos. *Ann. Math. Monthly* **82**, 985–92.

May, R. M. 1974. *Stability and complexity in model ecosystems.* Englewood Cliffs, NJ: Princeton University Press.

May, R. M. 1976. Models for single populations. In *Theoretical ecology: principles and applications*, R. M. May (ed.), 4–25. Oxford: Blackwell.

Parker, G. 1976. On the cause and characteristic scales of meandering and braiding in rivers. *J. Fluid Mech.* **76**, 457–80.

Parzen, E. 1974. Some recent advances in time series. *IEEE Trans Automatic Control*, AC-19, **6**, 723–30.

Schumm, S. A. 1979. Geomorphic thresholds: the concept and its applications. *Trans. Inst. Br. Geogs.* **4**, 485–515.

Thornes, J. B. 1973. Markov chains and slope series: the scale problem, *Geogr. Annal.* **5**, 322–8.

Thornes, J. B. 1980. Structural instability and ephemeral channel behavior, *Z. Geomorph.* **36**, 233–44.

Thornes, J. B. and M. W. Clark 1976. *The effects of missing data on autocorrelation and partial autocorrelation functions.* Unpublished paper of the Non-Sequential Water Quality Records Project. London School of Economics.

Valdares Tavares, L. 1975. Continuous hydrological time series discretization. *Am. Soc. Civ. Engrs Proc., J. Hydraul. Div.* **101**, HY1, 49–63.

Wallis, J. R. and P. E. O'Connell 1972. Small sample estimation of p_i. *Water Resources Res.* **8**, 707–13.

16

Geomorphic responses to climatic forcing during the Holocene

Wayne M. Wendland

INTRODUCTION

Climatic variability or climatic change is often invoked as the cause or partial cause of changes in the geomorphic, faunal, floral and archeological records. Although there are many possible causes for such discontinuities, when changes are areally synchronous or near synchronous, climatic change is difficult to dismiss and, in many cases, very easy to accept. Such associations by themselves do not prove a relationship. Indeed, they can only demonstrate temporal association. Increased credibility may only be derived from additional results found in independent data.

Without a specific cause–effect model relating climate change and a particular response, observed changes in an alluvial stratigraphy could be a lagged response to an initial shift upstream, or a change in vegetation, or perhaps natural succession. However, if botanical, archeological or geomorphological evidence from a large area suggests synchronous change, climatic change is a likely causal mechanism.

Changes in temperature and precipitation may be related to several causal mechanisms, e.g., change in the strength or mean latitude of the westerlies, change in the mean latitude of the semi-permanent pressure systems, etc. Another model which may be invoked to cause changes in surface climatic regimes is changing frequencies of dominating airmasses. When one airmass is replaced by another, temperature and humidity characteristics change. For example, the Mississippi River valley is typically dominated by warm, moist maritime tropical (mT) air from the Gulf of

Mexico in summer. There is ample evidence which shows that these characteristics have not always been present in the region during the past, e.g., during mid-Holocene the upper Mississippi River valley was drier and somewhat warmer, suggesting more frequent dominance by maritime polar (mP) air from the Pacific Ocean (Bryson & Wendland 1967, Webb & Bryson 1972, Wendland 1980a).

Similar arguments apply to other areas and other times. To understand possible and/or likely alterations in airmass frequencies, one must realize where the source regions for airmasses are located, and their area of influence.

GENERAL CLIMATIC BACKGROUND

An assemblage of meteorological events (climate) which yields the climatic character at a given location (e.g. type and intensity of precipitation, temperature and humidity, wind direction, thunderstorm frequency, blizzards, etc.) is largely the result of the assemblage of airstreams which dominate a given area. Airstreams originate in specific regions on the surface of the Earth, determined in part by the shape of the Earth, the relative locations of continents and oceans, the orbital characteristics of the Earth about the Sun, and location of major mountain chains. From the above, a mean general circulation pattern evolves, reaching equilibrium, with permanent or near-permanent feature locations, including (i) subtropical anticyclones located over low-latitude oceans, (ii) sub-polar cyclones over the Aleutians and Iceland, (iii) the jet stream, (iv) Hadley and Rossby circulation regimes of low- and mid-latitudes, respectively, (v) polar anticyclones, (vi) areas with high frequency of frontal activity, (vii) areas of preferred cyclone frequency, etc.

Several airstream source regions can be identified over the Earth's surface, each developing temperature, humidity and stability characteristics in accordance with the surface characteristics.

GENERAL CIRCULATION REGIMES OF THE NORTHERN HEMISPHERE

Several years ago, Borchert (1950) demonstrated that the prairie grassland, the natural vegetation of the Great Plains, is limited to that area of North America which is dominated by mP air from the Pacific Ocean during winter. Both to the north and south of the grasslands, the climate is sufficiently different, due to the dominance of other airmasses, that other vegetation assemblages dominate. In 1966, Bryson showed that the major natural vegetation areas of eastern North America are near-coincident with dominant areas of major airmasses during particular times of the year. The dominance of a given airmass during a specified season, or the frequencies of all airmasses which affect a given region during the mean year, may be used as an index of climate. Similarly, climatic change may be assessed by noting changes to airmass frequencies over time.

Airstream climatology

Source regions of 19 different airstreams of the Northern Hemisphere have recently been identified (Wendland & Bryson 1981). These sources (Fig. 16.1) are the origin of the different airmasses which thence migrate

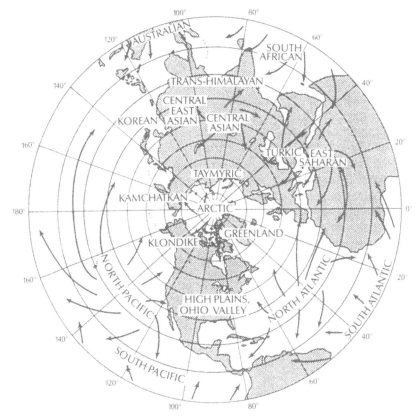

FIGURE 16.1
Source locations of the 19 airstreams, after Wendland and Bryson (1981). Generalized streamlines shown for several of the larger sources.

from place to place. Two principal features of the map are the large sub-tropical anticyclones located over the low-latitude Pacific and Atlantic Oceans. These two sources are annual, i.e., they are present during all 12 months of the year and spread the warm and humid characteristics of the low-latitude ocean to the outer margins of flow. Although smaller in areal extent, source regions located over eastern Asia, the Arctic and Turkey are also present during all months of the year. Their size expands and contracts seasonally to a much greater degree than that exhibited by the Atlantic and Pacific sources. The three sources are all of continental origin, therefore considerably drier than the maritime sources with temperatures of the Arctic (continental polar, cP) source being coldest and Saharan (continental tropical, cT) source being warmest. Each of the remaining airstream sources occupies a diminishing area of the Northern Hemisphere and is in existence for nine months or less.

Although meteorological parameters often exhibit mild gradients within an airstream, much stronger gradients are found across the boundaries between one airstream and another, i.e., along fronts. Similarly, certain meteorological qualities or events tend to be associated within airstreams, whereas others concentrate along fronts. This preferential association permits one to draw inferences about the regional climatological characteristics typically experienced within airstreams, or fronts identified in Figure 16.1.

The characteristics associated with different synoptic features of these climatic charts are essentially the same as those found on daily weather maps. Perhaps more importantly for this paper, climatic maps (indicating synoptic features) suggest aggregates of weather phenomena which can exert influence on surficial geomorphological processes. Relationships from recent data may be applied to reconstructed synoptic maps for earlier times (see e.g., Bryson & Wendland 1967, Wendland 1978) to infer likely forcing mechanisms which bring about surficial change.

METEOROLOGICAL CHARACTERISTICS ASSOCIATED WITH SYNOPTIC FEATURES

Anticyclonic source regions

Anticyclones are large areas of subsiding air and surface diffluence, motions which do not support widespread cloud cover or precipitation. Cumulus clouds, however, often form within anticyclones, i.e., if the air is sufficiently unstable. The cumulus clouds often grow to sufficient altitude to initiate short-duration, large-drop sized showers. Within areas covered by one airstream for the average year, frontal passages and associated frontal-type weather are essentially unknown, and wind direction is relatively constant. Of importance to surficial geomorphological processes are: (i) consistent wind direction in a given sector, (ii) frequent daily sun, (iii) precipitation restricted to short-lived showers with accumulations of the order of 1 cm or so, (iv) precipitation rates of a few cm per hour, and (v) relatively constant temperatures and humidities. All of these characteristics diminish in intensity and consistency as one proceeds further away from the center of the anticyclone.

Areas of persistent cyclones

The wind direction within a sector of a persistent cyclone is relatively constant in direction through the year, except when/if migrating lows move through the area. The wind speeds are stronger within cyclones than in anticyclones, averaging perhaps 8–15 mps. Because of the surface convergence within and around cyclones, widespread sheet-like clouds develop which may easily cover 100 000 km^2. Systematic cloud cover of this type most often supports continuous precipitation.

The impacts on surficial processes are much more complex than those experienced within an anticyclone: (i) the force of the wind is greater and may change with migrating cyclones, (ii) cloudy episodes often continue for days without any direct solar radiation, (iii) precipitation tends to continue for one or two days (particularly on the polar side of cyclones) with only short intervals of no precipitation, and accumulations are typically only a few cm, (iv) precipitation rates are usually a few tenths to 1 cm per day, and (v) temperature tends to be constant, although strong spatial gradients may exist.

Areas with frequent fronts

Areas with persistent fronts are identified in Figure 16.2 and result from persistent anticyclones adjacent to each other. Clouds associated with persistent fronts tend to be linear and aligned along the front. The form of precipitation associated with frontal activity can be either continuous or convective showers. Precipitation rates are often of the order of a few cm per hour and total accumulations can be as much as 5 to 10 cm.

Migrating fronts, anticyclones and cyclones

A significant portion of middle and high latitudes experience effects of migrating cyclones, anticyclones and frontal activity. These locations experience rather large day-to-day variations in weather, i.e., one airmass dominates for a few days, yielding to another with the passage of a front. The associated weather, therefore, includes the wide variability of meteorological events associated with anticyclones *and* cyclones. Areas which experience the greatest frequency of migrating fronts, cyclones and anticyclones can be identified by those areas not dominated by one airstream (Fig. 16.3) or a persistent front (Fig. 16.2).

CLIMATIC CHANGE PARADIGM

The Holocene climatic record may be viewed as a warming trend following late-glacial time, reaching a maximum during the middle Holocene, followed by a cooling trend. There is a growing body of evidence

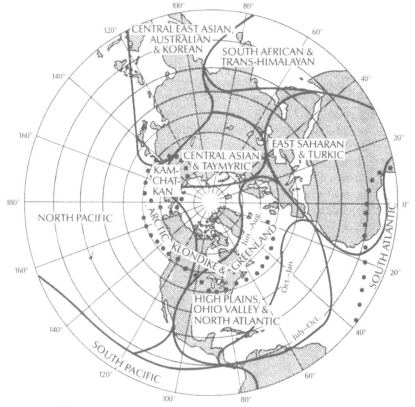

FIGURE 16.2
Lines indicate areas of preferred frontal locations. Fronts identified from strongest gradients of airstream frequencies by principal components analysis of mean monthly streamline analyses used in Wendland and Bryson (1981).

from botanical and archeological records which suggests that the climatic changes occurred abruptly as a step function, i.e., a relatively stable climatic episode was separated from another by an abrupt short-term transition. The step function model of climatic change, proposed by Bryson and Wendland (1967), was supported by discontinuities identified from several hundred radiocarbon dates indicative of (i) cultural associations and (ii) those marking times of vegetation change (Wendland & Bryson 1974). Although the cause of each date of change is not known to be the result of climatic influences, that a few preferred times were found in two independent data sets (the times of which compared favorably with each other) suggested that these synchronous changes were caused by

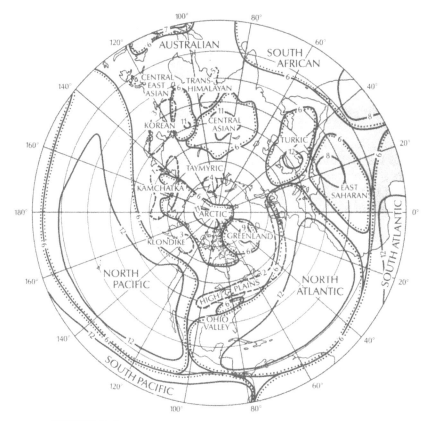

FIGURE 16.3
Outer limits of airstream dominance regions (months per year). Innermost contour shows mean maximum number of months airstream is in existence (Wendland & Bryson 1981). (Reproduced with the permission of the American Meterological Society, J. Appl. Meteor.)

some large-scale forcing function, i.e., a change in climate. Eight synchronous discontinuities were identified in the two independent records, and the authors concluded that these times marked large-scale climatic change of sufficient magnitude to be evidenced in the botanical and archeological response records.

Knox (1972) suggested that geomorphic evidence should also tend to exhibit discontinuities in process on or about the times of climatic change. He identified several discontinuities from a sample of 102 radiocarbon dates from the periphery of the Great Plains, indicative of alluvial changes.

The research which follows also analyzes discontinuities in several surficial, geomorphological records where the chronology was established by radiocarbon evidence. This is an attempt to determine if changes in geomorphological processes occurred near-synchronously over relatively large areas, the synchroneity of which suggests climate change, perhaps realized by changing frequencies of airmasses through a year. Synchronous climatic changes should not be expected to be of a similar type (i.e., warming or cooling) because, when dominating circulation patterns change, the change may be experienced over most of the hemisphere, the new emerging anomaly fields exhibit both negative and positive change areas, as well as an area between where little or no change occurred.

Radiocarbon evidence (primarily from North America) purported to indicate the beginning or ending of a geomorphic process episode was obtained from the literature (marked with an asterisk in the references), in which the author presented either individual radiocarbon assays representative of such changes, or interpreted the radiocarbon evidence in terms of beginnings and/or endings of process episodes. The 122 data in this study primarily marked times of discontinuity in alluvial sequences, terrace cut–fill sequences, changes in sedimentation rates, and dates associated with soil creep or solifluction events. Sites within the North American Great Plains (including Alberta, Wyoming, Kansas, Missouri, Iowa, Illinois, Wisconsin, Oklahoma and New Mexico), Yukon, Australia and northern Europe formed the data base.

A temporal histogram of these data was constructed (Fig. 16.4). Because of the relatively small sample size, the raw data were filtered by means of a binomial filter:

$$\hat{x}_i = 0.25\, x_{i-1} + 0.5\, x_i + 0.25\, x_{i+1} \qquad (16.1)$$

where $\hat{x}_i$ is the filtered value for the i-th cell.

Dates associated with changes in sedimentation rate of soil creep and solifluction events are shown separately in Figure 16.4. Since there were so few data of this type, they were not analyzed further, nor added into the fluvial data histogram.

Results

Several prominent peaks may be identified within the fluvial histogram, notably at about 2000, 3000, 4000, 4500, 5000, 6000, 7500, 8000 and 11 000 yr before present (BP). The relatively few data associated with changes in sedimentation rates and soil creep and solifluction tend to reinforce the peaks indicated on the filtered fluvial histogram.

The arrows beneath the histogram on Figure 16.4 represent times of discontinuity within a botanical and archeological data base, analyzed by Wendland and Bryson (1974). One cannot help but notice the similarity between the times of change identified between the botanical, archeological and fluvial records. However, not every peak in the fluvial record was also identified in the botanical and/or archeological study(ies). These differences may be the result of insufficient data in either of the three records, or may simply demonstrate that the botanical, archeological, and fluvial responses are each sensitive to either different events associated with climate change, or to climatic change of different magnitudes.

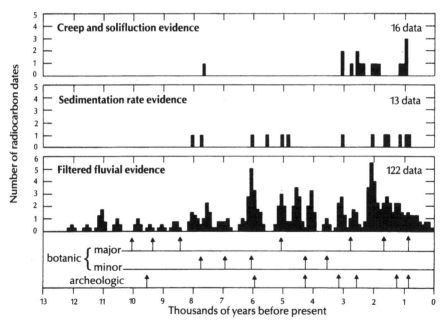

FIGURE 16.4
Histogram of ^{14}C dates indicative of discontinuities in soil creep and solifluction, sedimentation rates, and fluvial evidence.
Arrows mark times of discontinuity identified by Wendland and Bryson (1974) in botanical and archeological data.

Knox (1976) reported on discontinuities in the alluvial record from radiocarbon evidence from sites along the periphery of the Great Plains, from which a frequency histogram was constructed. His analysis indicated discontinuities at about 4600, 5900, 7000, 8500 and 9800 yr BP. Four of those five discontinuities are closely aligned with frequency peaks identified in the present data sample. This is not surprising, since some data may have been shared between the two studies.

Knox identified discontinuities in his study by displaying accumulative percentage change in the number of radiocarbon dates from one class interval to the next. Estimates of the discontinuity made by this method were identified at those times when the *change* in the number of dates abruptly increased from one class interval to the next. However, Wendland and Bryson (1974) argued that discontinuities are expected to be best marked when the number of radiocarbon dates maximize in a class interval. They further assumed that dates of discontinuity would tend to be normally distributed on either side of the actual change. On the other hand, the accumulative percentage change method marks the discontinuity somewhat earlier than the time indicated by maxima of radiocarbon dates.

To determine whether a similar lag relationship would be found in the fluvial data of the present study, a cumulative percentage change curve was prepared (Fig. 16.5). Relatively fewer discontinuities were identified by this method than were suggested by the histogram (Fig. 16.4). The four major discontinuities indicated by calculating cumulative percentage change are, in each case, earlier by 200–300 yr than the time suggested by the modes of Figure 16.4.

To investigate the differences in discontinuities identified by the two

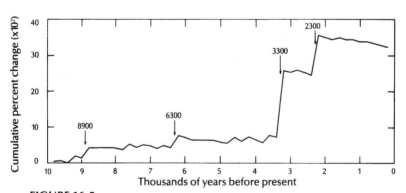

FIGURE 16.5
Cumulative percentage change in radiocarbon evidence of fluvial activity (this study) in one 200-yr class interval relative to the interval immediately preceding.

methods further, filtered botanical data shown in Figure 1 of Wendland and Bryson (1974) were subjected to cumulative percentage change analysis (Fig. 16.6). In all cases, the discontinuities identified from the cumulative percentage change curve of Figure 16.6 also precede the modal dates of Figure 1 in Wendland and Bryson (1974).

In light of the above results, it appears that discontinuities identified by cumulative percentage change precede those identified by the method which assumes that the modal class interval best marks the discontinuity. To render the results of the two analysis methods more comparable, discontinuities identified by the cumulative percentage change method should be 'adjusted' by subtracting 300 yr from the BP date. With that correction, Knox's (1976) results, those from Wendland and Bryson (1974), and discontinuities within the fluvial data (this study) are compared (Table 16.1). Discontinuities from the different studies were paired when temporal differences were less than 300 yr. Using that arbitrary placement, 75% of the 18 total discontinuities were identified in two or more studies.

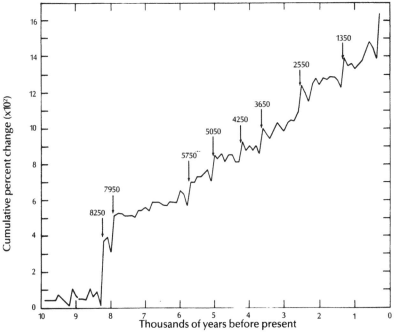

FIGURE 16.6
Cumulative percentage change in radiocarbon evidence of botanical history (Wendland & Bryson 1974) from one 100-yr class interval relative to the interval immediately preceding.

CLIMATOLOGICAL AND RESPONSE VARIATIONS
WITHIN EPISODES

The comments which follow refer to large-scale climatic episodes which typically exhibit durations from several hundred to one or two thousand years. These scale sizes represent changes brought about by atmospherically synchronous changes in upper air wind regimes and related airmass frequency changes. Transitions between episodes appear to be on the order of 1/10th the length of episodes. Relative lags of one response function to another may be estimated from the data presented and reviewed in this paper. In eight instances where alluvial discontinuities identified in this research were paired with the botanical record (see Table 16.1), the fluvial discontinuities were later than the botanic discontinuities in seven cases by an average of 105 yr. Wendland and Bryson (1974) claimed that botanical discontinuities precede those in the cultural record by c. 50 yr. This is not an unexpected result, since fluvial systems respond not only

TABLE 16.1
Comparison of inferred climate discontinuities from four studies, determined from radiocarbon chronologies.

Botanical record[a]	Archeological record[b]	Alluvial record[c]	Alluvial record[d]
850 BP	830 BP		
	1260		1200 BP
1680			1500
			2000
2760	2510		2700
	3110		3000
3570			
4240	4230	4300 BP	4000
			4500
5060			5000
		5600	
6050	5900		6000
6910		6700	6800
7740			7500
		8200	8000
			8600
8490			
9300	9530	9500	
10 030			

[a,b]Wendland and Bryson (1974).
[c]Knox (1976).
[d]This paper.

to changes in precipitation and temperature (two parameters associated with the climatic transition), but also to changes in vegetation type and cover, a change which apparently occurs over a few decades. After a climatic change is initiated, it is followed by vegetation, cultural and fluvial changes, each lagging the former.

Although apparently fewer geomorphological events occur within a stable climatic episode, the events are of the same type as those which mark transitions. Climatological evidence, on the other hand, suggests that the components of climate, i.e., thunderstorms, rainfall events, mean and extreme temperatures, etc., have occurred throughout the Holocene, but that the assemblage of these components is significantly different from one episode to another. For example, as the frontal frequency across a given location changes with time, so does the frequency and magnitude of precipitation received at that site.

Information gathered from 49 recording raingauges and 1344 rainstorms in east central Illinois (Huff 1969) showed that 20% of storms lasting fewer than 3 hours yielded about 70% of the cumulative precipitation from all storms, whereas 20% of storms lasting longer than 12 hours yielded only about 45% of the cumulative precipitation. Therefore, as cold frontal frequency decreases, precipitation *rate* also decreases, and vice-versa, and hence, the potential work done by the storm is decreased.

Meteorological–climatological events of two different episodes may be estimated by comparing historical data from the middle- and late-1800s (end of the neo-Boreal, also called the Little Ice Age), with that of the recent few decades. According to Wahl and Lawson (1970), the upper Midwest experienced mean temperatures about 1–2° F lower, and precipitation about 10% more during the mid-1800s, compared to recent normals. Although these magnitudes are less than several inferred for other times during the Holocene, the climate change of the last 100 yr exerted significant influence on human occupation, vegetation and geomorphological processes. For example, glacial expansion occurred in high peaks of the southern Rockies during the late 1800s, whereas today they are ice-free. In southern Wisconsin, there were a few reports of frost and snowfalls during summer months during the mid-1800s. Although the frost occurrences and snowfalls were very infrequent, the mere fact of occurrence represents a significant change from our recollections of the last few decades.

In addition, return frequencies of specified rainfall amounts would also have been significantly different from those of recent years, as would the frequency and distribution of thunderstorms and severe weather through the year. Hurricanes, for example, would have been much less frequent and severe during cooler episodes than those which we experience today, and more frequent during warmer episodes (Wendland 1977).

CONCLUSIONS

Climatic episodes of the Holocene, identified from botanical and archeological evidence, are suggested to have been the result of changes in the strength or pattern of the upper air circulation and surface airstream dominance frequencies. Evidence strongly suggests that the locations of airstream sources, and the characteristics of the airmasses of recent decades, were relatively constant throughout the Holocene (except perhaps the characteristics of arctic air which were probably different up to about 7000 yr BP, due to the existence of the Laurentide Ice Sheet – and perhaps also arctic air in Eurasia, although the area with continental ice in Europe was much smaller (Wendland 1980b).

Radiocarbon evidence associated with alluvial activity shows preferred times of process discontinuity which agree with those identified from botanical and archeological evidence. Although the data are still too few to draw concrete conclusions, archeological discontinuities appear to follow those found in the botanical record by about 50 yr, and fluvial discontinuities lag the botanical by about 100 yr.

ACKNOWLEDGEMENTS

J. C. Knox (University of Wisconsin-Madison) and C. S. Alexander (University of Illinois-Urbana) offered helpful suggestions for much of the geomorphic literature referenced in this article. J. Brother (Illinois State Water Survey, Champaign) drafted the figures and J. Lewis typed the manuscript. I thank all for their assistance.

REFERENCES

Borchert, J. R. 1950. The climate of the central North American grassland. *Assn Am. Geog. Ann.* **40,** 1-39.
Bryson, R. A. 1966. Airmasses, streamlines and the boreal forest. *Geog. Bull.* **8,** 223–69.
Bryson, R. A. and W. M. Wendland 1967. Tentative climatic patterns for some late-glacial and post-glacial episodes in central North America. In *Life, land and water,* W. J. Mayer-Oakes (ed.), 271–98. Winnipeg: Univ. Manitoba Press.
Huff, F. A. 1969. Climatological assessment of natural precipitation characteristics for use in weather modification. *J. Appl. Meteorol.* **8,** 401–10.
Knox, J. C. 1972. Valley alluviation in southwestern Wisconsin. *Assn. Am. Geog. Ann.* **62,** 401–10.
Knox, J. C. 1976. Concept of the graded stream. In *Theories of landform development,* W. N. Melhorn & R. C. Flemal (eds), 169–98. London: George Allen & Unwin.

Wahl, E. W. and T. L. Lawson 1970. The climate of the midnineteenth century United States compared to the current normals. *Mon. Wea. Rev.* **98**, 259–65.

Webb, T. and R. A. Bryson 1972. Late and postglacial climatic changes in the Northern Midwest, USA. *Quat. Res.* **2**, 70–115.

Wendland, W. M. 1977. Tropical storm frequencies related to sea surface temperatures. *J. Appl. Meteorol.* **16**, 478–81.

Wendland, W. M. 1978. Holocene man in North America: the ecological setting and climatic background. *Plains Anthropol.* **23–82** (1), 273–87.

Wendland, W. M. 1980a. Holocene climatic reconstructions on the Prairie Peninsula. In *The Cherokee excavations*, D. C. Anderson & H. A. Semken, Jr. (eds), 139–48. New York: Academic Press.

Wendland, W. M. 1980b. *Late Wisconsin ice ablation in western Europe*. AMQUA Abstracts 201. Orono: University of Maine.

Wendland, W. M. and R. A. Bryson 1974. Dating climatic episodes of the Holocene. *Quat. Res.* **4**, 9–24.

Wendland, W. M. and R. A. Bryson 1981. Northern hemisphere airstream sources. *Mon. Wea. Rev. 109*, 255 – 70.

SOURCES OF RADIOCARBON DATES NOT CITED IN TEXT

Ahler, S. A. 1973. Post-Pleistocene depositional change at Rodgers Shelter, MO. *Plains Anthropol.* **18**, 1–26.

Ahler, S. A. 1976. Sedimentary processes at Rodgers Shelter. In *Prehistoric man and his environments*, W. R. Wood & R. B. McMillan (eds), 123–39. New York: Academic Press.

Albanese, J. P. 1980. *Geology of the Laddie Creek archeological site area, Big Horn Co., Wyoming.* Casper. M.S., 22p.

Albanese, J. P. and M. Wilson 1974. *Holocene alluvial chronology and climate change on the Northwestern Plains.* AMQUA Abstract 88. Madison: University of Wisconsin.

Albanese, J. P. and M. Wilson 1974. Preliminary description of the terraces of the North Platte River at Casper, WY. In *Applied geology and archeology: the Holocene history of·Wyoming*, M. Wilson (ed.), 8–18. Laramie: Geol. Survey of Wyoming.

Alexander, C. S. and L. W. Price 1980. Radiocarbon dating of the rate of movement of two solifluction lobes in the Ruby Range, Yukon Territory. *Quat. Res.* **13**, 365–79.

Artz, J. A. 1981. *Soil–geomorphic evidence for environmental change on the southeastern periphery of the Central Plains.* Museum of Anthropology, University of Kansas, Lawrence, M. S.

Benedict, J. B. 1970. Downslope soil movement in a Colorado alpine region: rates, processes and climatic significance. In *Periglacial processes*, C. A. M. King (ed.), 181–226. Stroudsburg, PA: Dowden, Hutchinson and Ross.

Benedict, J. B. 1976. Frost creep and gelifluction features: a review. *Quat. Res.* **6**, 55–76.

Butzer, K. W. 1977. *Geomorphology of the lower Illinois Valley as a spatial–temporal context for the Koster Archaic site.* Springfield, IL: Illinois State Museum Rept. of Invest. 34.

Costin, A. B. 1972. Carbon-14 dates from the Snowy Mts. area, Southeastern Australia and their interpretation. *Quat. Res.* **2**, 579–90.

Costin, A. B., B. G. Thom, D. J. Wimbush and M. Stuiver 1967. Nonsorted steps in the Mt. Kosciusko area, Australia. *Geol. Soc. Am. Bull.* **78**, 979–92.

Daniels, R. B. and R. H. Jordan 1966. *Physiographic history and the soils entrenched stream system and gullies, Harrison County, Iowa.* USDA Tech. Bull. 1348.

Giles, L. H. 1974. *Holocene soils and soil–geomorphic relations in an arid region of southern New Mexico.* AMQUA Abstract 30–39. Madison: University of Wisconsin.

Hall, S. A. 1978. *Late Holocene alluvial chronology from Northeastern Kansas.* AMQUA Abstract. 207. University of Alberta.

Hall, S. A. 1980. *Corresponding geomorphic, archeologic and climatic change in the southern Plains: new evidence from Oklahoma.* AMQUA Abstract 89. Orono: University of Maine.

Haynes, C. V., Jr 1966. *Geochronology of late Quaternary alluvium.* Geochron. Lab., Univ. Arizona. Interim Res. Rept. 10.

Haynes, C. V., Jr. 1968. *Geochronology of late Quaternary alluvium.* INQUA, VII Congress Proc. **8**, 591–631.

Haynes, C. V., Jr. 1976. Late Quaternary geology of the Lower Pomme de Terre Valley. In *Prehistoric man and his environments*, W. R. Wood & R. B. McMillan (eds), 47–63. New York: Academic Press.

Henry, D. O. 1978. *The prehistory and paleoenvironment of Hominy Creek Valley.* Lab of Archeology, Univ. of Tulsa, Okla. Skiatook Lake Project, DACW56-77-C-0222.

Hoyer, B. E. 1980. The geology of the Cherokee site. In *The Cherokee excavations*, D. C. Anderson & H. A. Semken, Jr (eds), 21–66. New York: Academic Press.

Johnson, W. C. 1978. *Intensified fluvial activity in response to Holocene climatic variations.* AMQUA Abstract 216. University of Alberta.

Johnson, W. C. 1980. *An episode of late Holocene soil development on flood-plains on the Central Plains.* AMQUA Abstract 116. Orono: University of Maine.

Johnson, W. C. 1980. *Stream response to climatic change during the Holocene in the midwestern United States.* Dept. of Geography, University of Kansas, Lawrence, M.S.

King, R. H. 1980. *Holocene paleo-environmental record preserved in the soils of Jasper National Park, Alberta, Canada.* AMQUA Abstract 120. Orono: University of Maine.

Knox, J. C. 1976. *Impact of fluvial erosion on the Great Plains altithermal cultural hiatus.* Paper presented at Plains Conf., Minneapolis, Oct. M.S.

Knox, J. C. and W. C. Johnson 1974. Late Quaternary valley alluviation in the driftless area of southwestern Wisconsin. In *Late Quaternary environments of Wisconsin*, J. C. Knox & D. M. Mickelson (eds), 134–51. Field Guide to AMQUA meeting. Madison: University of Wisconsin.

Knox, J. C., P. J. Bartlein, K. K. Herschboeck and J. R. Muckenhirn 1975. *The response of floods and sediment yields to climatic variation and land use in the Upper Mississippi Valley.* Rept. 52, Inst. of Environ. Studies, Univ. Wisconsin, Madison.

Knox, J. C., P. F. McDowell and W. C. Johnson 1981. Holocene fluvial stratigraphy and climatic change in the driftless area, Wisconsin. *Quaternary Paleoclimate.* Norwich: Geo Abstracts. In press.

McDowell, P. F. 1980. *Holocene fluvial activity in Brush Creek watershed in the driftless area of Wisconsin.* AMQUA Abstracts 135. Orono: University of Maine.

McMillan, R. B. and W. R. Wood 1975. Man and environment in the western Ozarks. *Quat. Paleoenvironment history of the Western Missouri Ozarks.* Field book of Midwest Friends of the Pleistocene, 30–32. University of Missouri, Columbia.

McMillan, R. B. 1976. The dynamics of cultural and environmental change at Rodgers Shelter, MO. In *Prehistoric man and his environments.* W. R. Wood & R. B. McMillan (eds), 211–32. New York: Academic Press.

Starkel, L. 1966. Postglacial climate and the moulding of European relief. In *World climate from 8000 to 0 BC*, 15–32. London: R. Meteorol. Soc.

Wendland, W. M. 1981. *Preliminary estimates of recent frontal position variance and implications for climatic interpretation during the Holocene.* Paper given at AMS First Conf. on Climate Variations. San Diego, M.S.

Geographical Index

Alberta, 59, 64-78, 154: Bow River Valley, 173; Brooks, 66, 74-75; Cathedral Crags, 185-90; Dinosaur Provincial Park, 59-78; Elbow Valley, 173; Frank, 174; Kananaskis Valley, 173; Lake Louise area, 171-90; Mt. Elpoca, 185-90; Mt. Hector, 185-90; Mt. Rae, 171-90; North Saskatchewan River, 162; Red Deer River, 64-65; Rocky Mountains, 154; southern, 66, 164; Surprise Valley, 183
Aleutians, 356
Appalachians, 97, 100, 111, 114
Arctic, 358
Arizona, 278
Asia, 358
Atlantic Ocean, 358
Australia, 362

Bahamas, 150: Andros Is., 150
Bermuda, 150
British Columbia: Fraser River, 1; Lillooet River, 1; Selkirk Mts., 160; Vancouver Is., 155-56, 164

California, 272, 279, 285-313: Baja, 286; Belinas Point, 286; Cloverdale, 319, 322; Coast Ranges, 322; Duxbury Point, 286; Gulf of, 286; Oxnard alluvial plain, 288, 310; Point Hueneme, 287-88; Point Mugu, 287-88; San Clemente Is., 269-78, 281-82; San Francisco, 286; Santa Clara, 289; southern, 269; Transverse Ranges, 286, 310
Canada, 161, 164: prairies, 65; Rocky Mts., 160, 171-90, 243

Colorado, 40, 194: Front Range, 7, 25, 31, 38; Indian Peaks, see Front Range; northwestern, 63; San Juan Mts., 50-56; western, 2

East Pacific Rise, 286
Eire, 156
England: Lake District, 50-56; Lea Marston, 337-38; Mendip Hills, 150; north, 148; River Thame, 340; River Trent, 337; southwest, 150; Stour River, 339-40, 349-50; Yorkshire, 148, 155
Europe, 368: northern, 362

France (southern), 159

Giza (Egypt), 148
Golan Heights, 280
Great Plains (USA), 356, 362, 364
Gulf of Mexico, 355-56

Iceland, 356
Illinois, 120, 123-24, 131, 137, 367; Casey, 125-26, 130-31, 138, 142, 144; Clark County, 126-29, 142; Cumberland County, 126-27; northeastern, 124; south-central, 131-32, 135, 142; Springfield Plain, 125
Iowa, 120, 123-24, 155, 228, 236-37; Cherokee Sewer Archeological Site, 228-38; Little Sioux River Valley, 228-38; Missouri River, 228-29
Israel, 279: coastal plain, 279; Sharon Plain, 279

Lebanon (southern), 280
Libyan desert, 279

Mackenzie Mts. (N.W.T., Canada), 153
Mississippi River Valley, 355-56
Mojave desert, 272
Midwest (USA), 117, 118, 122, 367

Negev desert, 279
Newfoundland, 160
North America, 356-57, 362
North Pole, 167
Northern Hemisphere, 357-58

Ontario (southern), 245-46

Pacific Ocean, 356
Pennsylvania (central), 97, 100-01, 111

Rocky Mountains, 173: Alberta, 154;
    Canadian, 160, 171-90, 243;
    southern, 38, 367

Sahara desert, 279, 358
Sinai desert, 279
Spain, 226, 280: Andalusia, 280

Tasmania, 48-56: Ben Lomond, 50-56
Turkey, 358

Utah, 194, 319: Dirty Devil River, 196, 205;
    Fremont River, 196-214; Henry Mts.,
    195-214; Fishlake Plateau, 319-21;
    Sandy Creek, 196, 204-14;
    Sweetwater Creek, 196, 204-14; Town
    Wash, 196, 204-14

Wales (south), 227
West Virginia, 151

Subject Index

Accretion-gley, 118-19, 132-33, 144
Aeric Albaqualf, 279
Airmass, 355
Airstreams, 356-68
Alfisol, 38, 280
Allometry, 212: models, 203-06
Alluvium, 316
Altithermal, 36
Analysis: correlation, 294-95; cluster, 4;
        factor, 294; multiple regression, 264;
        spectral, 294, 303-05, 339; trend, 294;
        variance, 4, 10-22, 69-70, 102
Anhydrite, 147-68
Aquifer, 150
Archeologic record, 361, 363
Arcsin transform, 317, 323
ARIMA (Autoregressive-Integrated-
        Moving-Average), 86-114, 337-52
Arroyo, 6, 194-214
Atomic absorption spectrometry, 270
Auburn soil (California), 250
Autocorrelation, 86-114, 343-52;
        coefficient, 88-114, 343-52; partial
        coefficient, 88-114, 343-352

Badlands, 6, 59-78, 196
Berry Clay, 134, 138
Blair soil (Illinois), 130, 141, 143
BOD (biochemical oxygen demand),
        327-52
Botanic record, 361, 363, 365
Box-Jenkins model, see mixed
        autoregressive-moving ave.

Calcite, 147-68
Canmore Advance, 173
Carbonate solution, 147

Carbon dioxide, 147-68; partial pressure,
        147-68
Catastrophe theory, 210-11
Catena, 26-7, 127, 269, 274-82
Cary Event, 229-30
Cavell Advance, 173, 184
Central limit theorem, 157, 316
Channel network, 197-214
Chapin Soil, 124
Chi-square, 97
Chlorite, 133-34, 136, 276-78, 133-34, 136,
        276-78
Chromoxerert, 280
Chronology, 225: geomorphic, 172
Chronosequence, 27, 40
Cisne-Shiloh catena, 139
Cisne soil (Illinois), 130, 133, 135, 137-43
Clay: expanding, 133; mineralogy, 276-78
Climatology, 355-68; synoptic, 358-60
Climatic change, 355, 360-68; forcing,
        355-68; variability, 355
Clinomorphic Ratio, 71
Coastal change, 285-313
Coefficient: correlation, 8, 257-58, 262,
        264; debris diffusion, 82;
        determination, 201-02; diffusion, 82;
        recessional, 81; retreat, 81; sorting,
        10; subduing, 82
Correlogram (spatial), 46, 49, 52
Corrington Fan (Iowa), 230-38
Creep, 81, 82, 96
Critical power, 61
Critical shear stress, 62
Crowfoot Advance, 173, 184
Cumulic Haplaquoll, 141
Cycle: Davisian, 112, 223-24, 235;
        diffusion, 106; first-order, 236; fifth-

order, 236; fourth-order, 236; mega, 234, 236; seasonal, 11; second-order, 236; third-order, 236

Dating: chronometric, 238; radiocarbon, 150, 228, 231, 233, 236, 316-62; magnetic, 152-68; U-series, 150-68
Debris: accretion, 171-90; shift, 171-90
Dendrochronology, 186-88
Devon soil (Alberta), 243-67
Distribution: beta, 317, 324; beta-binominal, 316; binominal, 316; normal generated, 315-24; gamma, 335; spatial, 25-41
Dolomite, 147-68
Dolostone, 147-68
Drainage network, 60, 193
Drift plain (Illinois), 125, 128
Dumfries loam (Ontario), 245-67
Dystric Pergelic Cryochrept, 36
Dystric Cryochrept, 35, 36

Ebbert soil (Illinois), 130, 139-40, 143
Edmonton rainfall simulator, 243-67
Eisenhower Junction Advance, 173
Eigenvalu, 331-33
Entisol, 38
Environment: alluvial, 227; alpine, 45-56; bog, 227-28; colluvial, 227; eolian, 227; homogeneous, 243; lacustrine, 223, 227; marine, 223; paleo-, 227; subaerial, 227
Equilibrium, 60-61, 148, 193, 287, 311, 315, 327, 329, 356: behaviour, 327, 329; chemical, 157; disequilibrium, 61, 64; dynamic, 61, 311, 327, 329, 340; dynamic metastable, 27-28, 62, 78, 150, 152, 171-2, 189, 281; metastable, 148-51, 311, 312; static, 62, 220, 311-2, 327; steady state, 37-9, 46, 77, 194, 220, 311-2, 327; thermodynamic, 161
Ergodic: assumption, 55; hypothesis, 114; principle, 81-114; procedure, 46
Erosion: episodic, 172, 234; rainwash, 243-67

Fan, 187, 189: alluvial, 82, 173, 270; paraglacial, 188
Farmdale Soil (Illinois), 131, 134, 138-9, 141, 144
Farmdale surface, 139-40
Farmdalian time, 40

Feedback, 294, 333: mechanism, 285; negative, 62, 335; positive, 62, 206
Flandrian transgression, 286, 310
Flood-debris Flow, 171-90
Flowstones, 149
Fluvial, 193-214; record, 363, 365
Force differentials, 206-9, 212
Foreshore: lower, 299-300, 311; middle, 300-1, 311; upper, 300-3, 311
Fourier series, 51, 52, 303, 337
Franciscan melange, 322-4
Front, 358-9

Gerlach trough, 2, 6
Glacial Lake Hitchcock (Mass.-Conn.), 225
Glaciation: continental, 117
Glossaqualf, 279
Graded time, see time
Gravity law, 200
Great Cordilleran Advance, 172
Guelph loam (Ontario), 245-67
Gumbotil, 118, 132
Gypsum, 147-68

Hadley circulation, 356
Halite, 147-68
Haplaquoll, 141
Haploxeralf, 279-80
Hapludalf, 141
Hapludult, 141
Histic Pergelic Cryaquept, 36
Histic Pergelic Cryaquoll, 36
Holocene, 30, 151, 156-7, 167, 172-4, 184, 189, 221, 227, 236-7, 310, 355-68
Hoyleton soil (Illinois), 180, 189, 141, 143
Huey soil (Illinois), 130, 139
Humic gley, 141
Humic Pergelic Cryaquept, 36
Hurst phenomenon, 224
Hydraulic geometry, 193-4, 329
Hydraulic radius, 206-7
Hypsometric integral, 71

Illinoian, 117-9, 125, 131, 139-40, 172; till, 124, 131, 134, 144; till plain, 128, 132, 137-8, 141-4
Illite, 133-6, 280
Inceptisol, 38
Isostatic, 219

Jokulhlaup, 185-6

Kaolinite, 133-6, 276-8, 280
Karst, 147-68; minerals, 147
K-cycles, 27-41
Kinetic energy, 244-5
King clay loam (Ontario), 245-67
Köppen classification, 65-6, 197, 270
Kurtosis, 97

Lag time, 304, 333
Landform, stable, 109-14
Landslide, 318-24
Laurentide Ice Sheet, 367
Least squares, 201, 337
Limestone, 147-68
Lithic Cryorthent, 36
Lithic Haploxeralf, 270
Lithic Xerorthent, 271-2
Lithology, 96-114
Little Ice Age, 173-4, 186, 189, 367
Lockport clay (Ontario), 245-67
Loess, 27, 31, 35, 37, 39, 118-44, 230
Lognormal, 5, 48
Log transformation, 11, 12
Long shore current: direction 291-313; velocity, 291-313

Magnitude and frequency, 63, 219
Marine terraces, 270, 278
Markov process, 105, 344, 348
Mass movement, 61, 320
Mass-wasting, 30, 46, 48, 55, 171-90; controls, 46; rate, 45
Mazama: ash, 189; tephra, 173
Mediterranean climate, 269-82
Mica, 276-8
Microrelief, 244: gauge, 249, 259-60
Microtopography, 249-63
Milliken loam (Ontario), 245-67
Model: allometric, 203-6; cause-effect, 355; explanatory, 32; geomorphic, 219-39; morpho-dynamic, 285; noise, 329; observational, 32; predictive, 32; process-response, 285, 294; soil geomorphic, 25-41; structural, 335
Modern soil, 133
Moisture (antecedent), 62, 257
Montane forest, 4, 6-22

Natrixeralf, 297
Neoglacial Advance, 173, 184, 189
Newberry soil (Illinois), 130

Normal generated distribution, see distribution

Ochraqualf, 141
Offshore zone, 295-7
Orchard soil (Alberta), 243-67
Oxygen: isotope, 227; $O^{18/16}$ ratio, 155-6, 168; O^{18}, 221

Pachic Cryumbrept, 36
Paleotemperature, 155
Paraglacial, 173
Pedisediment, 122
Pelloxerert, 272, 279-80
Peoria Loess, 125, 131-4, 139
Pergelic Cryaquoll, 36
Pergelic Cryaquept, 36
Pergelic Cryoboralf, 36
Pergelic Cryochrept, 36
Pergelic Cryohemist, 36
Pergelic Cryumbrept, 36, 40
Permafrost, 150
Photography: aerial, 185, 200; terrestrial, 261
Plagioclase, 276, 278
Pleistocene, 30, 122, 172-3, 227, 320
Plio-Pleistocene boundary, 221
Podzolic soil, 141, 243
Pollen, 228
Pontypool sand (Ontario), 245-67
Power spectrum, 49, 303, 336
Pre-Illinoian Till, 229-30
Processes: fluvial, 192-214; geomorphic, 45, 219-39

Quartz, 276, 278, 291
Quaternary, 27, 118, 121-3, 221-39, 287

Rainsplash, 5, 21-2
Rainwash, see erosion
Rate: erosion, 71-2, 233-8; retreat, 286; runoff, 252-3; sedimentation, 233-8; subsidence, 310
Rendollic Xerochrept, 280
Response (complex), 62, 224, 234-8
Rhodoxeralf, 279-80
Rock glacier, 320
Rockfall, 171-90
Roxana Silt, 124-5, 132, 134, 139-41, 144
Rossby circulation, 356
Roughness, 71-73, 259
Runoff, 62

Sangamon Soil, 117-44: accretion-gley, 118-9; gumbotil, 118-9; in-situ, 118-9, 132, 141, 143
Sangamon surface, 117-44
Sangamonian, 118-9, 124-5, 139: time, 123, 144
Santa Ana wind, 272
Scale, 219-39: analysis, 205; catchment, 21-2; log, 172; plot, 2-4, 22; regional, 172; spatial, 45, 48-9, 59, 77
Sediment delivery: rate, 2; ratio, 45
Semi-arid lands, 59, 65, 192-214
Shiloh soil (Illinois), 130, 133-5, 137-41, 143
Shorezone, 235-313: summer, 285-313
Skewness, 5, 21, 97
Slope: angle, 2, 71, 259; backslope, 279; bedrock, 270; cross-slope angle, 259; downslope angle, 259; energy, 206; evolution, 81; footslope, 279; microtopography, 264; mountain, 196; original, 106; parallel retreat, 81; recession, 81; shape, 243, 259-61; stable, 103-5, 109; steepness, 264
Slopewash, 92, 122
Smectite, 276-9
Soil: alpine, 25-41; Andalusian black, 280; buried, 26, 38, 119, 122-3; exhumed, 28; relief, 28; truncated, 122-3
Soil creep, 362-3
Soil formation, 234, 269
Soil loss, 2-22, 243-67
Solifluction, 34, 320, 362-3
Solution rates, 147-68
Spatial interaction law, 198-202, 212
Spectographic analysis, 133
Spodosol, 38
Springvale sandy loam (Ontario), 245-67
Stability, 330-52: analysis, 330-52
Stalactites, 149-68
Stalagmites, 149-68
State factors, 26, 37, 40
Step function, 361
Stoneline, 123
Stratigraphy, 65, 172: alluvial, 355; bio-, 238; chrono-, 238; litho-, 230-8; micro-, 230-8; Midwest, 117-8, 122-3, 367; record, 219-39; Quaternary, 118, 123
Sub-alpine forest, 6-22
Surface: depositional, 120-1, 123; erosion, 119-21, 123; geomorphic, 26, 28, 120-3, 142, 269; ground, 26, 121; erosion, 23, 119-21; pedomorphic, 26; soil, 119; stable, 121; wash, 82, 96.
Surf zone, 290-313
Synthetic Alpine Slope model, 25-41
System: cascading, 149; closed, 163, 269; deterministic, 327; geomorphic, 147; natural, 327; open, 163, 269; oscillating, 327-52; transfer function, 328

Tamalco soil (Illinois), 130, 143
Tazewell Event, 229-30
Tectonic, 219, 221
Temporal data series, 327-52
Threshold, 61-3, 76, 148, 172, 193, 198, 211, 282, 287, 311, 315-6, 324: between, 60, 178, 189, 194, 198, 282; climatic, 237; dynamic, 147, 156, 160; event, 190; extrinsic, 61, 64, 78, 156, 178, 224, 236-7, 287, 311, 315; geomorphic, 28, 31, 39, 41, 171, 193; intrinsic, 61-2, 64, 77-8, 156, 167, 178, 224, 236-7, 287, 315; limiting, 60; magnitude and frequency, 63-4, 311; micro-, 77; pedologic, 281; qualitative, 76; subthreshold, 59-78
Time, 69: cyclic, 64, 220, 238-9; graded, 64, 219-20, 224-38; steady, 64, 220
Time series, 294
Toronto simulator, 245-67
Tractive force, 206-9, 212
Transition zone, 297-8
Transport-limited landform, 60, 64
Tundra: alpine, 4, 6-22, 29; ridge-top province, 29-41; valley-bottom province, 29-31, 40; valley-side province, 29-31, 40
Typic chromoxerert, 282
Typic Albaqualf, 279
Typic Cryumbrept, 35, 36
Typic Natrixeralf, 282
Typic Pelloxerert, 271

Unit: soil geomorphic, 26; soil landscape, 26; spatial, 46
Universal Soil Loss Equation, 48

Variance, 3, 46-7, 49-56, 96, 102, 104, 109, 114, 158, 161, 318, 338-52; antocovariance, 303, 336; co-, 161;

error, 346-51; log-, 51, 54; raw
spectrum, 52
Variation: coeffficient of, 5, 11, 46, 175-90,
243-67; spatial, 3, 11, 27, 40, 45, 55, 59-
60, 147-68, 193-5, 202-6, 210-11, 219-20,
225, 243, 269-82, 315-24; temporal, 27,
40, 55, 59-60, 147-68, 214, 219, 285-313,
315-24
Variogram (spatial), 46, 49
Varve, 224-8
Vertic Rendollic Xerochrept, 280
Vertisol, 272, 278, 281: proto-, 272-3, 276

Wave: angle, 290-313; height, 290-313;
period, 290-313; type, 290-313
Wisconsinan, 117-9, 124-5, 131, 139, 141,
144, 165, 228-9

Wisconsin, 172-3, 189: ice sheet, 65
Woburn loam (Ontario), 245-67

X-ray diffraction, 135, 270-82
Xerert, 280
Xerochrept, 279
Xerorthent, 280

Yarmouth Soil, 123
Yarmouth-Sangamon surface, 120

Zone: alternating, 28-9; coastal, 285-313;
deposition, 28-9; erosion, 28-9; one
(Zone 1), 1; persistent, 28-9

Ingram Content Group UK Ltd.
Milton Keynes UK
UKHW021944050423
419675UK00023B/175

9 780367 278182